M000049996

Cell–Cell Interactions in Early Development

The Forty-Ninth Annual Symposium of
the Society for Developmental Biology
Washington, D.C., June 27–30, 1990

Cell–Cell Interactions in Early Development

Editor
John Gerhart
Department of Molecular and Cell Biology
University of California
Berkeley, California

WILEY-LISS
A JOHN WILEY & SONS, INC., PUBLICATION
New York • Chichester • Brisbane • Toronto • Singapore

Address all Inquiries to the Publisher
Wiley-Liss, Inc., 605 Third Avenue, New York, NY 10158-0012

Printed in United States of America

Library of Congress Cataloging-in-Publication Data

Society for Developmental Biology. Symposium (49th : 1990 :
Washington, D.C.)
Cell–cell interactions in early development : the 49th Symposium of the Society for Developmental Biology / editor, John Gerhart.
p. cm.
Includes index.
ISBN 0-471-56123-1
1. Embryology—Congresses. 2. Cell interaction—Congresses. 3. Morphogenesis—Congresses. I. Gerhart, John, 1936– . II. Title.
QH491.S63 1990 91-22485
574.3′3—dc20 CIP

Contents

VI. MINISYMPOSIUM SUMMARY

Contributors

Roger N. Beachy, Department of Biology, Washington University, St. Louis, MO 63130 **[273]**

Robert J. Clifford, Department of Molecular Biology, Princeton University, Princeton, NJ 08544 **[163]**

Martyn Cook, Division of Eukaryotic Molecular Genetics, National Institute for Medical Research, Mill Hill, London NW7 1AA, England **[129]**

Michael Costa, Department of Molecular Biology, Princeton University, Princeton, NJ 08544 **[1]**

Carl M. Deom, Department of Biology, Washington University, St. Louis, MO 63130 **[273]**

A.J. Durston, Hubrecht Laboratorium, Uppsalalaan 8, 3584 CT Utrecht, The Netherlands **[109]**

Richard P. Elinson, Department of Zoology, University of Toronto, Toronto, Ontario, Canada **[297]**

Charles A. Ettensohn, Department of Biological Sciences, Carnegie Mellon University, Pittsburgh, PA 15213 **[175]**

Janice P. Evans, Department of Biology, University of North Carolina, Chapel Hill, NC 27599-3280 **[297]**

David L. Gard, Department of Biology, University of Utah, Salt Lake City, UT 84112 **[297]**

Jeff Hardin, Department of Zoology, Duke University, Durham, NC 27707 **[15]**

Robert K. Ho, Institute of Neuroscience, University of Oregon, Eugene, OR 97403 **[203]**

Jon M. Holy, University of Wisconsin-Madison, Madison, WI 14627 **[297]**

Paul Hunt, Division of Eukaryotic Molecular Genetics, National Institute for Medical Research, Mill Hill, London NW7 1AA, England **[129]**

Keith A. Jermyn, The Imperial Cancer Research Fund, Clare Hall Laboratory, South Mimms, Herts EN6 3LD, England **[261]**

Donald A. Kane, Institute of Neuroscience, University of Oregon, Eugene, OR 97403 **[203]**

Jon Karpilow, Institute of Molecular Biology, University of Oregon, Eugene, OR 97403 **[227]**

Brian K. Kay, Department of Biology, University of North Carolina, Chapel Hill, NC 27599-3280 **[297]**

Ray Keller, Department of Molecular and Cell Biology, University of California, Berkeley, CA 94720 **[31]**

Judith Kimble, Department of Biochemistry, College of Agriculture and Life Sciences, Graduate School, Laboratory of Molecular Biology, University of Wisconsin-Madison, Madison, WI 53706 **[283]**

Charles B. Kimmel, Institute of Neuroscience, University of Oregon, Eugene, OR 97403 **[203]**

The numbers in brackets are the opening page numbers of the contributors' articles.

Mary Lou King, Department of Anatomy and Cell Biology, University of Miami, Miami, FL 33101 **[297]**

Michael W. Klymkowsky, Department of MCDB, University of Colorado, Boulder, CO 80309-0347 **[297]**

Robb Krumlauf, Division of Eukaryotic Molecular Genetics, National Institute for Medical Research, Mill Hill, London NW7 1AA, England **[129]**

Eric J. Lambie, Department of Biochemistry, College of Agriculture and Life Sciences, Graduate School, Laboratory of Molecular Biology, University of Wisconsin-Madison, Madison, WI 53706 **[283]**

Lynn J. Manseau, Department of Molecular Biology, Princeton University, Princeton, NJ 08544 **[163]**

Heather Marshall, Division of Eukaryotic Molecular Genetics, National Institute for Medical Research, Mill Hill, London NW7 1AA, England **[129]**

David R. McClay, Department of Zoology, Duke University, Durham, NC 27707 **[15]**

Douglas A. Melton, Department of Biochemistry and Molecular Biology, Harvard University, Cambridge, MA 02138 **[79]**

Patricia J. Moore, Department of Agronomy, University of Kentucky, Lexington, KY 40546 **[273]**

John Morrill, Department of Biology, New College, Sarasota, FL 33580 **[15]**

Ian Muchamore, Division of Eukaryotic Molecular Genetics, National Institute for Medical Research, Mill Hill, London NW7 1AA, England **[129]**

Stefan Nonchev, Division of Eukaryotic Molecular Genetics, National Institute for Medical Research, Mill Hill, London NW7 1AA, England **[129]**

Christiane Nüsslein-Volhard, Max-Planck-Institut für Entwicklungsbiologie, Spemannstrasse 35/III, D-7400 Tübingen, Germany **[145]**

A.P. Otte, Hubrecht Laboratorium, Uppsalalaan 8, 3584 CT Utrecht, The Netherlands **[109]**

Nancy Papalopulu, Division of Eukaryotic Molecular Genetics, National Institute for Medical Research, Mill Hill, London NW7 1AA, England **[129]**

Suki Parks, Department of Molecular Biology, Princeton University, Princeton, NJ 08544 **[1]**

Carey R. Phillips, Department of Biology, Bowdoin College, Brunswick, ME 04011 **[93]**

James V. Price, Department of Molecular Biology, Princeton University, Princeton, NJ 08544 **[163]**

Elizabeth C. Raff, Department of Biology, Indiana University, Bloomington, IN 47405 **[297]**

Amy Sater, Department of Molecular and Cell Biology, University of California, Berkeley, CA 94720 **[31]**

Gary C. Schoenwolf, Department of Anatomy, University of Utah School of Medicine, Salt Lake City, UT 84132 **[63]**

Trudi Schüpbach, Department of Molecular Biology, Princeton University, Princeton, NJ 08544 **[163]**

Mai Har Sham, Division of Eukaryotic Molecular Genetics, National Institute for Medical Research, Mill Hill, London NW7 1AA, England **[129]**

John Shih, Department of Molecular and Cell Biology, University of California, Berkeley, CA 94720 **[31]**

Sergei Sokol, Department of Biochemistry and Molecular Biology, Harvard University, Cambridge, MA 02138 **[79]**

Edwin C. Stephenson, Department of Biology, University of Rochester, Rochester, NY 14627 **[297]**

Leslie M. Stevens, Max-Planck-Institut für Entwicklungsbiologie, Spemannstrasse 35/III, D-7400 Tübingen, Germany **[145]**

Susan Strome, Department of Biology, Indiana University, Bloomington, IN 47405 **[297]**

Dari Sweeton, Department of Molecular Biology, Princeton University, Princeton, NJ 08544 **[1]**

Ikuo Takeuchi, National Institute for Basic Biology, Okazaki 444, Japan **[249]**

Gerald Thomsen, Department of Biochemistry and Molecular Biology, Harvard University, Cambridge, MA 02138 **[79]**

Tadmiri Venkatesh, Institute of Neuroscience, University of Oregon, Eugene, OR 97403 **[227]**

Judith A. Verbeke, Department of Plant Sciences, College of Agriculture, University of Arizona, Tucson, AZ 85721 **[241]**

Jenny Whiting, Division of Eukaryotic Molecular Genetics, National Institute for Medical Research, Mill Hill, London NW7 1AA, England **[129]**

Malcolm Whitman, Department of Biochemistry and Molecular Biology, Harvard University, Cambridge, MA 02138 **[79]**

Eric Wieschaus, Department of Molecular Biology, Princeton University, Princeton, NJ 08544 **[1]**

Jeffrey G. Williams, The Imperial Cancer Research Fund, Clare Hall Laboratory, South Mimms, Herts EN6 3LD, England **[261]**

Paul Wilson, Department of Biochemistry and Molecular Biology, Harvard University, Cambridge, MA 02140 **[31]**

Tod Woolf, Department of Biochemistry and Molecular Biology, Harvard University, Cambridge, MA 02138 **[79]**

Preface

The 49th Symposium of the Society for Developmental Biology was held on June 27–30, 1990 at the Thomas and Dorothy Leavey Conference Center of Georgetown University, Washington, D.C. Professors David Nishioka and Ellen Henderson of that University served as local organizers of the meeting. On behalf of the Society members and symposium participants, I thank them for their inspired planning and direction. Furthermore, we thank the National Science Foundation, and especially Dr. Judith Plesset, program administrator for developmental biology, for partial support of the travel expenses of the speakers.

Two minisymposia were held prior to the meeting, one on "Meristem Organization and Function," arranged by Ian Sussex (University of California, Berkeley), Carl McDaniel (Rensselaer Polytechnic Institute), and Holly Schauer (Society for Developmental Biology, Washington, D.C.), and another on "Cytoskeletons of Oocytes and Embryos," arranged by Brian Kay and Janice Evans (University of North Carolina). The second topic is rarely reviewed, so I have asked Drs. Kay and Evans to prepare a summary of the minisymposium for inclusion in this volume as a service to interested members and readers.

The topic of the main symposium was "Cell–cell interactions in early development." The 27 speakers discussed such interactions in a wide variety of developing organisms and presented a range of experimental approaches to this venerable and thought-provoking subject. Since cell and developmental biology overlap extensively on the subject of cell interactions, researchers in both fields were included in the program.

It has become apparent that all metazoan and metaphytan organisms make extensive use of cell–cell interactions in the course of developing their complex multicellular organization. The only alternative to interactions is that of mosaic development based on the autonomous action of localized cytoplasmic materials of the oocyte and egg. While this alternative has been shown to be essential and sufficient in establishing a small number of initial regional differences within early embryos of many organisms (*Drosophila* being the best analyzed example), it is soon supplanted by patterning processes based on intercellular signals, for example, by induction and morphogenesis.

One of the earliest and most striking examples of embryonic patterning by cell interactions is provided by the vertebrate organizer. Although first discovered in amphibian embryos in 1924 by Spemann and Mangold, the organizer is far from understood to this day and serves as the subject of several chapters in this volume. The amphibian organizer consists of a group of approximately 1,000 of the 20,000 cells of the early gastrula. If these are removed at the start of gastrulation, the embryo develops merely as a ventralized ball without a nervous system or skeletal muscle, even though these normally arise entirely from the residual cells. The embryo fails to develop a body axis.

Organizer cells organize the anteroposterior and dorsoventral dimensions of the vertebrate body axis by engaging in three kinds of interactions among themselves and with their neighbors: 1) they undertake specialized morphogenesis based on interdigitation with one another, with the consequence that their population changes from a multilayered square array into a single- or double-layered elongated array, thereby lengthening the body axis; 2) they signal lateral neighboring cells to rearrange and differentiate as somites, heart, and kidney (an inductive event called "dorsalization of the mesoderm"); and 3) they signal the neighbors near the animal pole to repack and differentiate as neural tissue ("neural induction"). Eventually, organizer cells differentiate as the notochord, surrounded by their induction products, the neural tube and the segmented rows of somites.

In the years since the organizer's discovery, information on cell interactions has, of course, increased greatly. The present volume summarizes recent progress not only on vertebrate axis formation but also on axis formation in *Drosophila* and *Dictyostelium*, and on the patterning of local differentiations of cell types in many kinds of embryo. In the course of these interactions, cells of the embryo must not only provide signals in a coherent way related to their position, but must also respond appropriately to signals from other cells. Recent molecular genetic analyses, especially of *Drosophila* development, have allowed a glimpse into the inner workings of signaling and responding. Genes expressed at specific places and times in early development often have large, complex promoter-enhancer regions, and the activity of any one transcription factor binding to one of these regions depends on the presence or absence of other factors and on posttranslational modifications subject to various conditions. At the same time, work on receptors and second messengers, especially in vertebrate cells, has led to an appreciation of specific intercellular signals and their effects. Inductive signals now seem likely to act as selective, rather than as instructive, agents, and the major share of specificity in the outcome of an interaction is probably defined by the responding partner. Embryonic cells seem to have the double burden of generating the variety of responses as well as the selective conditions, both needed for development progress. It remains to be learned how many signals are used during a major patterning event such as dorsalization or neural induction during amphibian gastrulation, and how many responses are simultaneously available to a cell. Only recently has the first metazoan example been identified of a receptor and its signaling ligand used in an interaction of cells (the steel:c-kit case in mice). While hypotheses over the years have favored the use of one or few signals to carry positional information in a graded distribution, recent molecular genetic results raise the possibility of many families of receptors, each with many members, and, by implication, of an equally large number of signals.

Finally, signals and responses must be related by a set of rules, an intercellular circuitry or "language". It is significant that such rules have recently been deduced for the first time from studies of the patterning of small cell groups in vulval and gonad development in *C. elegans* and in eye development in *Drosophila*, thanks to the incisiveness provided by genetic analysis in these organisms. Rules are also well understood for the cellular slime mold *Dictyostelium* in terms of the mutual cross-inhibitions and cross-activations among the few interconverting cell types of the slug. The complexity of these rules is impressive in these seemingly simple cases.

Although this volume will show that significant progress has been made in recent years in the understanding of cell interactions in development, it will also demonstrate that we are still in the early stages of appreciating the immensity of the problem and can look forward to the exploration of this enduring subject for years to come.

John Gerhart

Young Investigators Awards, 1990

First Place

Donald A. Kane
Institute of Neuroscience
University of Oregon
Eugene, OR 97403

Gastrulation in Zebrafish *spt-1* Mutants: Extension Without Convergence?

Second Place

Susan M.E. Smith
Plant Science Group
Department of Biology
Rensselaer Institute of Technology
Troy, NY 12180

Floral Determination Assayed In Vitro

Carol Garvin
Department of Biology
Indiana University
Bloomington, IN 47405

Analysis of a Grandchildless Mutation in *C. elegans*

Abstract of the First Place Young Investigator Award

Gastrulation in Zebrafish *spt-1* Mutants: Extension Without Convergence?

Donald A. Kane, Robert K. Ho, Rachel M. Warga, and Charles B. Kimmel
Institute of Neuroscience, University of Oregon, Eugene, OR 97403

The wild-type function of *spt-1* is necessary for the correct dorsalwards convergence of lateral mesoderm during embryogenesis. The lack of *spt-1* function results in the cell-autonomous reduction of trunk myotomes. In *spt-1* mutants, a transient swelling appears above the dorsal margin an hour after the start of gastrulation. This swelling may be the result of the failure of two movements at the beginning of gastrulation. First, using transplants of mutant cells into wild-type embryos, we see failure of convergence of lateral mesoderm just after cells start to involute. Second, using time lapse microscopy, we see a slight early slowing of the migration of prospective prechordal plate toward the animal pole of the egg. The early deficiency of both these groups of cells delineates, respectively, the posterior and anterior edges of the swelling.

Remarkably, in spite of the early misplacement of the prechordal plate, axial mesoderm still extends along the dorsal midline in the mutant. The head forms normally but must then recruit cells from a different area of the fate map than in the wild type. Consistent with this idea, we found a changed fate map position for the rostral end of the mutant. The animal pole of the blastoderm, which is normally fated to become nose and eyes in wild type, mapped ventral to the mouth in the mutant. But this 12° ventralwards change is only minor. We have proposed that the *spt-1* mutation stops convergence. If extension-convergence is one movement, how then does extension continue without convergence?

Cell-Cell Interactions in Early Development, pages 1–14

1. Genes Controlling Cell Shape Changes During Gastrulation in *Drosophila melanogaster*

Eric Wieschaus, Suki Parks, Michael Costa, and Dari Sweeton
Department of Molecular Biology, Princeton University, Princeton, New Jersey 08544

INTRODUCTION

Cleavage stage embryos from a variety of organisms have long provided attractive experimental systems for cell biologists interested in various functional aspects of cellular structure and function (Wilson, 1925). The availability of synchronously developing gastrulae and the reproducibility of morphogenetic movements may offer a similar potential for understanding the basis of cell shape changes and morphogenetic movements. In most organisms, the cells are large at this stage and the morphogenetic movements are extremely rapid and reproducible from one embryo to the next. In spite of their large scale, however, the cell shape changes of gastrulation probably involve the same cytoskeletal and cellular components utilized by cells at later points in the life cycle. In addition to its general cell biological interest, gastrulation is of great importance to developmental biologists because the localized cell shape changes which occur at that time provide one of the earliest indications that the cells of the blastula have become programmed to different fates. The mechanisms that underlie this programming have been the subject of intense investigation in the past 10 years in a variety of different organisms. While many of the steps are now fairly well understood, it is not clear in any of these organisms exactly how this spatial programming is linked to the morphogenetic movements and cellular transformations that follow from it.

This chapter focuses on the role of apical constriction in two of the major morphogenetic events in *Drosophila* gastrulation, ventral furrow formation and posterior midgut invagination. These invaginations bring mesodermal and endodermal precursors into the embryo's interior and are thus of fundamental importance for the development of the organism. Although they occur in different regions of the embryo and are governed by different systems of maternal positional determinants, the sequence of morphological changes that occur during the formation of both invaginations is very similar. This morphological similarity suggests that the underlying cellular events may be controlled by identical molecules and genetic mechanisms.

DROSOPHILA GASTRULATION: APICAL CONSTRICTION IN THE FORMATION OF THE VENTRAL FURROW AND POSTERIOR MIDGUT INVAGINATIONS

Drosophila embryos develop very rapidly (Poulson, 1950; Campos-Ortega and Hartenstein, 1985). Fertilization is followed by rapid nuclear replications with no intervening cytokinesis. Initially, these nuclei are located in the central yolky cytoplasm of the egg cell, but after eight rounds of division, they migrate to the surface (Fig. 1). When the number of nuclei has reached 6,000, the replications pause and the embryo is subdivided into cells. Membrane invaginates from the surface between adjacent nuclei and cellularization occurs simultaneously over the entire surface of the embryo. After about 50 min, the invaginating membrane has reached the central yolk mass and cells are pinched off. The embryo then begins to gastrulate immediately.

Ventral furrow formation begins when cells in a domain approximately 20 cells wide and 80 cells long on the ventral side of the embryo flatten on their apical side. Subsequently, cells in a 12-cell-wide central stripe within this flattened domain begin to constrict their apices (Leptin and Grünewald, 1990; Sweeton et al., 1991), ultimately causing a shift in their shape from columnar to trapezoidal. During the constriction, the apical surface of these cells becomes ruffled and thrown into membranous blebs, as if being pulled together by contractions in the apical cortex below the plasma membrane (Turner and Mahowald, 1977; Fig. 2). The blebbing morphology makes it unlikely that the cell shape change is produced by internalization of apical membrane, or on a zipping up of lateral adhesions (Gustafson and Wolpert, 1962). The initial constricting cells appear overall to be randomly placed within the primordium, although cells closest to the ventral midline have a slightly greater tendency to constrict early. As more cells constrict, a shallow furrow forms along the ventral side of the embryo (Fig. 2b). After further constriction and cell shortening, the shallow furrow deepens and the whole region buckles into the interior of the embryo.

Similar apical constrictions appear at the posterior pole of the embryo and are responsible for the formation of the posterior midgut. The posterior midgut primordia can first be identified as a round plate of cells at the posterior pole shortly after the onset of gastrulation. As the apical surfaces of these cells constrict, blebs are observed that are very reminiscent of those seen in

---

Fig. 1. Schematic diagram of early development in *Drosophila*. The drawings show the development of *Drosophila* from the late syncytial stages through cellularization and early, middle, and late stages of gastrulation. In the early gastrula, the ventral furrow (VF) and posterior midgut (PMG) primordium are indicated. In the midgastrula, cephalic furrow (CF) and transverse dorsal folds (TDF) are indicated. In the lower left-hand corner, the approximate timing of each stage is given, measured in hours and minutes from fertilization.

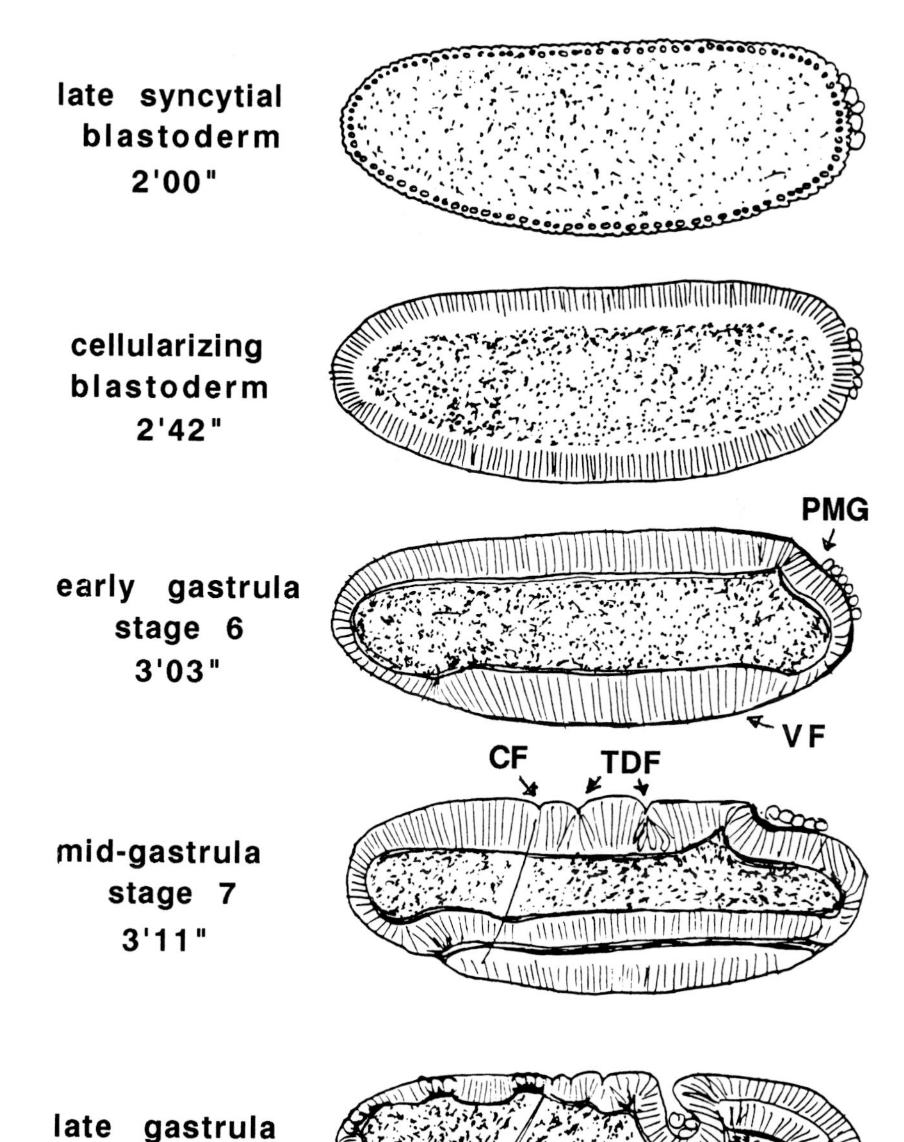
late syncytial
blastoderm
2'00"
cellularizing
blastoderm
2'42"
PMG
early gastrula
stage 6
3'03"
VF
CF
TDF
mid-gastrula
stage 7
3'11"
late gastrula
stage 8
3'26"

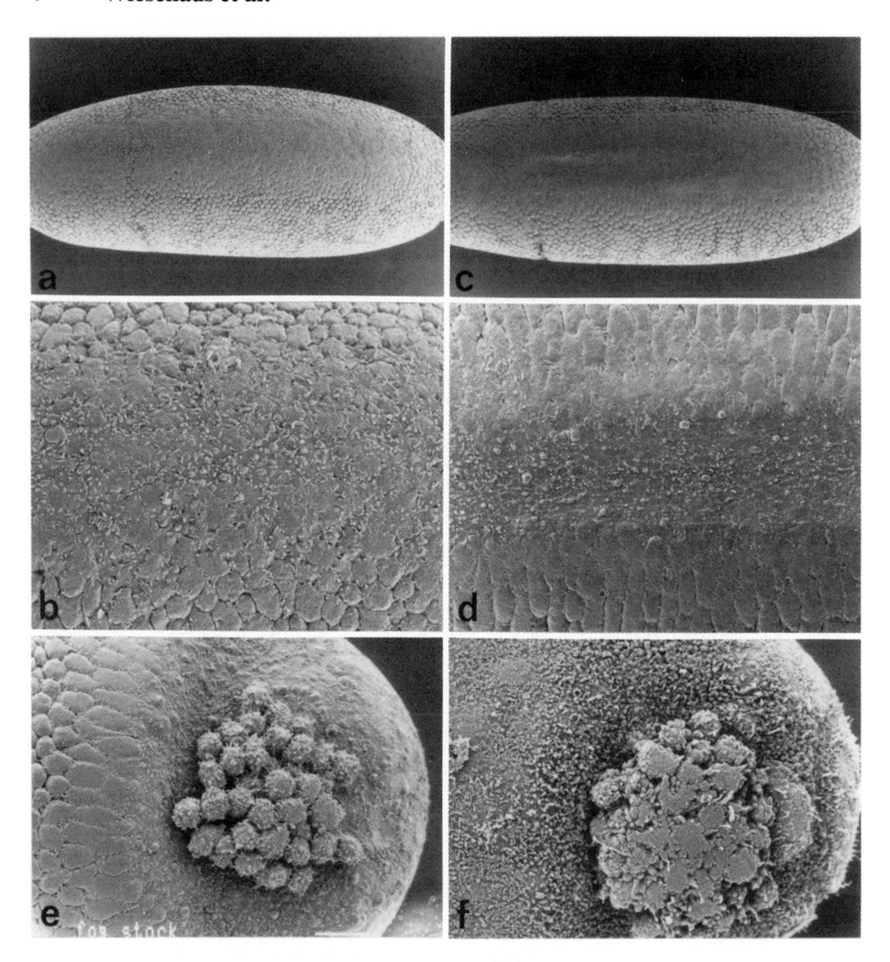

Fig. 2. Scanning electron micrographs of VF and PMG formation. In the initial stage of VF formation (**a**), the cells which will give rise to mesoderm are recognized as a flattened zone on the ventral side of the embryo. Within this zone (**b**), random isolated cells constrict their apices, initiating changes in cell shape from columnar to trapezoidal. Associated with regions of constriction is a ruffling or blebbing of the surface membrane. As more cells constrict, a shallow furrow forms on the ventral side of the embryo (**c**). The apical surface of cells in this furrow are covered with blebs (**d**). Similar membranous blebbing is observed on the surface of the PMG invagination (**e**). Embryos that lack the entire left arm of the second chromosome do not subdivide the syncytial blastoderm into cells. In spite of their syncytial nature, the posterior ends of such 2L-embryos do form blebs characteristic of apical constriction (**f**).

the ventral furrow (Fig. 2c). Ultimately, these constrictions cause the formation of a shallow concave primordium, which shifts dorsally to form a cup around the pole cells (Fig. 1). The posterior midgut closes over once the germband elongates around the posterior tip of the embryo and moves anteriorly along the dorsal side.

Formation of the ventral furrow (VF) and posterior midgut (PMG) occurs concurrently in the *Drosophila* gastrula with other morphogenetic movements. A fold (the cephalic furrow) forms in the lateral ectoderm about one-third of the distance from the anterior end of the embryo (Fig. 1). In addition, germband expands on the dorsal side of the embryo and the dorsal and lateral sides are drawn into deep transverse folds at defined positions along the anterior–posterior axis. In contrast to the VF and PMG invaginations, these furrows and folds are transient. While they may have structural significance, they do not result in a permanent displacement of cells into the interior. These folds also appear to arise by a different cellular mechanism: They are not associated with the formation of a flattened plate of cells, nor with apical constriction or the formation of membranous blebs.

APICAL CONSTRICTION IN THE VF AND PMG PRIMORDIA CAN OCCUR IN THE ABSENCE OF CELLS

In normal embryos, the initial cytoskeletal rearrangements of gastrulation appear to be tightly coupled to the completion of cellularization. VF cells, for example, never begin to flatten or constrict before cell membranes have reached the yolk sac, yet those cell shape changes normally begin within a minute of the completion of cellularization (Sweeton et al., 1991). It is possible that gastrulation is triggered by the cell membranes reaching the yolk sac. Alternatively, the cytoskeletal changes associated with gastrulation may not be possible while the cellularization process is going on, and thus the completion of cellularization may only provide the permissive circumstances for gastrulation.

Cellularization per se is not necessary for the initiation of gastrulation. We have previously identified two zygotically active genes required for cellularization on the left arm of the second chromosome (Merrill et al., 1988). Homozygous $2L^-$ embryos, which lack that entire chromosomal arm, develop normally to cycle 14. During the 50 min that follow, cell membranes do not invaginate between nuclei and the embryos remain a nuclear syncytium. When normal embryos would begin to form a PMG invagination at the posterior pole, the plasma membrane overlying posterior ends of $2L^-$ embryos begins to ruffle and is thrown into blebs (Fig. 2d), indicating that the underlying cortex has begun to constrict. This constriction is restricted to an area in the syncytial embryo that would normally give rise to the PMG. Moreover, the syncytial area does in fact make a shallow invagination which mimics the morphology of the normal PMG. Similar observations have been made on a variety of other acellular embryos (Rice and Garen, 1975; Swanson and Poodry, 1981). The limited spatial domain of constriction in $2L^-$ embryos indicates that subdividing the embryo into cells is not necessary to establish embryonic domains programmed to different fates. Cellularization is also evidently not required for the mechanical aspects of apical constriction, providing further evidence that the initial morphological changes occur through a restructuring of cytoskeleton rather than changes in cell surface properties such as adhesion. Lastly, because the apical constriction occurs at an approximately normal time, a repressive effect of cellularization on gastrulation movement cannot be the sole factor timing the onset of gastrulation.

GASTRULATION MOVEMENTS ARE CONROLLED BY A SYSTEM OF MATERNAL AND ZYGOTIC TRANSCRIPTION FACTORS WHICH DEFINE SPATIAL POSITIONS AND CONTROL CELL FATE

The behavior of cells in the VF and PMG primordia reflect their programming to mesodermal and endodermal fates. Much work on pattern formation in *Drosophila* has focused on the genes and mechanisms that underlie this programming. Results from a number of different labs have shown that the pattern of determination at the blastoderm embryo reflects the distribution of certain maternal gene products deposited in the egg during oogenesis (Nüsslein-Volhard, 1979; Anderson and Nüsslein-Volhard, 1984; Schüpbach and Wieschaus, 1986; Nüsslein-Volhard et al., 1987). Dorso-ventral polarity, for example, depends on a gradient of nuclear *dorsal* protein (Steward et al., 1988; Steward, 1989; Roth et al., 1989). At the blastoderm stage, this protein is found in highest concentration in nuclei on the ventral side of embryo, in the mesodermal precursors that will form the VF (Fig. 3). Changes in the distribution or activity of *dorsal* product shift the programming of cells and thus the size of the domain which undergoes apical constriction of gastrulation. Mutations that increase the level of dorsal protein in the lateral cells (e.g., $Toll^{10b}$, Anderson et al., 1985; Leptin and Grünewald, 1990; *torpedo*, Schüpbach, 1987) cause those cells to enter mesodermal fates, extending the apical constrictions further laterally and enlarging the VF region. Mutations that reduce concentrations of dorsal protein in the nuclei of the ventral cells prevent them from entering the mesodermal pathway or forming a VF (Nüsslein-Volhard, 1979; Anderson et al., 1985).

It is unlikely that *dorsal* protein directly interacts with the cytoplasmic components to effect the cytoskeletal and cellular changes responsible for gastrulation. Because it is localized to the nucleus and is likely to act as a transcription factor, *dorsal* probably affects gastrulation by controlling transcription of other

---

Fig. 3. Hypothetical genetic pathway for formation of the VF. Formation of the VF depends on a sequence of maternal and zygotic gene activities. Positional cues provided during oogenesis ultimately result in a gradient of nuclear localization of maternal *dorsal* protein. Cells with high levels of *dorsal* protein will form the VF. High levels of *dorsal* protein are thought to activate zygotic transcription of other genes that direct the cells into the mesodermal pathway. These zygotic gene products may then direct the synthesis of cytoplasmic factors, which interact with maternally supplied cytoskeletal components, bringing about the apical constriction and resultant cell shape change (see text).

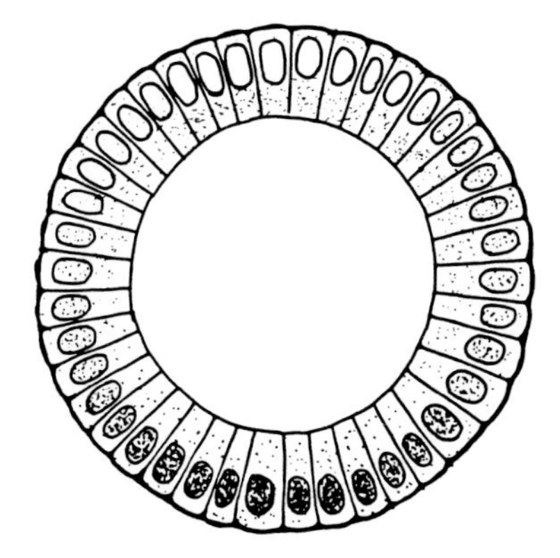

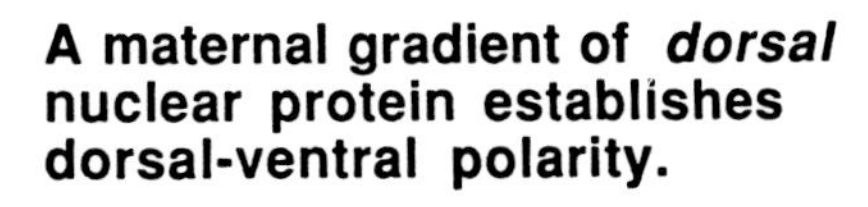

A maternal gradient of *dorsal* nuclear protein establishes dorsal-ventral polarity.

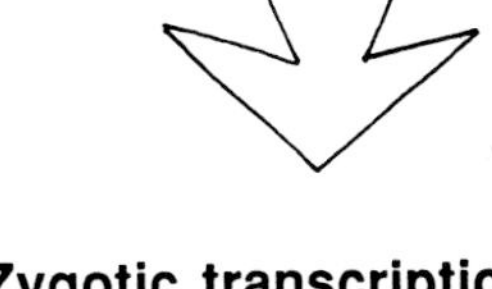

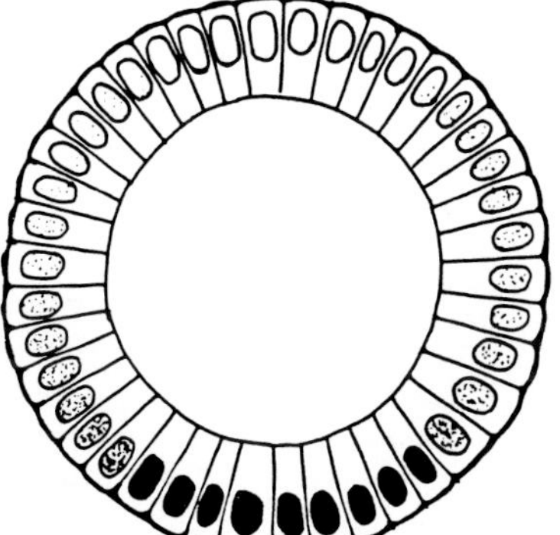

Zygotic transcription of certain genes occurs only in those cells with high concentrations of *dorsal*

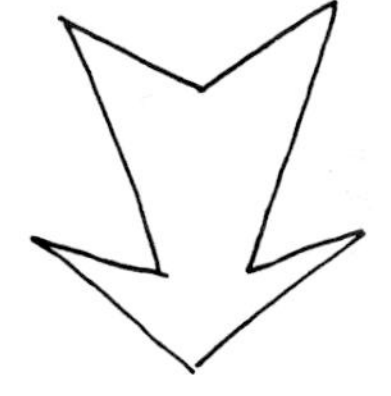

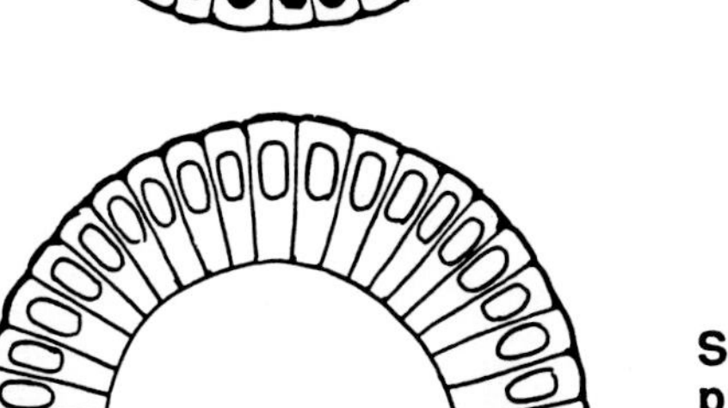

Some of those zygotic gene products elicit the cell shape changes responsible for ventral furrow formation.

genes during early development. The zygotic genes immediately downstream of *dorsal* are probably involved in subdividing the embryo into regions and establishing broadly defined fates (e.g., mesoderm, lateral ectoderm, amnion-serosa). Like mutations in *dorsal* itself, mutations in these downstream genes should affect gastrulation, but possibly again only as a secondary consequence of their alterations in cell fates. Candidates for zygotically active genes that function in the VF are *twist* and *snail* (Simpson, 1983; Thisse et al., 1988; Boulay et al., 1987; Leptin and Grünewald, 1990). Both are normally expressed in a band of cells along the ventral side of the embryo, i.e., roughly in the cells that would form the VF. Mutations in these genes block VF formation (Fig. 4), presumably because the cells that would normally have formed a VF are shifted to more lateral fates. At the blastoderm stage, ventral cells in such mutants express gene products (e.g., *single minded*), found in wild-type embryos only lateral to the VF (Nambu et al., 1990) and probably give rise to larval cuticle, rather than forming mesodermal derivatives (Wieschaus, unpublished).

Both *twist* and *snail* are homologous to vertebrate transcription factors (Boulay et al., 1987; Thisse et al., 1988). It is possible that among their downstream targets are the genes actually responsible for flattening, apical constriction, and cell shape change. Mutations in both genes specifically affect the ventral region

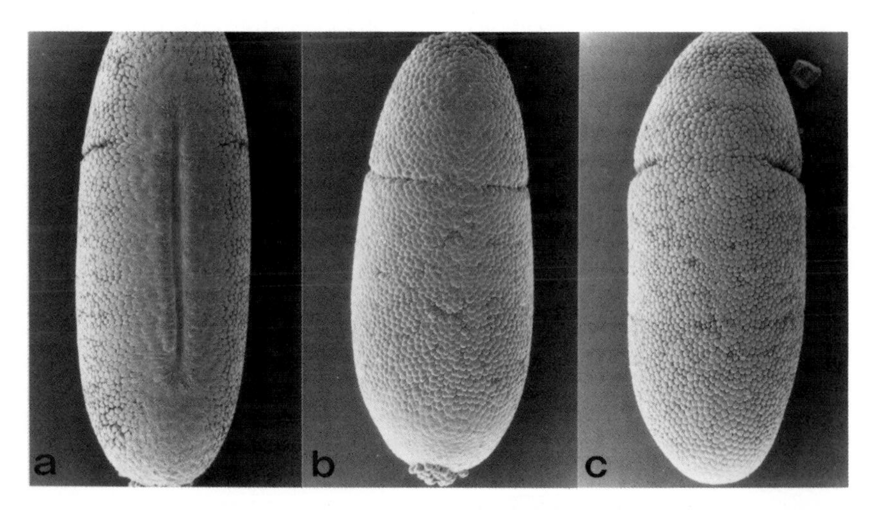

Fig. 4. Embryos homozygous for mutations in *twist* or *snail* fail to make a normal VF. In wild-type embryos, the mesodermal precursors invaginate via a clearly visible furrow on the ventral side of the embryo (**a**). In embryos homozygous for *twist* (**b**) or *snail* (**c**), a normal VF fails to form, presumably because at least some of these cells have been programmed to more lateral fates. In many *twist* embryos, folds and furrows do eventually form on the ventral side of the embryo. These have been interpreted as abortive VF (Leptin and Grünewald, 1990), but because of their delayed appearance and abnormal morphology, we have not regarded them as such.

of the embryo, and although *twist* expression does in fact extend into the PMG primordia, neither gene when mutated affects apical constrictions involved in making the PMG. On the other hand, genes that affect posterior determination have been identified (for example, *torso*, Schüpbach and Wieschaus, 1986; *tailless* and *hückebein*, Weigel et al., 1990). Mutations in these genes do affect PMG formation. Based on the morphological similarities in the VF and PMG, we have argued that the cellular genes actually responsible for the apical constriction may be the same in both processes. If this is the case, then those genes must be activated by two different sets of control genes, one controlling ventral determination, the other controlling posterior determination.

THE UNIDENTIFIED TARGET GENES THAT PARTICIPATE IN THE CELL SHAPE CHANGES AND RESTRUCTURING OF THE CYTOSKELETON

It has proven very difficult to get a genetic handle on the target genes that function downstream of *twist* and *snail* and may actually induce the apical constriction. If their activities are controlled by transcription factors like *twist* and *snail*, they must be supplied by zygotic transcription in the embryo itself. When mutated, they should produce defects in homozygous embryos and as such should be represented in the large collection of zygotic lethal mutations isolated in *Drosophila* in the past 10 years (Nüsslein-Volhard et al., 1984; Jürgens et al., 1984; Wieschaus et al., 1984). *twist* and *snail*, however, are the only zygotic loci in these collections that block VF formation. The reasons behind the failure to detect genes potentially downstream of *twist* and *snail* are unclear. One possibility is that there are multiple downstream genes, each of which contributes quantitatively to the constriction, but none of which is absolutely essential. When only one such gene is mutated in a particular stock, ventral furrowing may still occur normally, or show only minor phenotypic defects that would escape detection in most screens.

Because of the similarities at the cellular level between epithelial invaginations in a wide variety of species, it is possible to extrapolate from what is known from other systems to the cellular components that may function in *Drosophila* (Ettensohn, 1985). In many other species where this kind of invagination has been observed, the apical surface of the cells involved contains a rich matrix of actin and myosin, which is thought to provide the contractile force that constricts the apical surface (Baker and Schroeder, 1967; Burnside 1971; Lee and Naegle, 1985). Actin and myosin and other cytoskeletal proteins are localized to the apex of cells in the PMG and VF primordia of *Drosophila* (Pesacreta et al., 1989; Young et al., 1990). One would therefore anticipate a role for these apically localized cytoskeletal components, as well as actin-binding proteins and regulators of myosin activity. In *Drosophila*, large maternal stores of actin, cytoplasmic myosin, and many other major cellular proteins have been demonstrated in the egg. In the simplest model, their organization and activity would be controlled by the downstream targets of *twist* and *snail* in the VF, or *tailless* and *hückebein* in the PMG invagination. These major cytoskeletal components would thus represent the final downstream target in the hierarchy of cell shape changes.

Although a number of these abundant cytoskeletal genes have been cloned, no mutations at these loci have been isolated that specifically block gastrulation. Because the gene products are supplied maternally, homozygous embryos gastrulate normally, using products made by their heterozygous mother. The mutant individual does eventually die, when those maternal supplies are used up, or become insufficient due to the embryo's growth. On the other hand, it is probably reasonable to assume that these components function as they do in other cell types. Our genetic analyses have therefore focused on the few components that are specific for the process and can be identified genetically, in the expectation that these gene products produce their effects on morphology by regulating or modifying the configuration of the more ubiquitous components.

COORDINATION OF CELL SHAPE CHANGES—AN UNANTICIPATED SIGNAL TRANSDUCTION PATHWAY

After the initial flattening, the subsequent constriction and cell shape changes associated with VF or PMG formation do not occur simultaneously. Instead, individual isolated cells within the primordia begin to constrict. Although the cells closest to the ventral midline are slightly more likely to initiate constriction than the more laterally situated cells, time-lapse videos and scanning electron microscopes do not reveal any large-scale pattern or waves that spread from a single well-defined center. Instead, the initial apical constrictions appear to occur in cells at random positions within the primordium. The initial stage of random constriction continues until about 30–40% of the cells have begun to constrict. It is followed by an extremely rapid transition phase in which the remaining cells in the midventral zone constrict. It is this rapid second phase of constrictions that results in the formation of a flattened groove on the ventral side of the embryo. Mutations in two genes, *concertina* and *folded gastrulation*, appear to block this second, rapid phase of constriction (Zusman and Wieschaus, 1985; Schüpbach and Wieschaus, 1989; Parks and Wieschaus, 1991; Sweeton et al., 1991). Mutant embryos make normal cellular blastoderms and initiate gastrulation normally. The cells on the ventral side form a flattened zone and random cells within that domain begin to constrict apically. This initial phase, however, is not followed by a more rapid phase in which the cells that have not yet constricted constrict (Fig. 5). Constrictions continue to begin in random isolated cells at the slow rate characteristic of the initial phase or they cease entirely. In most cases, an abnormal VF eventually forms, still containing a large fraction of unconstricted cells. Both *concertina*

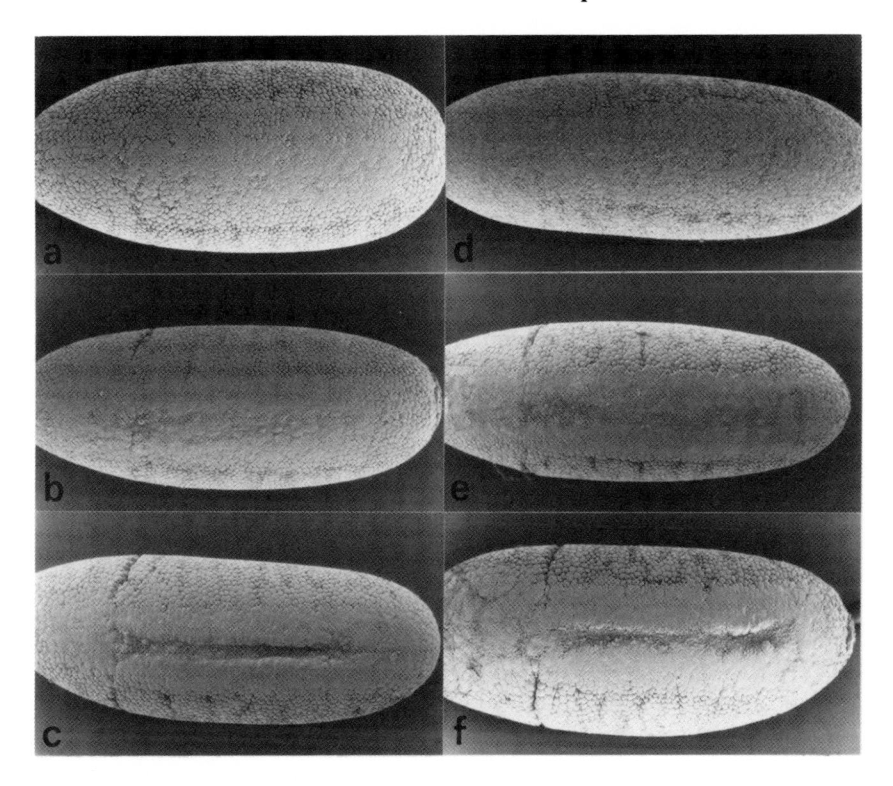

Fig. 5. Mutations in *concertina* and *folded gastrulation* cause the formation of abnormal VF. Ventral views of early gastrula homozygous for *folded gastrulation* (**a,b,c**) or derived from mothers homozygous for *concertina* (**d,e,f**). In both mutants, only a fraction of the cells in the VF primordium undergo apical constriction. Although a furrow is eventually formed, it is morphologically abnormal. Similar defects in apical constriction are observed during the formation of the PMG invagination.

and *folded gastrulation* also affect cell shape changes in the PMG invagination, and in a manner similar to their effect on the VF (Sweeton et al., 1991). Because other cell movements of gastrulation (cephalic furrow, dorsal transverse folds) appear normal in the mutant embryos, the observation of effects in both the VF and PMG provides further evidence that the two invaginations share certain cellular mechanisms.

One difference between *concertina* and *folded gastrulation* is that *concertina* is a maternal effect mutation, and embryos from a homozygous mother are abnormal regardless of their own genotype. Because homozygous embryos derived from heterozygous mothers survive normally to adult stages, it is likely that *concertina* product is specifically required only for gastrulation (Parks and Wieschaus, 1991). In contrast, transcription of *folded gastrulation* is absolutely required in the embryo itself and can be eliminated during oogenesis with no effects on embryonic viability (Zusman and Wieschaus, 1985).

concertina has been cloned (Parks and Wieschaus, 1991) and shown to be a highly homologous to alpha subunits of G-proteins used in signal transduction pathways (Gilman, 1987). This suggests that the transition between the early and the rapid phases of constriction might involve a cell signaling process. In one model, the cells in the VF primordia that have already constricted would signal to those that have not yet constricted, inducing them to constrict as well. This signaling would presumably serve to coordinate the cell shape changes in the VF and would ensure that in spite of the stochastic variability in constriction, all the cells will have begun to constrict by the time the furrow begins to form. Since Gα proteins are normally activated by a receptor, the *concertina* product might represent a ubiquitous factor in the egg that would only be activated in cells of the VF and PMG primordia at gastrulation. The factors involved in its activation are not known. One obvious candidate for a component involved in the process would be *folded gastrulation*. Because the level of *folded gastrulation* in the embryo depends on zygotic transcription, it might be transcribed in a temporally restricted manner that would limit the ability of the cells to respond to the signaling pathway. On the other hand, the spatial domain, which can respond to the signaling pathway and in which apical constriction can occur, would be defined in the VF by *twist* and *snail* and their downstream targets.

ACKNOWLEDGMENTS

We thank numerous members of the *Drosophila* labs at Princeton for advice, encouragement, discussions, and critical readings of the manuscript. These experiments were supported by National Institutes of Health grant 5RO1HD15587 to E.W.

REFERENCES

Anderson KV, Nüsslein-Volhard C (1984): Information for the dorsal-ventral pattern of the *Drosophila* embryo is stored as maternal mRNA. Nature 311:223–227.

Anderson KV, Jürgens, G, Nüsslein-Volhard C (1985): Establishment of dorsal-ventral polarity in the *Drosophila* embryo: Genetic studies on the role of the Toll gene product. Cell 42:779–789.

Baker, PC, Schroeder TE (1967): Cytoplasmic filaments and morphogenetic movement in the amphibian neural tube. Dev Biol 15:432–450.

Boulay JL, Dennefeld C, Alberga A (1987): The *Drosophila* developmental gene snail encodes a protein with nucleic acid binding fingers. Nature 330:395–398.

Burnside, B (1971): Microtubules and microfilaments in early amphibian neurulation. Dev Biol 26:416–441.

Campos-Ortega JA, Hartenstein V (1985): "The Embryonic Development of *Drosophila melanogaster*." Berlin, Heidelberg: Springer Verlag.

Ettensohn, CA (1985): Mechanisms of epithelial invaginations. Q Rev Biol 60:289–307.

Gilman AG (1987): G proteins: Transducers of receptor-generated signals. Ann Rev Biochem 56:615–649.

Gustafson T, Wolpert L (1962): Cellular mechanisms in the morphogenesis of the sea urchin larva. Changes in shape of cell sheets. Exp Cell Res 27:260–279.

Jürgens G, Wieschaus E, Nüsslein-Volhard C, Kluding H (1984): Mutations affecting the pattern of the larval cuticle in *Drosophila melanogaster*. II. Zygotic loci on the third chromosome. Roux Arch Dev Biol 193:283–295.

Lee H-Y, Naegle RG (1985): Studies on the mechanism of neurulation in the chick: Interrelationships of contractile proteins, microfilaments and the shape of the neuroepithelial cells. J Exp Zool 235:205–215.

Leptin M, Grünewald B (1990): Cell shape changes during gastrulation in *Drosophila*. Development 110:73–84.

Merrill PT, Sweeton D, Wieschaus E (1988): Requirements for autosomal gene activity during precellular stages of *Drosophila melanogaster*. Development 104:495–509.

Nambu JR, Franks RG, Hu S, Crews ST (1990): *The single-minded* gene of *Drosophila* is required for the expression of genes important for the development of CNS midline cells. Cell 63:63–75.

Nüsslein-Volhard C (1979): Maternal effect mutations that alter the spatial coordinates of the embryo of *Drosophila melanogaster*. Symp Soc Dev Biol 37:185–211.

Nüsslein-Volhard C, Frohnhofer HG, Lehmann R (1987): Determination of anteroposterior polarity in *Drosophila*. Science 238:1675–1681.

Nüsslein-Volhard C, Wieschaus E, Kluding H (1984): Mutations affecting the pattern of the larval cuticle in *Drosophila melanogaster*. I. Zygotic loci on the second chromosome. Roux Arch Dev Biol 193:267–282.

Parks S, Wieschaus E (1991): The *Drosophila* gastrulation gene concertina encodes a G alpha subunit. Cell (in press).

Pesacreta TC, Byers TJ, Debreuil R, Kiehart DP, Branton D (1989): *Drosophila* spectrin: The membrane skeleton during embryogenesis. J Cell Biol 108:1697–1709.

Poulson, DF (1950): Histogenesis, organogenesis, and differentiation in the embryo of *Drosophila melanogaster* Meigen. In Demerec M (ed): "Biology of *Drosophila*." New York: Wiley, pp 168–274.

Rice TB, Garen A (1975): Localized defects in blastoderm formation in maternal effect mutants of *Drosophila melanogaster*. Dev Biol 32:304–310.

Roth S, Stein D, Nüsslein-Volhard C (1989): A gradient of nuclear localization of the *dorsal* protein determines dorsoventral pattern in the *Drosophila* embryo. Cell 59:1189–1202.

Schüpbach T (1987): Germ line and soma cooperate during oogenesis to establish the dorsoventral pattern of egg shell and embryo in *Drosophila melanogaster*. Cell 49:699–707.

Schüpbach GM, Wieschaus E (1986): Maternal-effect mutations altering the anterior-posterior pattern of the *Drosophila* embryo. Roux Arch Dev Biol 195:302–317.

Schüpbach GM, Wieschaus E (1989): Female sterile mutations on the second chromosome of *Drosophila melanogaster*. I. Maternal effect mutations. Genetics 121:101–117.

Simpson P (1983): Maternal-zygotic gene interactions during formation of the dorsoventral pattern in *Drosophila* embryos. Genetics 105:615–632.

Steward R (1989): Relocalization of the *dorsal* protein from the cytoplasm to the nucleus correlates with its function. Cell 59:1179–1188.

Steward R, Zusman SB, Huang LH, Schedl P (1988): The *dorsal* protein is distributed in a gradient in early *Drosophila* embryos. Cell 55:487–495.

Swanson MM, Poodry CA (1981): The shibire^{ts} mutant of *Drosophila:* A probe for the study of embryonic development. Dev Biol 84:465–470.

Sweeton D, Parks S, Costa M, Wieschaus E (1991): Gastrulation in *Drosophila*: The formation of the ventral furrow and posterior midgut invaginations. Development (in press).

Thisse B, Stoetzel C, Gorostiza-Thisse C, Perrin-Schmitt F (1988): Sequence of the *twist* gene and nuclear localization of its protein in endomesodermal cells of early *Drosophila* embryos. EMBO J 7:2175–2183.

Turner FR, Mahowald AP (1977): Scanning electron microscopy of *Drosophila melanogaster* embryogenesis. II. Gastrulation and segmentation. Dev Biol 57:403–416.

Weigel DG, Jürgens G, Klingler M, Jaeckle H (1990): Two gap genes mediate maternal terminal pattern information in *Drosophila*. Science 248:495–498.

Wieschaus E, Nusslein-Volhard C, Jürgens G (1984): Mutations affecting the pattern of the larval cuticle in *Drosophila melanogaster*. III. Zygotic loci on the X chromosome and fourth chromosome. Roux Arch Dev Biol 193:296–307.

Wilson EB (1925): ''The Cell in Development and Heredity.'' 3rd ed. New York: Macmillan.

Young P, Pesacreta TC, Kiehart DP (1990): Dynamic changes in the distribution of cytoplasmic myosin during *Drosophila* embryogenesis. Development (in press).

Zusman S, Wieschaus E (1985): Requirements for zygotic gene activity during gastrulation in *Drosophila melanogaster*. Dev Biol 111:359–371.

Cell-Cell Interactions in Early Development, pages 15–29

2. Archenteron Morphogenesis in the Sea Urchin

David R. McClay, John Morrill, and Jeff Hardin

Department of Zoology, Duke University, Durham, North Carolina 27707 (D.R.M., J.H.); Department of Biology, New College, Sarasota, Florida 33580 (J.M.)

INTRODUCTION

The progression of development involves an impressive array of morphogenetic rearrangements, each of which involves the coordination of multiple cellular functions and molecular events. We have been studying gastrulation in the sea urchin embryo as a relatively simple model system in an attempt to understand the sorts of rules observed by an embryo as it performs a single morphogenetic event. To the observer, gastrulation in this embryo involves two major cell movements. First, primary mesenchyme cells ingress and display a series of migratory behaviors leading to the assembly of the larval skeleton. Second, invagination of the archenteron leads to the formation of the primitive gut. The ingression and subsequent behavior of primary mesenchyme cells is examined elsewhere in this volume (Ettensohn, Chapter 11). Here we review events associated with formation of the archenteron.

In echinoderm embryos, archenteron formation begins with an indentation of the vegetal plate, followed by elongation of the indented area until a tubular archenteron forms that extends into the blastocoel. Elongation continues until the tube reaches a defined, anatomically specific region on the wall of the blastocoel (Fig. 1). All through invagination, secondary mesenchyme cells at the tip of the archenteron extend filopodia that make contact with the wall of the blastocoel. This behavior of secondary mesenchyme cells continues until final contact is made with the anatomical target for archenteron extension. Later, the stomodeum forms, usually just ventral to the site of attachment of the archenteron.

More than twenty years ago, Gustafson and Wolpert (1963,1967) published extensive reviews based on their time-lapse observations. These provide a descriptive background against which experiments can be performed. They also proposed a number of models suggesting how invagination might work. They proposed, for example, that the initial inward bending of the vegetal plate could be brought about by a cell shape change. The cells at the vegetal plate can be observed to change shape from cuboidal to columnar, and then

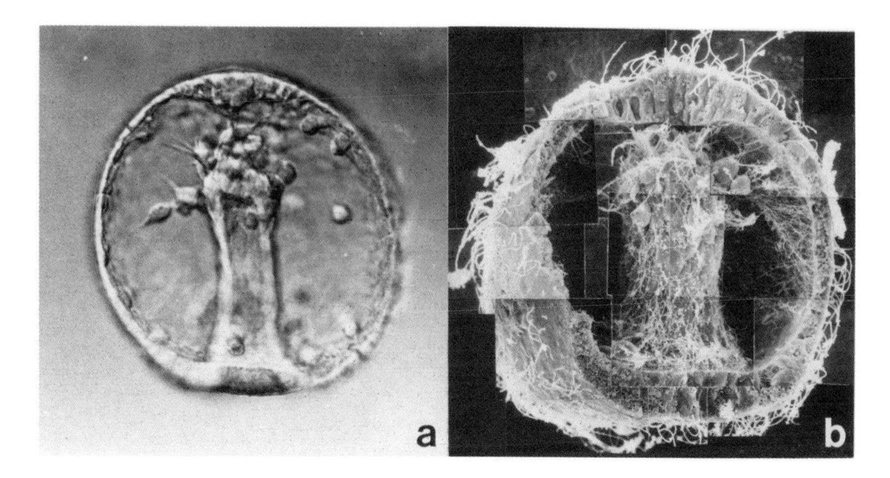

Fig. 1. Archenteron invagination in the sea urchin embryo. Cells of the endoderm rearrange as the archenteron extends from the vegetal plate toward the animal pole region. (**a**) In a micrograph using Nomarski optics, numerous filopodia can be seen to be extending from secondary mesenchyme cells at the tip of the archenteron before it reaches its target near the animal pole. (**b**) A scanning electron micrograph of an embryo at a similar stage to that shown in a. Numerous extracellular matrix fibers can be seen around the archenteron and lining the blastocoel wall.

some cells become keystone-shaped in profile (Fig. 2). Coincident with these shape changes, the vegetal plate begins to bend inward. Mechanistically, the shape change could be driven intrinsically by the cells of the vegetal plate, or there could be external forces imposing the cell shape changes. One intrinsic mechanism that has been proposed is apical constriction of cells (Rhumbler, 1902). Gustafson and Wolpert (1963,1967) pointed out that changes in adhesion of cells in the vegetal plate could be the mechanism leading to the apical constriction; if cells lost adhesive affinity for the hyaline layer and gained affinity for the lateral surfaces of neighboring cells, the keystone shape might result. Alternatively, apical constriction was hypothesized to result from contraction of circumapical microfilament bundles during amphibian neurulation by Burnside (1971) and Baker and Schroeder (1967; see reviews by Ettensohn, 1985b; Fristrom, 1988 for further discussion). Later, stretch-activated constriction of apical microfilaments was presented as a model for the initial invagination of the archenteron in the sea urchin by Odell et al. (1981). The major requirement for these models is that the cell shape changes should be driven locally. Moore and Burt (1939) and later Ettensohn (1984) showed that isolated vegetal plates would invaginate on their own. This meant that global forces from the remaining embryo, or from sort of negative pressure within the blastocoel, were not crucial for invagination. In addition, the vital staining experiments of Hörstadius (1935) showed that there is no large global epibolous movement of cells toward the vegetal plate as it bends inward. When the veg_2 layer of cells was stained, it was found that only the veg_2 layer contributed to the vegetal plate, and cells lateral to the inbending region do not move substantially toward it. These marking experiments were later refined by Ettensohn (1984) to show that while there is some movement of marked regions of the vegetal plate during the early phase of invagination, only the vegetal plate contributes to the archenteron. Furthermore, dye marking and time-lapse videomicroscopy indicate that there is no involution during the subsequent elongation of the archenteron (Hardin, 1989). Thus, the forces of invagination are located near the vegetal plate and only cells of the vegetal plate contribute to the archenteron. Does this indicate that the forces for invagination are generated within each cell? Perhaps, although there has been no formal test of that hypothesis. Furthermore, it is not known what those forces might be. Are they cytoskeletal changes? Or, as suggested by Gustafson and Wolpert, are there changes in cell adhesion? Beyond these unanswered questions about the mechanism of invagination itself, there is no understanding of the controlling influences that help to initiate the invagination. Invagination begins after the primary mesenchyme cells (PMCs) have ingressed in most species, though removal of the PMCs from the blastocoel has no effect on the beginning of the invagination movements (Ettensohn and McClay, 1988). Thus, while the descriptive information on the beginning phases of invagination is fairly complete, there are many local forces that are incompletely understood. Below, we discuss experiments that indicate a role for the synthesis of new embryonic proteins during gastrulation. This de novo synthesis shows that there is developmentally regulated, differential gene expression concurrent with gastrulation; however, the molecular regulatory apparatus that actually controls initiation of the inbending at the vegetal pole is completely unknown.

THE SIGNAL TO INVAGINATE

At the molecular level, recent studies have suggested that there are spatial and temporal regulatory molecules that activate genes at gastrulation (reviewed by Davidson, 1988, 1989, 1990). Presumably, DNA binding proteins and inductive signals are coordinated to activate the genes required for initiation of gastrulation. What are the signals and what genes are activated to initiate the process of invagination? One idea is that there is some kind of internal clock that directs presumptive endoderm cells at a certain time to begin their cell shape change (Spiegel and Spiegel, 1986). Such a clock might rely entirely upon *cis* regulatory elements that control entry into gastrulation as a result of certain rate constants intrinsic to the cells. In indirect support of such a model are the above experiments, in which only the presumptive vegetal plate region need be present for invagination to begin, implying localized signals. Other experiments, however, indicate that the timing model is probably incorrect

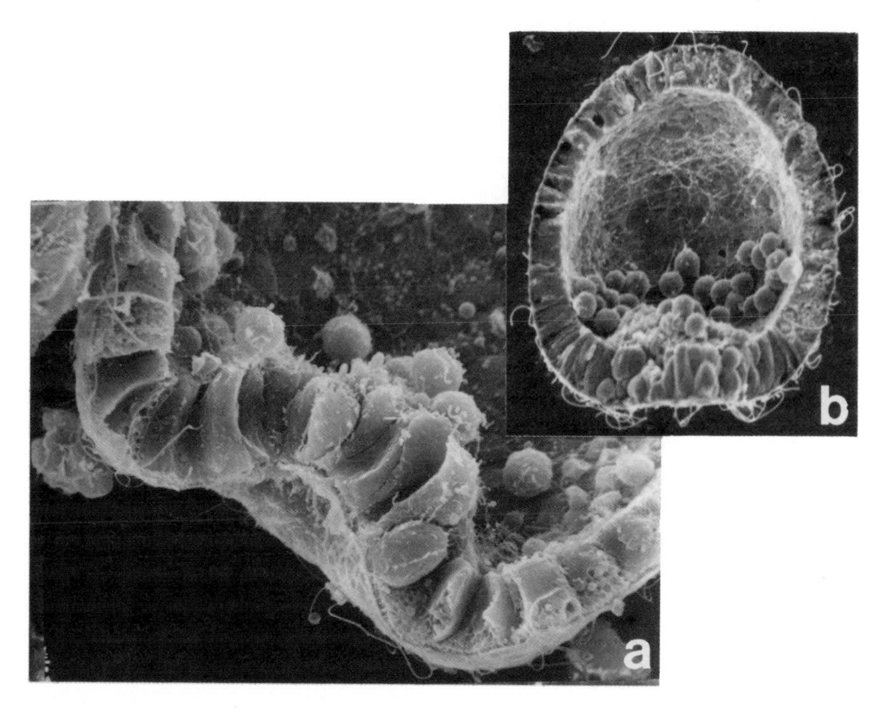

Fig. 2. (**a**) A scanning electron micrograph showing an embryo that has been split along a midsagittal plane. At the vegetal plate the early stages of invagination show a number of presumptive endoderm cells that have changed shape either causing or in response to the forces that initiate the invagination. (**b**) An embryo that has been cut lateral to the midsagittal plane to show the bulge in the vegetal plate and the primary mesenchyme cells surrounding the indentation and lining the blastocoel wall.

and that epigenetic signals are necessary to launch the invagination program. If one disrupts the interaction of cells with the hyaline layer (Adelson and Humphreys, 1988), or with the basal lamina (Wessel and McClay, 1987; Butler et al., 1987), gastrulation is totally blocked. In both cases, the blockage is not permanent since the effects are reversible. Once the block is removed, gastrulation begins and normal larvae result. These data indicate that timing cannot be an explanation since the embryos can be held at the mesenchyme blastula stage for extended periods of time and then allowed to continue with normal development. Instead, the data suggest that the embryos are responsive to epigenetic signals for continuation of development.

What is the block to gastrulation and what does it tell us about the putative epigenetic signals that appear to be necessary for gastrulation? Presumptive endoderm cells adhere to at least two hyaline layer proteins, hyalin (McClay and Fink, 1982; Fink and McClay, 1985) and echinonectin (Alliegro et al.,

1988; Burdsal et al., 1991). A monoclonal antibody to hyalin was produced that blocked the interaction of cells to hyalin substrates (Adelson and Humphreys, 1988). When the antibody was added to embryos at low concentrations (5–10 μg/ml), invagination movements failed to occur. Blastomeres pulled away from the hyalin layer and the embryos appeared to shrink. By a number of criteria, the cells continued to metabolize and many genes continued to be expressed at the pregastrula levels (Adelson and Humphreys, 1988). The inhibitory effect could be reversed if the antibody were washed out of the culture medium.

While inhibition of gastrulation by blocking cell adhesion to hyalin was dramatic, a second kind of inhibition suggests the epigenetic signal is not specific to the cell–hyalin interaction. Treatment of embryos with β-aminopropriononitrile (βAPN) also inhibits gastrulation (Wessel and McClay, 1987; Butler et al., 1987). This lathrytic agent inhibits lysyl oxidase, the enzyme required for cross-linking of collagen. It was found that embryos would grow to the mesenchyme blastula stage in the presence of the βAPN, but no further. If left in the drug, the embryos remained at the mesenchyme blastula stage (Fig. 3), and continued to express all genes measured, including collagen, at control mesenchyme blastula levels. If the embryos were removed from the drug, they resumed development to normal pluteus larvae. New transcription of endodermal genes failed to occur until the drug was removed. Of importance, embryos could be held in the arrested state for long periods of time (Fig. 3), yet when released they progressed through development normally from that point onward. The βAPN treatment was shown to block the cross-linking of collagen, and other basal laminar proteins also failed to be assembled into the basal lamina during the treatment. These data indicate that an intact basal lamina is somehow important for initiation of gastrulation. From other studies it was known that cells of the mesenchyme blastula adhere to the basal lamina (Fink and McClay, 1985; Katow and Solursh, 1981; Solursh, 1986), but it is not known whether the lack of adhesion directly, or some other signal indirectly dependent upon adhesion, is the critical element missing from the blocked embryos. An interaction with the basal lamina might be instructive, or, more likely, simply permissive for further development. Whatever the reason for the inhibition of gastrulation by βAPN, or for that matter with other inhibitors such as xylosides (Solursh et al., 1986; Lane and Solursh, 1988), the reversibility of the treatment indicates that a critical epigenetic event occurs at the beginning of gastrulation. Proper assembly of the basal lamina or the hyalin layer can be blocked at any stage of development up to gastrulation without any noticeable effect. Data have indicated that the sensitivity to these reagents exists for only about 2 hr. Addition of the reagents before the mesenchyme blastula stage has no effect until mesenchyme blastula stage, or, if added after the brief period of high sensitivity, the reagents have a reduced effect on further development (Fig. 3).

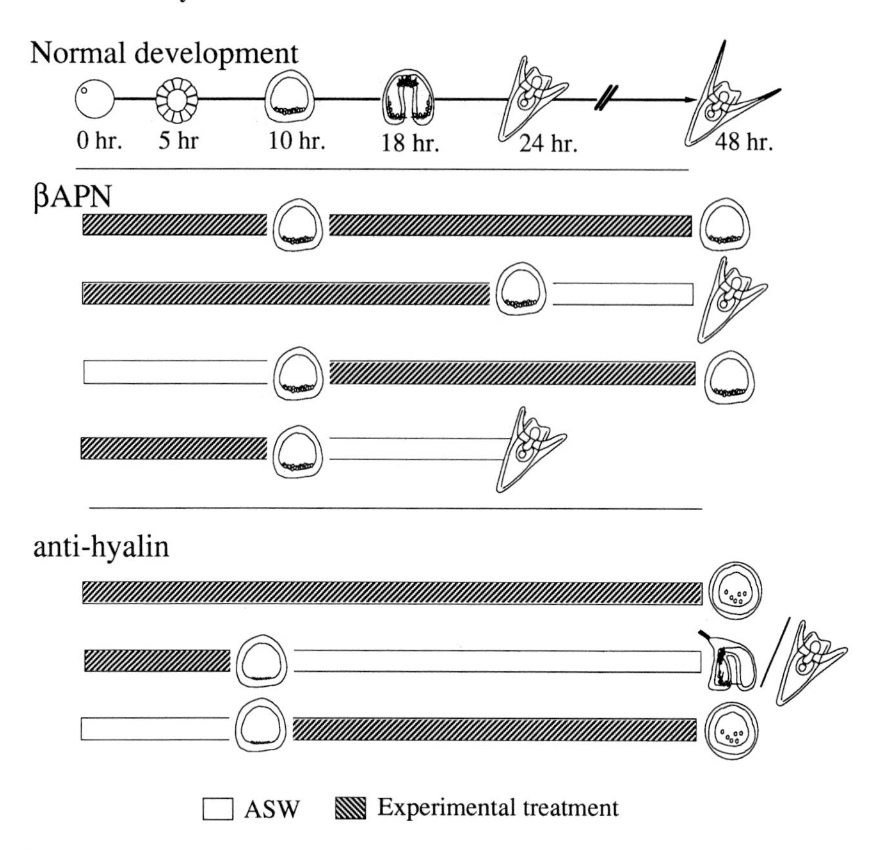

Fig. 3. Diagram summarizing treatments with β-aminoproprionitrile (βAPN) and antibody to hyalin. At the top, a time line shows several stages in development. The bars indicate embryos that were incubated in the presence (hash marks) or absence (open bar) of βAPN or antihyalin. When left in βAPN, the embryos reached the mesenchyme blastula stage on schedule but were arrested. The arrested behavior could be reversed (second bar) if the embryos were later washed in seawater without βAPN. The critical period of sensitivity was at the mesenchyme blastula stage since embryos were arrested if βAPN was added at the mesenchyme blastula stage (third bar), or they developed normally and on schedule if released from βAPN at the mesenchyme blastula stage. Similar results were seen with antihyalin.

A number of marker proteins have been identified that are expressed in the endoderm at the beginning of gastrulation (McClay et al., 1983; Wessel and McClay, 1985; Wessel et al., 1989a; Nocente-McGrath et al., 1989). All appear to be dependent for their expression on the βAPN-sensitive event. Among the genes expressed, presumably, are the genes required for the initiation of the invagination of the vegetal plate. Although the critical genes are not known, several properties of the invaginating tissue must be accounted for in the sequence of events that follow. First, the cells have apical–basal polarity (Schroeder,

1988; Nelson and McClay, 1988), and secrete matrix components toward both surfaces. The signal that launches the inward folding, if present inside each of the cells that change shape, must be responsive to that polarity. Second, as described below, the cells begin to rearrange within the plane of the cell sheet. This requires the onset of cell motility to drive the cell neighbor changes, and could require synthesis of new cell adhesion molecules. Third, directionality must somehow be present. The cells move in a net inward direction and so there must be some kind of signal imparted to the cell sheet that gives it that directionality. Finally, the movements must somehow be coordinated. Clearly, all of these processes are launched early in gastrulation, but the molecular identity of the components involved still are not known in any system.

DEEPENING THE INVAGINATION

The archenteron begins to elongate following the initial inward bending of the vegetal plate. The first half to two-thirds of the invagination are brought about largely by cell rearrangements described as "convergence and extension" in other morphogenetic systems (Fristrom, 1988; Keller, 1987) (Fig. 4a). Several kinds of experiments demonstrate that convergence-extension is a major mechanism of archenteron elongation. First, by counting cells in cross sections of the tube, it was observed that the number of cells in any given circumference steadily declined (Ettensohn, 1985a; Hardin and Cheng, 1986). Second, dye marking experiments showed that the elongation is local, i.e., there is no recruitment of cells from areas lateral to the site of invagination (Hardin, 1989). The only way to explain these observations is a rearrangement of the cells. Third, a patch of presumptive endoderm was marked with a fluorescent tag and observed throughout gastrulation. Cells in the patch shifted position relative to one another during gastrulation, resulting in a highly elongated patch of labeled cells (Fig. 4) (Hardin, 1989). Fourth, time-lapse films of the archenteron have detected shifts in cell position along the axis of elongation (Hardin, 1989). Finally, the one other major mechanism proposed for primary elongation of the archenteron, as suggested by Gustafson and Wolpert, was a pulling force due to traction of the secondary mesenchyme cells. The necessity for such traction is ruled out since exogastrulae extend archenterons without benefit of secondary mesenchyme pulling, and laser ablation experiments destroyed filopodia but the archenteron continued to extend until it reached up to two-thirds its final length (Hardin, 1988).

Cell adhesion changes also occur during the time of invagination as demonstrated by aggregation and sorting experiments (McClay et al., 1977; Bernacki and McClay, 1989). These changes readily show that endoderm cells develop the capacity to recognize one another relative to ectoderm and mesoderm. However, it is not known whether these experimentally observed adhesive changes might actually be important for the invagination process. The germ layers really

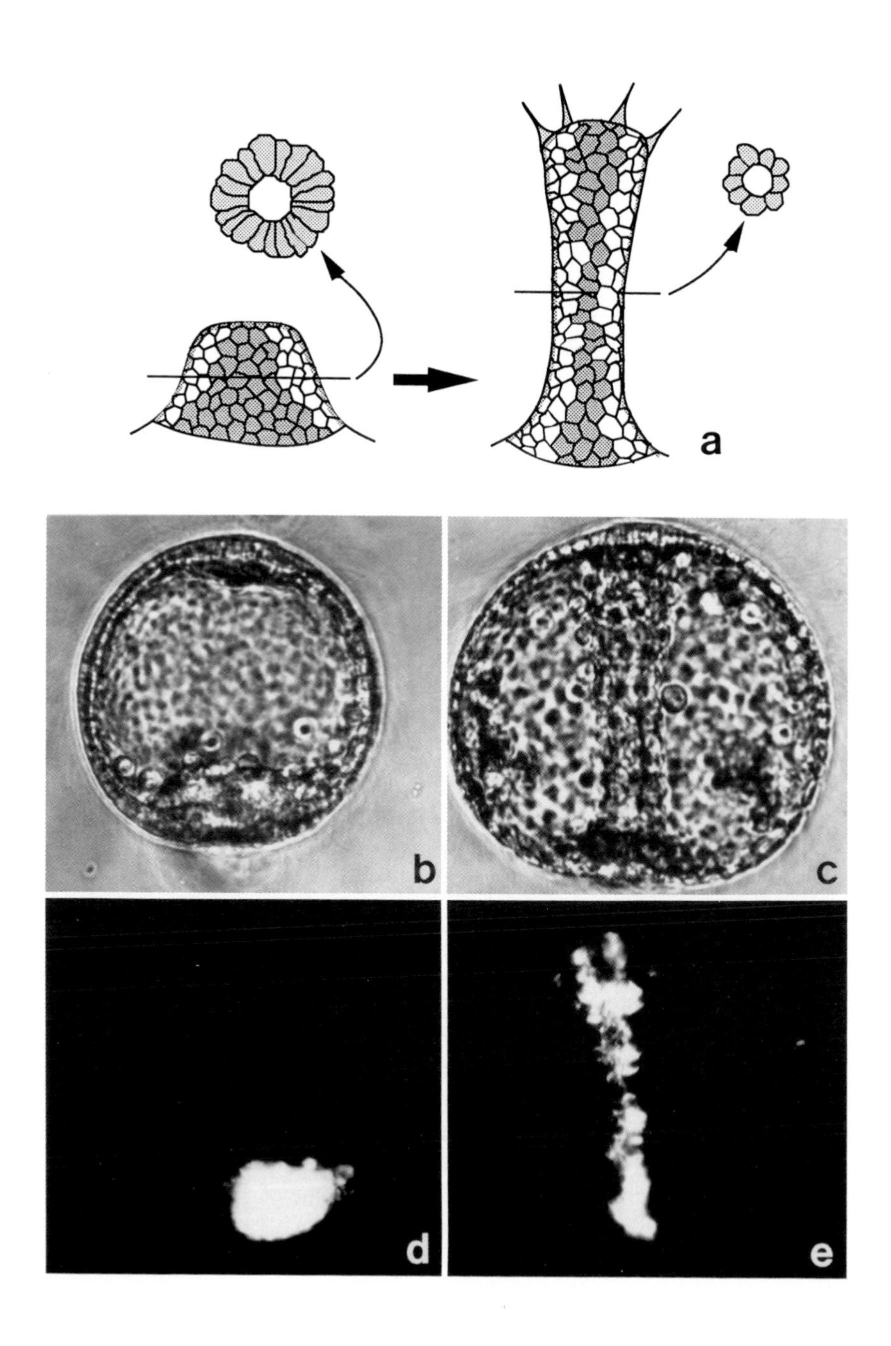
a
b
c
d
e

do not sort out from one another during normal invagination since the cells already are physically isolated from one another. The adhesion changes may reflect newly acquired properties that contribute to endodermal cell-cell rearrangements. In cell aggregates, endodermal cells quickly rearrange to form a tubular structure, demonstrating their tendency to interact with one another and to form a cell sheet one cell deep (Spiegel and Spiegel, 1975; Bernacki and McClay, 1989).

The most elusive property of gastrulation, and one of the most critical, is the directionality of invagination. If one examines the basal end of the endodermal cells during gastrulation, they can be seen to overlap adjacent cells in the direction of net movement of the archenteron (Hardin, 1989). How do the cells know which way is which? Apparently information along the animal–vegetal axis is all that is needed since the elongation continues when the dorsal–ventral axis has been destroyed (Hardin et al., in preparation). The signal probably is local (as opposed to global) since the archenteron continues to elongate in embryos that have had the animal hemisphere removed. This point of reference in space is an important element of all pattern-forming systems. Many experiments in the sea urchin embryo point to the existence of such directional signals; however, in this and in other embryonic systems, the identity of such signals remains obscure.

SECONDARY MESENCHYME CELL BEHAVIOR DURING ARCHENTERON FORMATION

Films of gastrulation show the dramatic sequence of filopodial extension and retraction exhibited by secondary mesenchyme cells (Fig. 5). Although it was shown experimentally that filopodial activity is not required for the first two-thirds of the extension of the archenteron (Hardin, 1988), completion of extension requires filopodia (Hardin, 1988, 1989). Indeed, toward the end of the extension process when filopodia finally reach the point at which the archenteron attaches to the wall of the blastocoel, one can observe a transient cell shape change in endodermal cells suggestive of a stretching in the direction of archenteron movement (Hardin, 1989). As indicated above, even though the filopodia are extended all through gastrulation, they apparently are not crucial until the last phases of the extension.

Fig. 4. Rearrangement of endoderm during archenteron elongation. (**a**) Diagrams of the cross section of the archenteron at an early and of a later stage of invagination. Many more cells line the lumen early in invagination compared with later. This observation enabled Ettensohn (1985a) to deduce that endoderm cells rearrange during invagination. (**b,c**) Phase micrographs of (**d**) and (**e**). A patch of presumptive endoderm was tagged with rhodamine isothiocyanate. Early in gastrulation (b,d) the patch was rounded and at the vegetal plate. Late in gastrulation (c,e) the patch had rearranged so that cells were found in a long thin linear array (Hardin, 1989), providing evidence for the cell rearrangement that occurs during invagination.

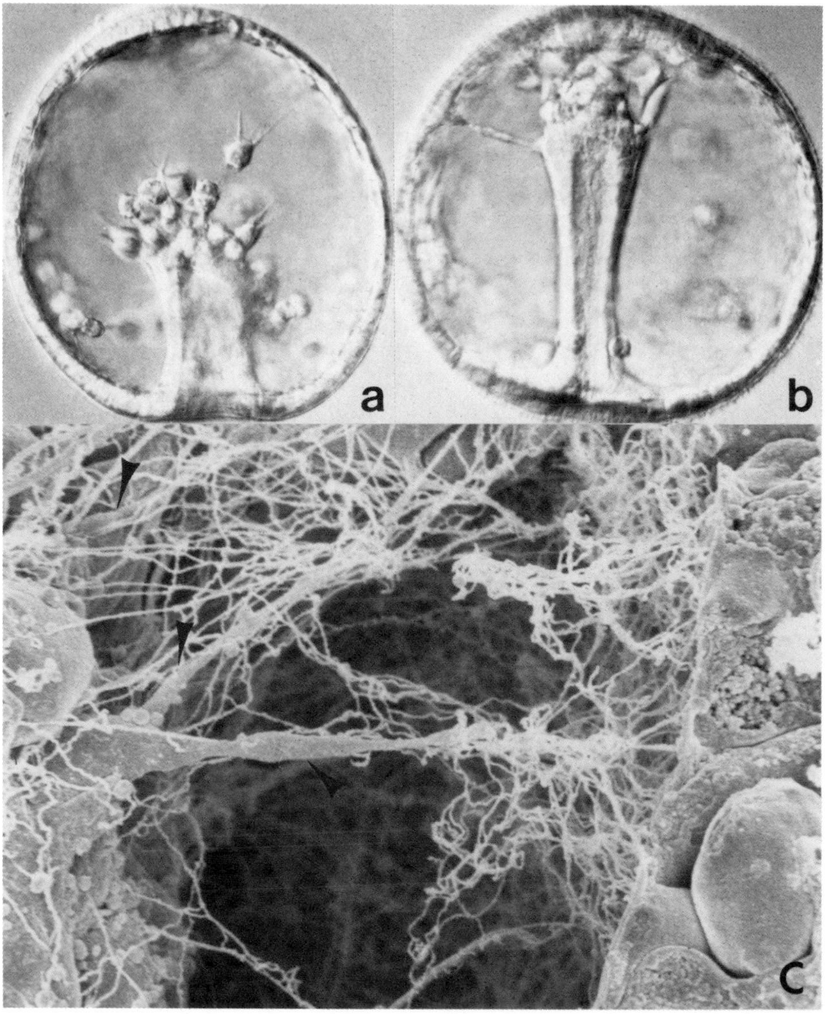

Fig. 5. Secondary mesenchyme cells extend filopodia for the duration of invagination. (**a**) A *Lytechinus pictus* midgastrula stage showing numerous filopodia. (**b**) A late gastrula embryo showing the secondary mesenchyme cells reaching the target, and one long filopodium at the left. (**c**) A scanning electron micrograph showing the extension of three filopodia (arrows) from the archenteron (at left) to the blastocoelic wall (at right). The tips of the filopodia are enmeshed in extracellular matrix fibers.

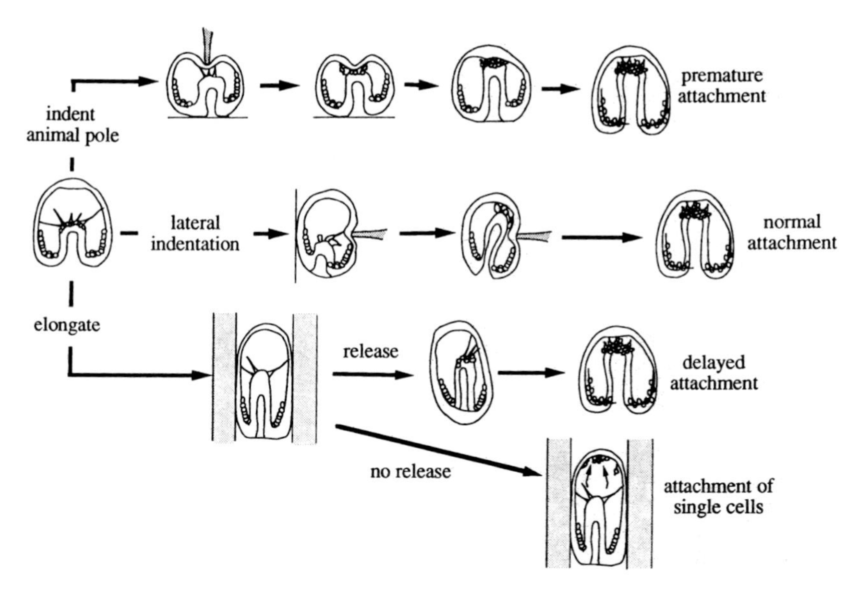

Fig. 6. Summary diagram of experiments demonstrating the existence of a target for archenteron extension (Hardin and McClay, 1990). If embryos are indented at the animal pole, there is a premature attachment to the target region. If the indentation is made laterally, the archenterons make contact but continue to move until the target region is reached. If the embryos are experimentally elongated, the archenteron fails to reach the target region, filopodial extension continues, and this behavior continues until embryos are released, allowing the archenteron to attach to its target region. Only when the target is reached do the secondary mesenchyme cells change their behavior by ending the repetitive extension of filopodia and assuming a flattened phenotype.

A curious characteristic of secondary mesenchyme cells was that many filopodia were extended without making contact with a substrate (Fig. 6) (Hardin and McClay, 1990). A careful study of filopodial behavior revealed that when contact was made with the wall of the blastocoel, the cells held that contact for intervals lasting from 2 to 10 min. If cells did not reach a substrate within about 35 μm, the filopodia were withdrawn. This behavior continued for the duration of gastrulation, until contact was made with a region of the blastocoel wall near the animal pole. When contact with this region was made, the behavior of secondary mesenchyme cells changed dramatically. The duration of contact was extended much beyond the 10 min maximum that was observed up until then, and in many cases the filopodia were never withdrawn from the animal pole region. In addition, the secondary mesenchyme cells changed phenotype and took on a more fibroblastic appearance.

Was there something special about contact with the animal pole region of the embryo that led to the behavior change? Or was the change in the secondary mesenchyme cells independent of a contact-mediated stimulus that might exist in the animal pole region? The experiments that follow show that contact with a specific target region is necessary for the phenotypic change to occur. Comparative studies in other species, described below, also indicate that for each species there is an anatomical target for attachment of the archenteron. The experiments that reveal the existence of the target and some of the properties of that target are described below.

A TARGET FOR ARCHENTERON EXTENSION

Since the filopodial behavior, described above, continues for the duration of gastrulation until contact with the blastocoelic wall is made, the question arises as to whether the change in secondary mesenchyme cell phenotype could be induced precociously by early contact with the animal pole region. By simply pushing the animal pole region into contact with the archenteron tip, the secondary mesenchyme cells changed phenotype much earlier in response (relative to control embryos of the same age). If contact with the animal pole region were prevented by preventing the secondary mesenchyme cells from reaching the animal pole region, one might predict (if contact were essential) that secondary mesenchyme cells would continue their filopodial extension behavior for an abnormally prolonged time. This prediction was tested by inserting embryos into narrow-bore glass tubes so that the embryos were elongated (Fig. 6). It was observed that filopodial extension continued for hours longer than the behavior occurred in the control embryos. If the embryos were released from the tubes, they rounded up, contact with the animal pole region was made, and the phenotypic change occurred. Thus, until contact with the animal pole was made, the secondary mesenchyme cells continued their filopodial extension behavior (Hardin and McClay, 1990).

The contact-induced change in phenotype occurred when the secondary mesenchyme cells reached the animal pole. Was that the only region of the blastocoelic wall capable of inducing the phenotypic change? To address this question, the tip of the archenteron was brought into contact with other regions of the blastocoelic wall by denting the wall from the side (Fig. 6). Although extensive filopodial contact was made with these indentations, the filopodial contact was brief (in the 2–10 min range), and the archenterons continued to extend until contact with the animal pole was made. Thus, only contact with the animal pole region brought about the phenotypic change.

Secondary mesoderm cells give rise to pigment cells, cells of the coelomic pouches, and muscle (Gibson and Burke, 1985). Normally these cells do not participate in the production of the larval skeleton, although experiments have shown that secondary mesenchyme cells retain the capacity to produce spicules until quite late in gastrulation (Ettensohn and McClay, 1988; Ettensohn, 1990). This lineage conversion is a property of the secondary mesenchyme cells that can be induced experimentally by depriving the embryo of cells of the primary mesenchyme cell lineage from very early in development (Hörstadius, 1939), through gastrulation (Ettensohn and McClay, 1988), until about the time of target contact (Ettensohn, 1990). In lineage mapping, studies have yet to determine when secondary mesenchyme cells are specified. The experiments of Ettensohn (1990) suggest that the lineage is finally committed irreversibly sometime around the end of gastrulation, or at about the time the secondary mesenchyme cells undergo the contact-mediated phenotypic change. The striking correlation between Ettensohn's experimental results showing loss of lineage conversion capacity and the phenotypic change observed in the present experiments suggest that at least one consequence of striking the target at the animal pole might be commitment toward the restricted fate of the secondary mesenchyme cell lineage.

The process of archenteron extension appears to have two critical control steps, based upon the experiments described above. At the beginning of invagination the cells are very sensitive to interactions with the extracellular matrix. Expression of a number of genes that normally are transcribed coincident with the beginning of invagination fails to occur unless, or until, contact with the extracellular matrix is established. At the end of gastrulation, target contact may be involved in cell lineage restriction of secondary mesenchyme cells (although this notion is entirely correlative at present). As the archenteron extends, cells appear to follow a simple set of instructions that may be given once at the beginning of gastrulation. The endoderm cells rearrange, and the secondary mesenchyme cells send out filopodia until further notice. The "further notice" appears to be contact with the target region. The accuracy of this model remains to be tested rigorously. Nevertheless, the behavioral data gathered, the inhibitor studies, the lineage conversion studies, and the gene expression studies that have been performed during gastrulation all support this idea.

What is meant by the phrase "set of instructions," as used above? Taking the behavior of the endodermal cells as an example, it is clear that the cells perform many behavioral subroutines, each of which must be coordinately regulated at the molecular level with the others. Indeed, at the beginning of gastrulation it is known that a group of genes is transcribed coordinately in the endoderm (McClay et al., 1983; Wessel et al., 1989a; Nocente-McGrath et al., 1989) (although expression of those genes has not been linked directly to the behaviors observed by the cells). Cells become motile and undergo circus movements that appear to jostle the cells as viewed by time lapse. They rearrange in an ordered sequence that involves initiating and breaking cell adhesions. They converge and extend the archenteron directionally. All through elongation the cells retain contact with the luminal matrix and the basal lamina. Each of these component cell behaviors must have a molecular basis, and that group of molecules must be under coordinate regulation. Thus, while it is easy to state that the behavior is a simple programmed sequence, this incomplete list of functions indicates that a complex network of molecular events underlies the process.

REFERENCES

Adelson DL, Humphreys T (1988): Sea urchin morphogenesis and cell-hyalin adhesion are perturbed by a monoclonal antibody specific for hyalin. Development 104:391–402.

Alliegro MC, Ettensohn CA, Burdsal CA, Erickson HP, McClay DR (1988): Echinonectin: A new embryonic substrate adhesion protein. J Cell Biol 107:2319–2327.

Baker PC, Schroeder TE (1967): Cytoplasmic filaments and morphogenetic movement in the amphibian neural tube. Dev Biol 15:432–450.

Bernacki SH, McClay DR (1989): Embryonic cellular organization: Differential restrictions of fates as revealed by cell aggregates and lineage markers. J Exp Zool 251:203–216.

Burdsal CA, Alliegro MC, McClay DR (1991): Tissue-specific temporal changes in cell adhesion to echinonectin in the sea urchin embryo. Dev Biol 144:327–344.

Burnside B (1971): Microtubules and microfilaments in newt neurulation. Dev Biol 26:416–441.

Butler E, Hardin J, Benson S (1987): The role of lysyl oxidase and collagen crosslinking during sea urchin development. Exp Cell Res 173:174–182.

Davidson E (1988): ''Gene Activity in Early Development,'' 3rd ed. New York: Academic Press.

Davidson EH (1989): Lineage-specific gene expression and the regulative capacities of the sea urchin embryo: A proposed mechanism. Development 105:421–445.

Davidson EH (1990): How embryos work: A comparative view of diverse modes of cell fate specification. Development 108:365–389.

Ettensohn CA (1984): Primary invagination of the vegetal plate during sea urchin gastrulation. Am Zool 24:571–588.

Ettensohn CA (1985a): Gastrulation in the sea urchin is accompanied by the rearrangement of invaginating epithelial cells. Dev Biol 112:383–390.

Ettensohn CA (1985b): Mechanisms of epithelial invagination. Q Rev Biol 60:289–307.

Ettensohn CA (1990): The regulation of primary mesenchyme cell patterning. Dev Biol 140:261–271.

Ettensohn CA, McClay DR (1988): Cell lineage conversion in the sea urchin embryo. Dev Biol 125:396–409.

Fink RD, McClay DR (1985): Three cell recognition changes accompany the ingression of sea urchin primary mesenchyme cells. Dev Biol 107:66–74.

Fristrom D (1988): The cellular basis of epithelial morphogenesis. A review. Tissue Cell 20:645–690.

Gibson AW, Burke RD (1985): The origin of pigment cells in embryos of the sea urchin *Strongylocentrotus purpuratus*. Dev Biol 107:414–419.

Gustafson T, Wolpert L (1963): The cellular basis of morphogenesis and sea urchin development. Int Rev Cyt 15:139–214.

Gustafson T, Wolpert L (1967): Cellular movement and contact in sea urchin morphogenesis. Biol Rev 42:442–498.

Hardin J (1988): The role of secondary mesenchyme cells during sea urchin gastrulation studied by laser ablation. Development 103:317–324.

Hardin J (1989): Local shifts in position and polarized motility drive cell rearrangement during sea urchin gastrulation. Dev Biol 136:430–445.

Hardin JD, Cheng LY (1986): The mechanisms and mechanics of archenteron elongation during sea urchin gastrulation. Dev Biol 115:490–501.

Hardin J, McClay DR (1990): Target recognition by the archenteron during sea urchin gastrulation. Dev Biol 142: in press.

Hörstadius S (1935): Über die Determination im Verlaufe der Eiachse bei Seeigeln. Pubbl. Staz. Zool. Napoli 14:251–429.

Hörstadius S (1939): The mechanics of sea urchin development, studied by operative methods. Bio Rev Cambridge Philos Soc 14:132–179.

Katow H, Solursh M (1981): Ultrastructural and time-lapse studies of primary mesenchyme cell behavior in normal and sulfate-deprived sea urchin embryos. Exp Cell Res 136:233–245.
Keller RE (1987): Cell rearrangement in morphogenesis. Zool Sci 4:763–779.
Lane MC, Solursh M (1988): Dependence of sea urchin primary cell migration on xyloside- and sulfate-sensitive cell surface-associated components. Dev Biol 127:78–87.
McClay DR, Fink RD (1982): Sea urchin hyalin: Appearance and function in development. Dev Biol 92:285–293.
McClay DR, Cannon GW, Wessel GM, Fink RD, Marchase RB (1983) Patterns of antigenic expression in early sea urchin development. In Jeffrey W, Raff R (eds): "Time, Space, and Pattern in Embryonic Development." New York: Alan R. Liss, pp 157–169.
McClay DR, Chambers AF, Warren RG (1977): Specificity of cell-cell interactions in sea urchin embryos. Appearance of new cell-surface determinants at gastrulation. Dev Biol 56:343–355.
Moore AR, Burt AS (1939): On the locus and nature of the forces causing gastrulation in the embryos of *Dendraster excentricus* . J Exp Zool 82:159–171.
Nelson SH, McClay DR (1988): Cell polarity in sea urchin embryos: Reorientation of cells occurs quickly in aggregates. Dev Biol 127:235–247.
Nocente-McGrath C, Brenner CA, Ernst SG (1989): Endo 16, a lineage-specific protein of the sea urchin embryo, is first expressed just prior to gastrulation. Dev Biol 136:264–272.
Odell GM, Oster G, Alberch P, Burnside B (1981): The mechanical basis of morphogenesis. I. Epithelial folding and invagination. Dev Biol 85:446–462.
Rhumbler L (1902): Zur Mechanik des Gastrulationsvorganges insbesondere der Invagination. Arch Entwicklungsmech Org 14:401–476.
Schroeder TE (1988): Contact-independent polarization of the cell surface and cortex of free sea urchin blastomeres. Dev Biol 125:255–264.
Solursh M (1986): Migration of sea urchin primary mesenchyme cells. In Browder L (ed): "Developmental Biology: A Comprehensive Synthesis: The Cellular Basis of Morphogenesis, Vol. 2" New York: Plenum Press, pp 391–431.
Solursh M, Lane MC (1988): Extracellular matrix triggers a directed cell migratory response in sea urchin primary mesenchyme cells. Dev Biol 130:397–401.
Solursh M, Mitchell SL, Katow H (1986): Inhibition of cell migration in sea urchin embryos by β-D-xyloside. Dev Biol 118:325–332.
Spiegel M, Spiegel E (1975): The reaggregation of dissociated embryonic sea urchin cells. Am Zool 15:583–606.
Spiegel M, Spiegel E (1986): Cell-cell interactions during sea urchin morphogenesis. In Browder L (ed): "Developmental Biology: A Comprehensive Synthesis." New York: Plenum Press, pp 195–240.
Wessel GM, McClay DR (1985): Sequential expression of germ-layer specific molecules in the sea urchin embryo. Dev Biol 111:451–463.
Wessel GM, McClay DR (1987): Gastrulation in the sea urchin embryo requires the deposition of cross linked collagen within the extracellular matrix. Dev Biol 121:149–165.
Wessel GM, Goldberg L, Lennarz WJ, Klein WH (1989a): Gastrulation in the sea urchin embryo is accompanied by the accumulation of an endoderm-specific mRNA. Dev Biol 136:526–536.
Wessel GM, Zhang W, Tomlinson CR, Lennarz WJ, Klein WH (1989b): Transcription of the Spec 1-like gene of Lytechinus is selectively inhibited in response to disruption of the extracellular matrix. Development 106:355–365.

Cell-Cell Interactions in Early Development, pages 31–62

3. Pattern and Function of Cell Motility and Cell Interactions During Convergence and Extension in *Xenopus*

Ray Keller, John Shih, Paul Wilson, and Amy Sater
Department of Molecular and Cell Biology, University of California, Berkeley, California 94720 (R.K., J.S., A.S.); Department of Biochemistry and Molecular Biology, Harvard University, Cambridge, Massachusetts 02140 (P.W.)

INTRODUCTION

The morphogenetic movements of convergence and extension have recently come to the attention of developmental biologists after being ignored since they were described in the heyday of *Entwicklungsmechanik* (developmental mechanics) in the early part of this century. These movements are deserving of renewed attention because they are the principal morphogenetic movements of the vertebrate organizer (see Spemann, 1938) and in this capacity are fundamental participants in nearly every aspect of early vertebrate embryogenesis. They establish the elongate shape and the dorsal position of the notochord, the somitic mesoderm, and the neural plate. In doing so, they play central roles in gastrulation (Keller, 1986), in neurulation (Jacobson and Gordon, 1976; Jacobson, 1981), and in establishing the anteroposterior and dorsoventral body axes of amphibians (Gerhart and Keller, 1986). The subsequent tissue interactions and movements of organogenesis are organized around the shapes generated by these primary movements. In the amphibian, their pattern and character are determined by information accumulated progressively through early development, beginning in oogenesis, passing through fertilization and cleavage, and continuing as these movements are carried out during gastrulation (see Gerhart et al., 1991). Lastly, convergence and extension represent a type of morphogenetic movement in which a population of cells distorts autonomously by forces generated within itself, a poorly understood process that has received little attention in cell and developmental biology. Movements of this type and convergence and extension in particular, are not unique to amphibians but play equally important roles throughout the metazoa (Keller, 1987). Thus a large gap exists in our understanding of how cells function in morphogenesis.

We begin with a brief review of the origin of the concepts of convergence and extension, since our current idea ". . . has not come out of a clear sky, but rests in turn on other work that has preceded it'' (p. vi, Morgan, 1927). We will then discuss types of cell motility involved in convergence and extension, their patterns of expression, and how they function in the gastrula and neurula of the African clawed frog, *Xenopus laevis*.

Early Descriptions of Convergence and Extension in the Amphibian

These movements were discovered near the beginning of experimental embryology and were important in the thinking of the time. It was known in the early 1900s that the prospective somites, notochord, and neural tube were arrayed in some fashion around both sides of the blastopore in the marginal zone of the amphibian gastrula. It was thought that during gastrulation the two halves fused or ``zipped up'' at the dorsal blastoporal lip without much change in shape, forming the elongated array of dorsal tissues and closing the blastopore at the same time (Fig. 1, top). The main evidence for this mechanism, called ``concrescence,'' was the fact that when the blastopore did not

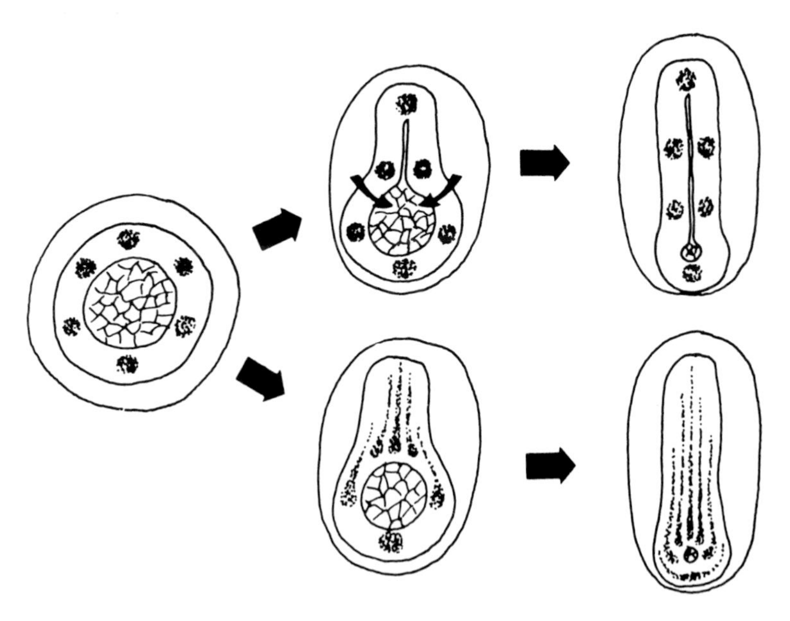

Fig. 1. Diagrams of the fates of dye marks placed in the marginal zone of the amphibian (**left**) illustrate schematically how the marginal zone would move and close the blastopore during gastrulation by *concrescence* **(top)** and by *convergence* and *extension* (**bottom**). These alternate mechanisms apply equally to the neural tissue that does not involute (shown here) and the mesodermal tissue that does involute (not shown).

close, these tissues differentiated, presumably in their places of origin, around both sides of the open blastopore (see Morgan, 1897; Bautzmann, 1933; Schechtman, 1942). But Vogt (1929), and Goodale (1911) before him, discovered that vital dye marks placed in the marginal zones of amphibian embryos narrowed toward the dorsal side (*convergence*) and elongated greatly along the future anterior–posterior axis (*extension*) during gastrulation and neurulation (Fig. 1, bottom). Convergence of the marginal zone around the yolk plug was thought to produce "constriction" of the circumblastoporal region, which resulted in blastopore closure. Thus the mesodermal and neural tissues move from their original location around the blastopore to their final elongated shape at the dorsal midline by *distortion* of the marginal zone rather than by *fusion* of its right and left halves. This finding was important because it directed subsequent thinking toward underlying cellular processes that would explain local distortion rather than fusion of tissues.

A variety of names was applied to these movements, often with somewhat different meanings (see Vogt, 1929; Schechtman, 1942). In our work, we have used *convergence* to describe the narrowing and *extension* to describe the commensurate elongation of tissue, in the same sense that one can narrow and lengthen bread dough by pulling on it (Keller, 1986). For the present, we will refer to these processes together, since they are often coupled. However, as we will discuss below, this is not always the case.

Convergence and Extension Are Autonomous, Region-Specific Processes

Convergence and extension of the marginal zone are just two of several regional movements comprising the unitary flow of tissues in the gastrula (Vogt, 1929). Early experimentation was directed at learning which movements are autonomous and actively driven by forces generated from within tissues, and which ones are passive responses to forces generated elsewhere. Explantation and microsurgical manipulations by Spemann, Vogt, O. Mangold, Holtfreter, and others (reviewed in Spemann, 1938; Keller, 1986; and Gerhart and Keller, 1986) demonstrated that convergence and extension are autonomous and specific to the dorsal marginal zone. Schechtman (1942) followed with a characterization of these movements and how they are integrated with other region-specific processes in *Hyla regilla*, the California tree frog.

Convergence and Extension Were Ignored and Attention Turned to Other Regional Processes

Since convergence and extension were shown to be autonomous, region-specific processes, further analysis and attention in the literature should have continued, up to the present, but in fact did not. Early texts presented the mistaken idea of concrescence, described above (see Morgan, 1897; McEwen, 1923; Shumway, 1927), and later ones often pointed out that the dorsal tissues "stretched" or "elongated." But convergence and extension were not usually recognized as autonomous processes, nor were they given formal names equivalent to those of the other region-specific processes, such as invagination, involution, ingression, and epiboly, with rare exceptions (see Hamburger, 1960) until recently (Gilbert, 1988).

There were several reasons for this neglect. No uniform terminology or clear characterization of these movements emerged from early work, and it was never clear exactly when and how they functioned in gastrulation or neurulation. Schechtman's insightful 1942 paper, which suggested new approaches, was not widely read. Also, there was no widely recognized paradigm relating the behavior of individual cells to the cell population behaviors of convergence and extension. Waddington (1940, p. 109) correctly concluded that since neither change in cell shape nor cell division could account for these movements, they must occur by cell rearrangement. But the idea of cells rearranging themselves, particularly in an active, force-producing fashion, is not easily visualized in terms of cellular properties or behaviors. Thus this notion did not flourish and no experiments or observations followed. In contrast, other regional processes could be thought about in terms of specific cell behaviors. Invagination was attributed to bottle cell invasiveness or change in shape (Rhumbler, 1902; Holtfreter, 1943a,b; Lewis, 1947). The arrangement of tissue layers at the end of gastrulation was explained by cell crawling (Holtfreter, 1943a,b, 1944, 1946), directed by selective affinities of tissues or cells for one another (Holtfreter, 1939; Townes and Holtfreter, 1955).

CONVERGENCE AND EXTENSION IN *XENOPUS*

Other Processes Do Not Account for the Bulk of Gastrulation Movements in *Xenopus*

Early experiments did not show convergence and extension to be robust and autonomous processes in *Xenopus*. Autonomous extension in explants was weak (Ikushima and Maruyama, 1971) and what did occur was rather late in development (Keller, unpublished results). On the other hand, interdiction of all the other identified regional processes did not stop the major movements of gastrulation in *Xenopus*. Removal of the bottle cells truncated the periphery of the archenteron, which is the tissue normally formed by the respread bottle cells, but gastrulation continued (Keller, 1981). Removal of the entire blastocoel roof (animal cap) stops both migration of mesodermal cells on the roof and the epiboly of the animal cap itself, but most of gastrulation continues, including involution of the marginal zone, closure of the blastopore, and extension of the axial mesodermal tissues in their proper pattern (Keller et al.,

1985a,b; also see Holtfreter, 1933). In contrast, manipulation of the marginal zone, particularly the deep mesodermal region, results in failure of all movements, including involution, blastopore closure, and elongation and narrowing of the axial tissues (Keller, 1981, 1984; Keller et al., 1985a,b). Thus the marginal zone plays an important role in accomplishing all these movements. But this role is not dependent on the presence of either the blastocoel roof or the bottle cells, and thus it must involve only behaviors of the marginal zone that were shown in the classical literature to be strictly autonomous—convergence and extension.

Demonstration of Two Regions of Convergence and Extension in Sandwich Explants of *Xenopus*

We returned to looking for autonomous convergence and extension in explants of the marginal zone of the early *Xenopus* gastrula, this time with improved technique. We made "sandwich" explants, by apposing the inner surfaces of two marginal zones of the early gastrula stage and allowing them to heal as shown (Fig. 2A). To our surprise, these explants showed not one but two regions of convergence and extension: one in the dorsal involuting marginal zone (IMZ), which differentiates into somites on both sides of a central notochord, and another in the dorsal noninvoluting marginal zone (NIMZ), which differentiates into neural tissue (Keller et al., 1985a,b; Keller and Danilchik, 1988) (Fig. 2A). These movements are specific to the dorsal marginal zone and provide an early indicator of experimental induction of this type of mesoderm (Smith and Symes, 1987). Because there are two regions in the marginal zone that converge around the blastopore, we distinguished between the one that involutes (IMZ) and the one that does not (NIMZ). It is uncertain to what degree convergence and extension of the NIMZ was represented among the numerous extensions reported for species other than *Xenopus*, mostly urodeles, in the classical literature.

Our earlier attempts, mentioned above, were unsuccessful because we did not remove all the involuted head mesoderm from the inner surface of the explant; its invasiveness resulted in partial reorientation and disorganization of the axial mesoderm prior to its attempted extension (see Keller and Danilchik, 1988). This experience warns that failure to demonstrate these complex movements in explants of any organism does not necessarily mean they do not occur.

The geometry and mechanics of how convergence and extension produce involution and blastopore closure in the spherical geometry of the gastrula are discussed elsewhere (Keller, 1986; Wilson and Keller, 1991; Keller et al., 1991). We now turn our attention to the mechanisms of these movements, first in the IMZ and then in the NIMZ.

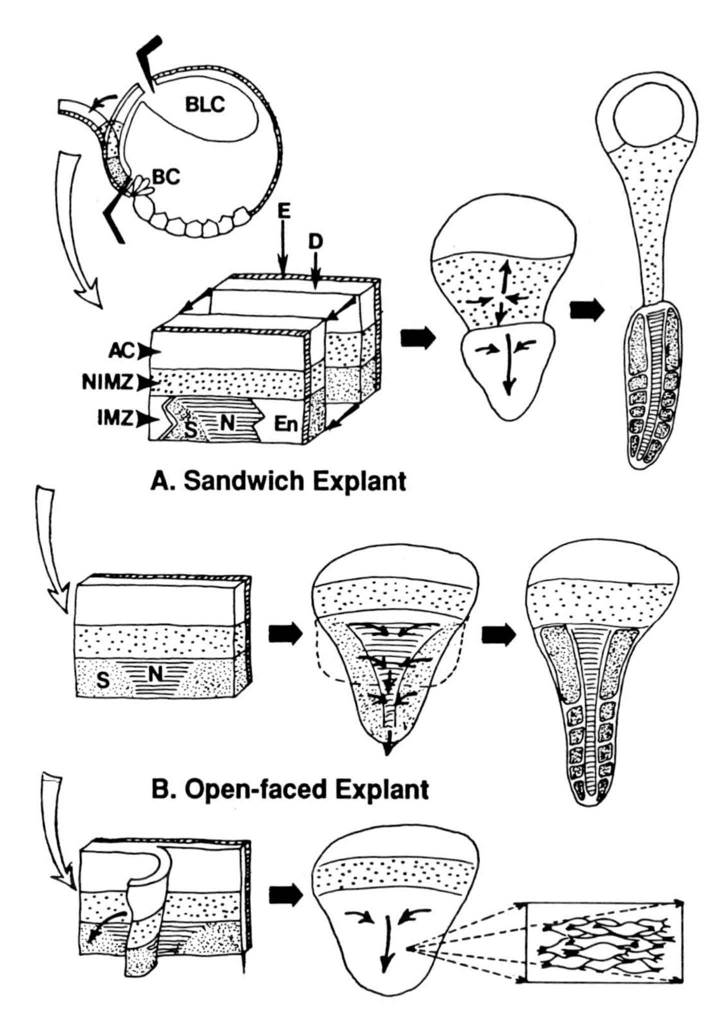

Fig. 2. A schematic diagram shows the development of sandwich explants (**A**), open-faced explants (**B**), and shaved, open-faced explants (**C**). The dorsal sector of the early gastrula, depicted in sagittal section at the upper left, is excised on both sides, 60 degrees from the midline, at the bottle cells (BC), and near the animal pole. This explant consists of a superficial epithelial region (E) and a deep, nonepithelial (mesenchymal) region (D), and is divided into an animal cap region (AC), a noninvoluting marginal zone (NIMZ), and an involuting marginal zone (IMZ). In normal development the AC and NIMZ have a uniform tissue fate in both layers, which is neural ectoderm, the AC contributing to the brain and the NIMZ contributing to the posterior hindbrain and spinal cord. The superficial and deep regions of the IMZ have different fates; the superficial layer forms endoderm (En) and the deep region forms mesoderm, specifically notochordal (N) and somitic (S) mesoderm. The sandwich explants show two regions of conver-

CONVERGENCE AND EXTENSION IN THE IMZ

The Mechanical Roles of the Epithelial Cells and the Deep Mesodermal Cells

The marginal zone of the early *Xenopus* gastrula has two components—a superficial, monolayered epithelium of prospective endoderm and an underlying deep region, consisting initially of four or five layers of mesenchymal cells, all of which are prospective mesoderm (Fig. 2) (Keller, 1975, 1976; Keller and Schoenwolf, 1977). In contrast to the other anurans (Vogt, 1929; Purcell, 1989) and urodeles (Vogt, 1929; Smith and Malacinski, 1983; Lundmark, 1986) studied thus far, *Xenopus* deep and superficial layers retain their integrity and do not exchange cells during gastrulation (Keller, 1975, 1976). The epithelial layer appears to be stretched passively by forces produced among the deep mesodermal cells. Replacement of the native epithelial layer of the dorsal marginal zone of the early gastrula with that of the animal cap does not disrupt movements of this region in whole embryos (Keller, 1981) or explants (Keller and Danilchik 1988), whereas the corresponding manipulation of the deep layer does block movement (Keller, 1981). Marginal zone epithelial cells elongate in the direction of extension and later return to a nearly isodiametric shape as they rearrange and divide (Keller, unpublished data), a behavior consistent with stretching and relaxation to the original shape during rearrangement. Lastly, deep cells can converge and extend without the superficial epithelium after the midgastrula stage (Wilson, 1990). In contrast, explants that include only the superficial layer are unstable and crumple up into a corrugated mass. We have not been able to see the relatively small amount of extension of epithelial sandwiches observed by Ikushima and Maruyama (1971).

Convergence and Extension of the IMZ Occurs by Radial and Mediolateral Intercalation

Time-lapse recordings reveal that the superficial epithelial cells accommodate convergence and extension by *mediolateral intercalation*: They intercalate between one another along the mediolateral axis to form a longer but narrower array (Keller, 1978) (Fig. 3A). The cells change neighbors even though

gence and extension, one in the NIMZ and one in the IMZ (arrows in A). The NIMZ develops into a tangle of neurons and the IMZ develops into notochord and somites. The open-faced explants are cultured in Danilchik's solution under a coverslip, conditions under which the IMZ will converge, extend, and differentiate into somites and notochord (B). Shaved explants (C) are made and cultured like open-faced explants but the innermost layers of deep cells are shaved or peeled off with an eyebrow knife, exposing the deep cells next to the endodermal epithelium to observation and videorecording for analysis of cell behavior at high resolution (box, lower right). From Keller et al., 1991.

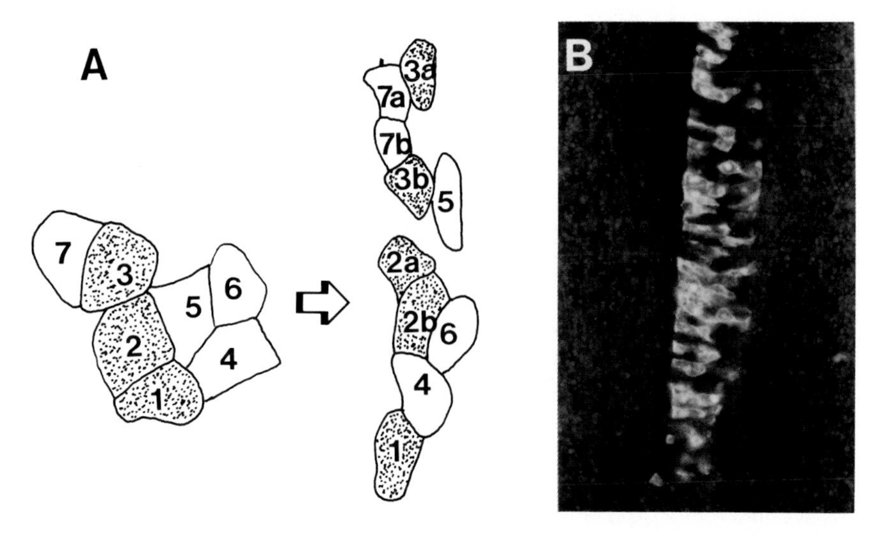

Fig. 3. Mediolateral intercalation of superficial, epithelial cells of the IMZ is illustrated by tracing of cells through gastrulation with time-lapse recordings (**A**). Mediolateral intercalation of deep cells is illustrated by the intercalation of unlabeled deep cells from lateral positions between cells of an originally cohesive block of labeled cells at the dorsal IMZ of a whole embryo (**B**). From Keller and Tibbetts (1989). The axis of extension is vertical in both diagrams.

they are bound into an epithelial organization by a circumapical junctional complex, including tight junctions and desmosomes. Epithelial cells commonly rearrange during morphogenesis (reviewed in Keller, 1987 and Fristrom, 1988), even in systems where a high-resistance physiological barrier is maintained throughout the rearrangement (Keller and Trinkaus, 1987). Deep cells also undergo mediolateral intercalation. A coherent block of deep cells, labeled with a lineage tracer, intercalate with unlabeled cells on both sides during convergence and extension of the dorsal marginal zone (Keller and Tibbetts, 1989) (Fig. 3B).

Since the deep cells are likely to generate much of the force for convergence and extension, we wanted to observe their behavior in more detail. We developed the open-faced explant system that allows normal development and morphogenesis of deep cells without the protection of an opaque epithelium on all sides (Keller et al., 1985a,b) (Fig. 2B). We have used this explant system with high resolution time-lapse videomicrography to reveal the cell behavior underlying convergence and extension during gastrulation (Wilson, 1990; Wilson and Keller, 1991; Shih and Keller, manuscript in preparation) and neurulation (Wilson et al., 1989; Keller et al., 1989a; Wilson, 1990).

The picture that emerges is complex, consisting of several specific types of cell behavior progressing in spatial and temporal patterns that must be visualized in terms of the mass movements of tissues in the spherical gastrula. To summarize, there are two principal cell behaviors, *radial intercalation* and the *mediolateral intercalation* already mentioned above, each driven by specific types of cell protrusive activity. In a given region, they occur sequentially, in this order, perhaps with some overlap. This sequence of behaviors begins in the dorsal midline at the vegetal end of the IMZ, which is prospective anterior notochord, and spreads progressively in the animal (prospective posterior) direction and also laterally and ventrally within the IMZ. Finally, mediolateral intercalation becomes specialized in the notochord and somitic mesoderm as the boundary forms between these tissues in the late midgastrula, occurring by new cell behaviors and producing different results in each tissue.

We will first briefly describe the sequence of radial and mediolateral intercalation as they occur in the dorsal IMZ explants (for details see Wilson, 1990, and Wilson and Keller, 1991) (Fig. 4). From the onset of gastrulation through the midgastrula stage, the IMZ thins and extends but converges very little, as deep cells intercalate between one another, along the radial aspect of the embryo, forming fewer layers of greater area (*radial intercalation*) (Fig. 4). The increased

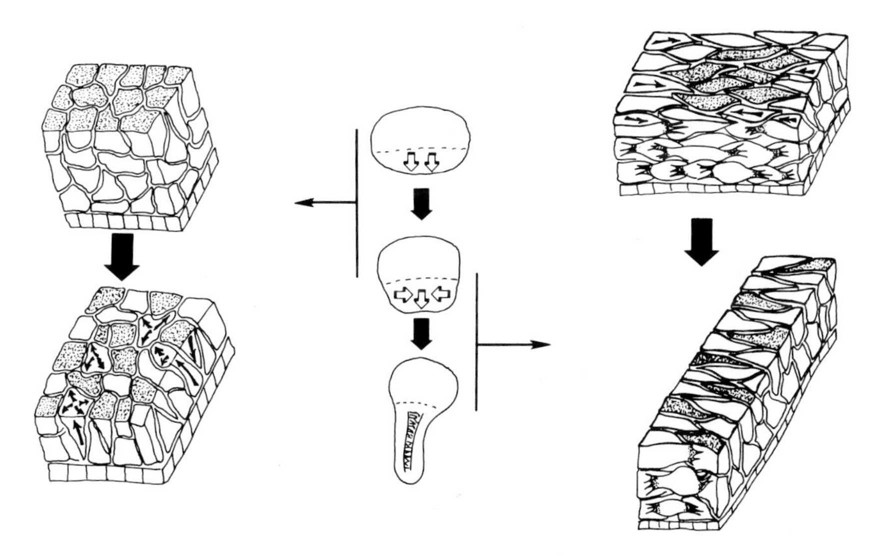

Fig. 4. A schematic diagram shows the behavior of cells during convergence and extension of the involuting marginal zone (IMZ) in open-faced explants. In the center column, change in the shape of open-faced explants are shown at early, middle, and late gastrula stages, and the cell behaviors producing these movements are shown on the sides. The inner, deepest surface is shown uppermost and the epithelial layer is at the bottom in all diagrams. Radial intercalation (**left**) produces thinning and extension in the first half of gastrulation, and mediolateral intercalation (**right**) produces convergence and extension in the second half. Modified from Keller et al. (1991).

area results in extension along the animal–vegetal (future anteroposterior) axis, rather than in all directions. At the midgastrula stage, radial intercalation has stopped in the vegetal (prospective anterior) end of the IMZ but continues at the animal (prospective posterior) end. At this point, the explant converges and extends as the deep cells at the vegetal end begin *mediolateral intercalation*, moving between one another along the mediolateral axis to produce a narrower but longer array (Fig. 4). Mediolateral intercalation involves bipolar protrusive activity and mediolateral alignment of elongated cells, discussed in more detail below. There is an anteroposterior progression of radial intercalation followed by mediolateral intercalation, but since the axis is only about 12 to 15 cells long at the early gastrula stage, these processes pass quickly from one end to the other. Note that in the early gastrula, radial intercalation produces extension as a result of thinning and mediolateral intercalation produces extension as a result of convergence.

We have named these cell behaviors "intercalations" (Keller et al., 1985a,b), meaning "to insert between or among existing elements" (*Webster's Third International Dictionary*) in preference to the previously used "interdigitation" (Keller, 1984), which has the inappropriate connotation of interlocking fingers, and "cell rearrangement," which is too general and is commonly used to describe global movements of cells during gastrulation, neurulation, and organogenesis.

Region-Specific Differences in Patterns and Types of Cell Intercalation

Radial intercalation progresses laterally into prospective anterior somitic mesoderm, and mediolateral intercalation follows to the extent that the anterior regions of both tissues are undergoing mediolateral intercalation by the time the boundary between the two tissues forms at the late midgastrula stage. At this point, the notochord and somitic mesoderm develop different spatial and temporal patterns of the radial/mediolateral intercalation sequence and different types of mediolateral intercalation. To compare the development of the notochordal and somitic mesoderm directly, we make explants at the late neurula stage. Since it is located mid-dorsally with its anteroposterior axis oriented in the vegetal–animal direction, the whole notochord develops in dorsal explants (Fig. 5A–C); by contrast, the anteroposterior axis of the somitic mesoderm passes from dorsal to ventral and thus only the anterior somites develop in dorsal explants (Fig. 5A–C). To visualize the development of the entire somitic mesoderm, along with the notochord, the dorsal structures and the circumblastoporal region are excised from the late gastrula stage (Fig. 5D). The endodermal roof of the archenteron is peeled off, revealing the deeper notochord and somitic mesoderm (Fig. 5E), which is cultured as an open-faced explant, where it continues to converge, extend, and differentiate (see Wilson et al., 1989; Wilson, 1990; Keller et al., 1989a) (Fig. 5F,G), having

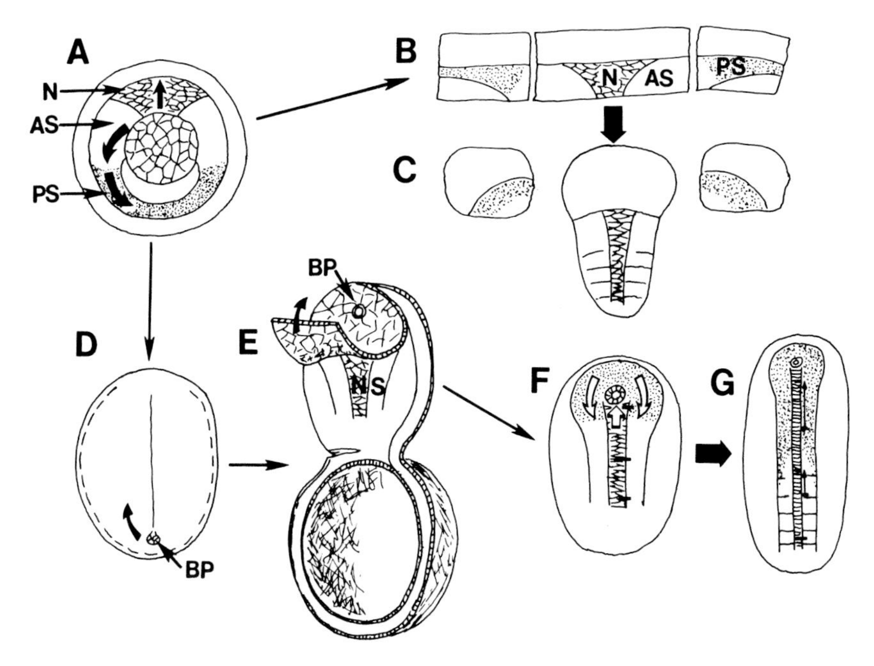

Fig. 5. The development of dorsal explants of the early gastrula (**B,C**) and explants of dorsal tissues of the late gastrula (**D–G**) is illustrated schematically. The prospective notochord (N), anterior somitic mesoderm (AS), and posterior somitic mesoderm (PS) of the early gastrula are shown as if the overlying endodermal epithelium were removed (**A**). Dorsal is at the top and the arrows indicate the prospective anteroposterior axes of the notochord and somitic mesoderm. In dorsal explants of the early gastrula (B), described in the previous figure, the posterior somitic mesoderm is cut off and does not participate in convergence or extension (C). Dorsal explants of the late gastrula are made by cutting off the dorsal half (D), folding it forward, and removing the endodermal archenteron roof, exposing the dorsal mesoderm (E). An open-faced explant of these tissues allows analysis of the cell behaviors underlying the coordinate convergence and extension of notochord and anterior and posterior somitic mesoderm (F–G). Shear between notochord and somitic mesoderm is indicated by arrows showing displacement of transverse bars (F,G).

lost its earlier dependence on the endodermal epithelium for development (Wilson, 1990). The following summary of cell behaviors is based on extensive investigation of the somitic mesoderm (Wilson et al., 1989; Wilson, 1990) and notochord (Keller et al., 1989a; Wilson, 1990) in these explants.

In late gastrula explants, the entire notochord and anterior somitic mesoderm have involuted to their mid-dorsal positions while the posterior somitic mesoderm remains in a thick ring around the lateral and ventral sides of the closed blastopore (Fig. 5F). Subsequently, the notochord extends posteriorly and pushes the blastopore region through the prospective posterior somitic mesoderm, which sweeps around both sides of the blastopore (arrows, Fig.

5F). As the posterior somitic mesoderm is swept alongside the posterior notochord, it increases in length by radial intercalation, as in the gastrula stage, and is added onto the posterior end of the anterior somitic mesoderm already in place (Fig. 5F,G). It then converges toward the midline by mediolateral intercalation, but this convergence is absorbed primarily as thickening rather as extension. Meanwhile, the notochord extends more or less uniformly along its length by mediolateral intercalation. But since the somitic mesoderm extends primarily at its posterior end, the notochord shears posteriorly with respect to the somitic mesoderm, the amount of shear increasing posteriorly (Fig. 5G).

What cell behaviors underlie the differences between movements in these two tissues? After its boundary forms in the late midgastrula stage, most of the notochord has completed radial intercalation and continues convergence and extension by mediolateral intercalation, using a modified cell behavior. Protrusive activity ceases at the boundary, making the boundary cells monopolar with all their invasive protrusive activity directed inward (Fig. 6). As a result, the notochord converges and extends by mediolateral intercalation but also thickens somewhat (Fig. 6). In the late neurula, protrusive activity spreads to the dorsal and ventral edges of the cells as well, and the cells spread circum-

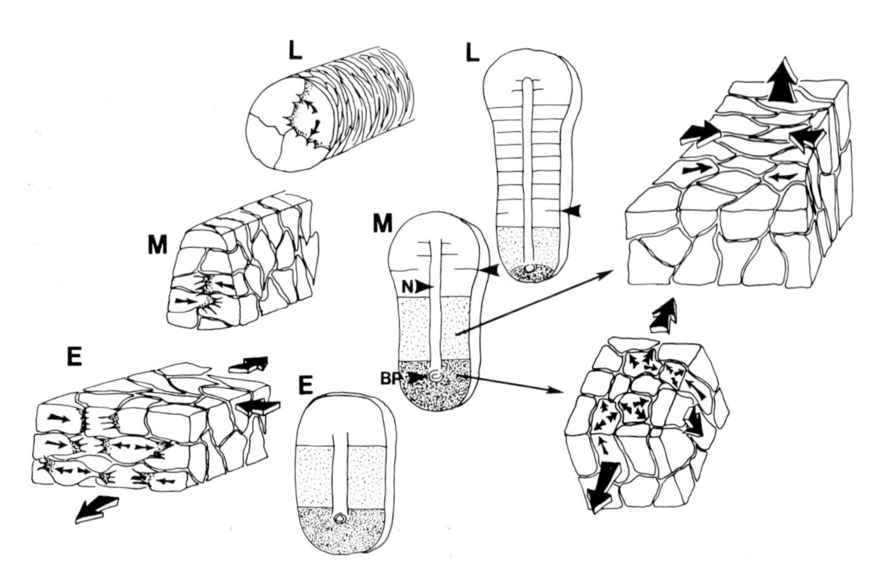

Fig. 6. A summary diagram illustrates the development of dorsal explants, made as in Figure 5D–G, at early (E), middle (M), and late (L) neurula stages (center column). The explants consist of somitic mesoderm on both sides of a central notochord (N). The cell behavior and arrangement in the notochord is shown in the left column. The somitic mesoderm cells pass through a sequence of cell behaviors, progressing anteroposteriorly: radial intercalation (dark shading), mediolateral intercalation (light shading), and segmentation (unshaded). Modified from Keller et al. (1989a) and Wilson et al. (1989).

ferentially, taking on shapes like pizza slices, and rounding the notochord (Fig. 6). Vacuolation and swelling of these cells follows in the late neurula stages and continues into the tailbud stages, lengthening and straightening the notochord in this period (see Adams et al., 1990; Koehl et al., 1990). Like the sequence of radial and mediolateral intercalations in the gastrula stage, these subsequent modifications of mediolateral intercalation progress from anterior to posterior but pass relatively quickly from end to end and thus show little difference between ends.

In contrast, the somitic mesoderm displays a dramatic anteroposterior progression of cell behaviors (Fig. 6, middle panel). Radial intercalation, which proceeded laterally and ventrally from its beginning mid-dorsally in the early gastrula, reaches the prospective posterior somitic mesoderm lying around the blastopore at the end of gastrulation (dark shading, Fig. 6). As radial intercalation produces more area at the expense of thickness, the area is absorbed almost entirely as extension of the posterior segmental plate on either side of the posterior notochord, mentioned earlier. Anterior to this region, mediolateral intercalation occurs (Fig. 6, light shaded zone), but the convergence produced by this intercalation yields only a little extension and considerable thickening of the somitic mesoderm. Mediolateral intercalation is accompanied by increase in cell height and elongation in the mediolateral direction (Fig. 6, right). Thus the somitic mesodermal cells form tall "buttresses" that appear to push upward on the lower surface of the neural folds, perhaps aiding neurulation (Schroeder, 1970). Most anteriorly, segmentation follows in the late neurula (Fig. 6, unshaded region), beginning anteriorly and proceeding posteriorly, the furrows forming from lateral to medial (arrowheads, Fig. 6). Shortly thereafter, the mediolaterally elongate cells rotate to bring themselves into their final anterior–posterior alignment (see Wilson et al., 1989).

The shearing of the notochord posteriorly with respect to the somitic mesoderm (Fig. 5G) is a function of the notochord extending throughout its length by mediolateral intercalation without significant addition of material posteriorly. By contrast, the somitic mesoderm lengthens at its posterior end by addition of cells from the circumblastoporal region and then extending the posterior region by radial intercalation. Mediolateral intercalation at anterior levels results mostly in thickening rather than lengthening.

Cell Interactions and Patterning of Cell Behavior

Differences in timing of behaviors along the anteroposterior axes of the notochord and somitic mesoderm may reflect the progress of the cell interactions setting them up (Wilson, 1990). Anteroposterior pattern in the notochord is probably determined by an inductive signal proceeding upward from the vegetal region (Sudarwati and Nieuwkoop, 1971; Stewart and Gerhart, 1991), whereas patterning in the somitic mesoderm may involve two signal geometries, if not two signals. The anterior somitic mesoderm, being alongside the notochord and having its axis more or less parallel to the notochord, may derive anteroposterior patterning from the same signal that patterns this aspect of the notochord. In contrast, the posterior somitic mesoderm is probably patterned by the dorsalizing signal from the organizer, acting over a longer time, throughout much if not all of gastrulation (Spemann, 1938; Dale and Slack, 1987). Organizer activity, and thus perhaps patterning of cell behaviors as well, does in fact proceed progressively from vegetal to animal on the dorsal side (Stewart, 1990; Stewart amd Gerhart, 1991; Gerhart et al., 1991). In addition, the progressive recruitment of ventral mesoderm to become somitic by interaction with the dorsal organizer has been characterized over some years (Spemann, 1938; Dale and Slack, 1987; Stewart, 1990; Stewart and Gerhart, 1990). In terms of the three-signal model of mesoderm induction (Smith et al., 1985; Dale and Slack, 1987), the anteroposterior axis of the notochord and anterior somitic mesoderm may both be organized by the second signal (the dorsal mesoderm inducing signal) and the posterior somitic mesoderm by the third (dorsalizing) signal. Since the morphogenesis resulting from the second and third signals seems to be the same, perhaps the signals themselves are the same.

Despite the anteroposterior progression of cell behaviors, no local, detailed pattern of intercalation and differentiation is specified. The original anteroposterior order of cells is expanded tremendously by an intercalation process that is not highly ordered; thus, neighboring cells wind up at very different anterior–posterior levels depending on the vicissitudes of the intercalation process (Wilson and Keller, 1991). Thus detailed patterning, such as determining what cells will be in which somite, must occur after or near the end of period of rapid intercalation (Wilson, 1990; Wilson et al., 1989). Neither convergence and extension nor anteroposterior patterning of the mesoderm is dependent on involution or contact with the overlying blastocoel roof (prospective neural plate), or the extracellular matrix there, since these processes occur in open-faced explants of the early gastrula that have not involuted and have had no contact with the blastocoel roof (Wilson, 1990; Wilson and Keller, 1991). They also occur in total absence of the blastocoel roof (Keller et al., 1985a,b; Holtfreter, 1933). Continued extension of the notochord in the neurula does appear to be dependent on somitic mesoderm, despite the shearing between the two tissues (Wilson et al., 1989).

The overlying endodermal epithelium may play a role in supporting and perhaps organizing deep cell behavior and differentiation (Shih and Keller, in preparation). Grafts of epithelium from the dorsal to the ventral marginal zones of early gastrulae induce ventral tissues to converge, extend, and differentiate a second set of somites and neural tube (Fig. 7A), as observed with whole organizer grafts (Gimlich and Cooke, 1983). Sandwich explants of ventral deep mesodermal cells with one or both superficial layers from the dorsal marginal zone converge, extend, and form somites, all events that do not occur

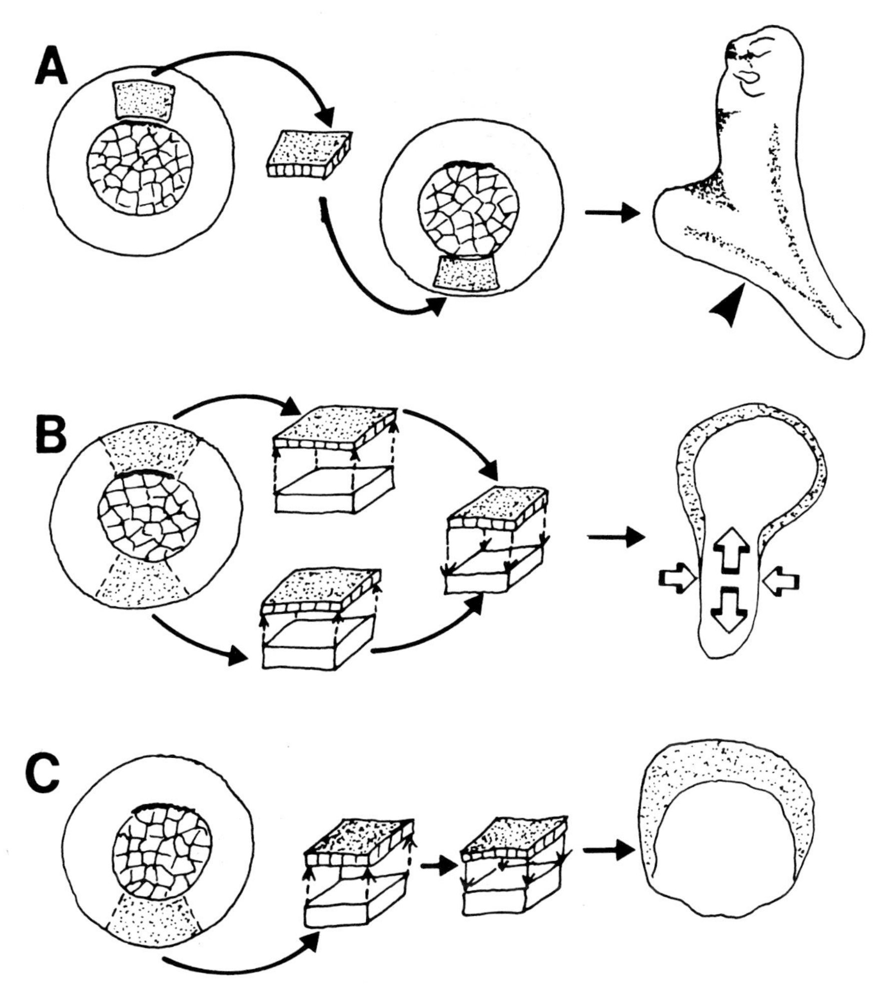

Fig. 7. Experiments showing the organizer function of the epithelial layer of the dorsal marginal zone are illustrated schematically. Grafts of dorsal marginal zone epithelium to the ventral side of another gastrula results in the development of a second axis that converges, extends, and develops a second set of somites and neural tube (arrowhead, **A**). An open-faced explant composed of the dorsal epithelial layer and a ventral deep region shows convergence and extension of the IMZ (**B**), whereas a ventral deep region with its own epithelial layer does neither (**C**).

when ventral explants are cultured with their native epithelium (Fig. 7B,C). Thus the epithelium of the dorsal marginal zone appears to be able to act as an organizer, inducing ventral tissues to adopt dorsal morphogenetic movements (convergence and extension) and tissue differentiation (as notochord and somites). It also appears essential up to the midgastrula stage for organizing convergence and extension of the dorsal IMZ, though not for its differentiation (Wilson, 1990). This conflicts with experiments cited above, which show that replacement of the epithelium of the IMZ in whole embryos or explants does not stop their movements. This is probably because in both cases there was opportunity for the deep cells beneath the grafted patch to be instructed by their better-informed neighbors to the sides.

Are the Different Results of Intercalation Due to Differences of Cell Behavior or Biomechanical Context?

A given type of cell intercalation produces different changes in tissue shape. Mediolateral intercalation produces convergence, but convergence can be absorbed by increase in length or thickness of the tissue. In the early notochord, convergence produces extension and a little thickening, which we have called "convergent extension" (see Keller et al., 1985a,b). But convergence of somitic mesoderm produces more thickening and less extension ("convergent thickening"), because the population of cells increases in height and width at the expense of length while cells intercalate mediolaterally (Wilson et al., 1989). Likewise, radial intercalation in the animal cap produces uniform spreading (Keller, 1980), whereas in the IMZ it produces extension only in the anteroposterior direction, whether it occurs in the early gastrula (Fig. 4) (Wilson and Keller, 1991) or in the neurula (Fig. 6) (Wilson et al., 1989).

We suggest that such differences in output could arise from changes in protrusive activity. Protrusive activity could be modified in a way that still brings about intercalation but also changes the shape or biases the arrangement of cells in another dimension. For example, changes in the shape of the notochord and the shape and arrangement of its cells in the late neurula follow the expansion of protrusive activity to the dorsal and ventral margins of its cells (Keller et al., 1989a) (Fig. 6).

Another possibility is that differences in biomechanical context could also modulate the effect of an *unchanged* cell behavior. For example, if there is greater resistance to extension, convergence might result in more thickening and less lengthening, the balance between the two being determined by relative resistance of the tissue or of adjacent tissues in two dimensions. For example, radial intercalation may be inherently unbiased, and may produce uniform expansion in the animal cap because the attached epithelial layer offers uniform resistance there, and may produce extension in the IMZ because the epithelium there stretches more easily along the anteroposterior axis. There are examples of cell behavior being dependent on the biomechanical context. The effect of the swelling of notochord cells by vacuolation is dependent on their attachments to one another and on the mechanical constraints of the notochordal sheath and surrounding tissues (Koehl et al., 1990). In the case of *Xenopus* bottle cells, both the local cell behavior (apical constriction) and the tissue-level outcome (rolling of the marginal zone) are dependent on the relative stiffness of adjacent tissues (Hardin and Keller, 1988).

CONVERGENCE AND EXTENSION IN THE NIMZ

Convergence and Extension of the NIMZ in Embryos and in Sandwich Explants

Vital dye mapping of whole embryos shows that the dorsal NIMZ of *Xenopus* converges and extends to form the posterior neural plate (Keller, 1975). The length of the spinal cord is generated by enormous convergence and extension of a zone that is only about five to seven cells in animal–vegetal extent at the early gastrula stage (Keller et al., in preparation) (Fig. 8). These movements are autonomous. The dorsal NIMZ converges and extends in sandwich explants where it lies end-to-end with the IMZ rather than on top of it (Keller et al., 1985a,b; Keller and Danilchik, 1988) (Fig. 2A). Convergence and extension begin abruptly and strongly at the midgastrula stage (stage 10.5–11) and have a character and pattern specific to the dorsal NIMZ (prospective posterior neural plate). However, the dorsal NIMZ of sandwich explants shows only the convergence and extension aspects of normal neural plate morphogenesis. The cells do not columnarize or become wedge-shaped, as they do in the whole embryo. Thus the NIMZ does not thicken and roll into an organized neural tube (Schroeder, 1970), but it does converge, extend, and differentiate into a disorganized tangle of neurons (Keller and Danilchik, 1988). Therefore, convergence and extension of the neural plate is independent of the normal columnarization and wedging of its cells (Keller and Danilchik, 1988). Convergence and extension of the dorsal NIMZ does not occur in open-faced

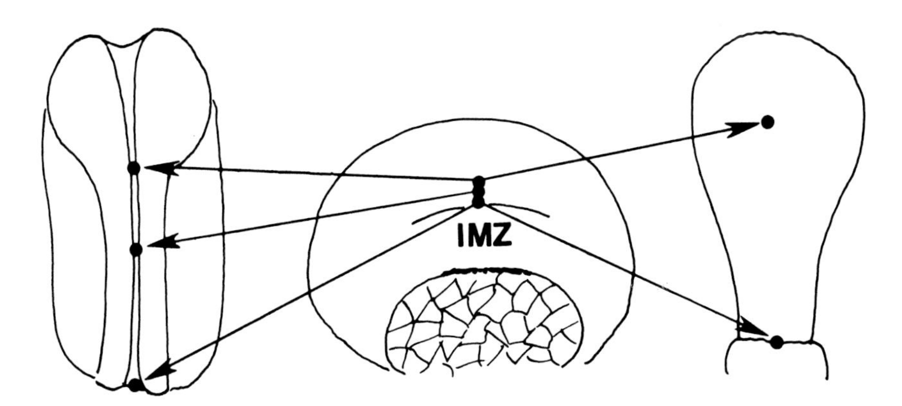

Fig. 8. Schematic diagrams show the graded convergence and extension of the dorsal NIMZ to form the length of the posterior hindbrain-spinal cord of a normal embryo (**left**) and the corresponding part of an extended NIMZ in a sandwich explant (**right**).

explants (Fig. 2B), most likely because the NIMZ extends too weakly to overcome friction in the open-faced explant (Shih, unpublished results).

Convergence and Extension of the NIMZ Occurs by Radial and Mediolateral Intercalation

Labeling deep cell populations shows that they undergo sequential radial and mediolateral intercalations, as in the IMZ. Radial intercalation first thins the NIMZ without much convergence, and then mediolateral intercalation produces convergence and extension, as in the IMZ (Keller et al., in preparation). Superficial cells of the NIMZ intercalate mediolaterally during convergence and extension (Keller, 1978).

The IMZ Induces Convergence and Extension of the NIMZ by an Early, Planar Interaction

Convergence and extension of the neural anlage represent an early morphogenetic response to a neural inducing signal moving from the IMZ through the plane of the tissue to the adjacent NIMZ (Keller and Danilchik, 1988). Dorsal NIMZ will extend and converge in sandwich explants only if left in contact with the IMZ (Fig. 9). Dorsal, lateral, and ventral NIMZ and animal cap tissue, which do not converge and extend alone, will do so when placed in

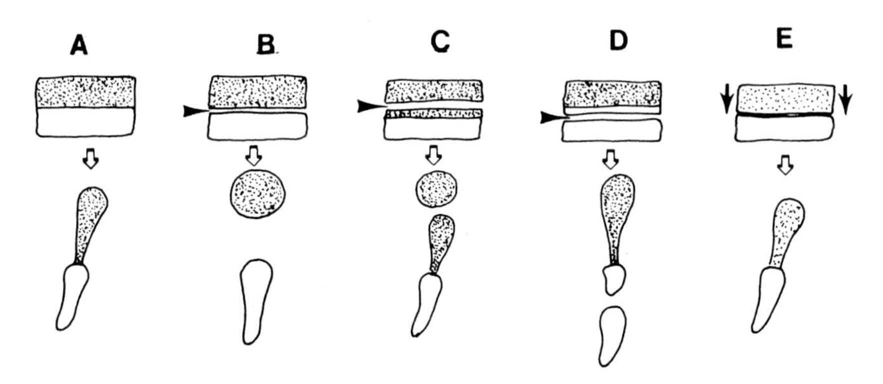

Fig. 9. Experiments showing the dependence of convergence and extension of the NIMZ (dark shading) on planar contact with IMZ (unshaded) are illustrated diagrammatically. Control explant shows extension of the NIMZ (**A**). The NIMZ separated from IMZ does not converge or extend (**B**). When a cut is made above the junction between the NIMZ and IMZ, only the NIMZ attached to the IMZ extends (**C**). When a cut is made below the junction, the NIMZ extends, provided IMZ is left on its vegetal end (**D**). Tissue from the animal cap, and the lateral NIMZ and the ventral NIMZ (light shading), which do not extend on their own in sandwich explants, will converge and extend when placed in planar or edgewise contact with IMZ (**E**). Based on Keller, Shih, Sater, and Moreno (in preparation).

planar apposition to the dorsal IMZ at the early gastrula stage (Fig. 9). Thus it appears that convergence and extension of NIMZ or animal cap tissues can be induced by planar interactions with the IMZ, and that normal extension of the NIMZ is dependent on this interaction.

Since convergence and extension of the NIMZ begins abruptly and strongly at the midgastrula stage (stage 10.5), this aspect of neural induction must occur in the first half of gastrulation. Although this is much earlier than had been thought in the traditional view, it is consistent with new evidence for early, planar signals and with Spemann's original view (see Phillips, Chapter 6, this volume). Phillips and his colleagues (Aker et al., 1986; London et al., 1988; Savage and Phillips, 1989) were the first to provide evidence for early planar interactions, by showing that early planar interaction was sufficient to repress expression of the epidermis-specific antigen Epi-1 in the prospective neural ectoderm. Kintner and Melton (1987) were able to visualize NCAM transcripts in the ectoderm of exogastrulated *Xenopus* embryos, which presumably have only planar interactions between mesoderm and ectoderm. NCAM is also expressed in "Keller" sandwiches of the dorsal marginal zone (Dixon and Kintner, 1989; Phillips and Doniach, submitted), in which planar signals alone are present. Planar signals alone are sufficient to organize anteroposterior pattern in the NIMZ of open-faced explants (see Gerhart et al., 1991) and to induce differentiation of neurons in open-faced explants without epithelia (Sater et al., 1991). This result shows that at least some planar signals, those responsible for differentiation of neurons, are transmitted through the deep layer of cells from the mesodermal component of the IMZ.

Vertical Signals Control Other Aspects of Neural Plate Morphogenesis

While planar interactions result in convergence and extension of the NIMZ, the columarization of cells that drives the formation of the neural plate, and the subsequent cell wedging that occurs during rolling of the plate into a tube (Schroeder, 1970; see Burnside, 1971; Jacobson and Gordon, 1976; Schoenwolf, 1985) do not occur in sandwich explants (Keller and Danilchik, 1988). These morphogenetic responses must depend on vertical signals. Holtfreter showed that any part of the neural tube adjacent to a notochord thinned to form a floorplate, whereas the regions adjacent to somitic mesoderm columnarized and thickened in the normal fashion of the lateral wall of the neural tube (Holtfreter and Hamburger, 1955). Wedging of median hinge-point (floorplate) cells is induced by the notochord in avian embryos (see Smith and Schoenwolf, 1989, and Schoenwolf, Chapter 4, this volume), and Jessell and his colleagues have demonstrated the role of the notochord in inducing the floorplate and other aspects of the ventrodorsal organization of the neural tube (see Jessell et al., 1989). Anterior notochord can induce neural markers by vertical interaction (Hemmati-Brivanlou et al., 1990). We do not yet know how the cell behaviors driving convergence and extension are integrated with those producing the thickening and rolling of the neural plate in *Xenopus*. It is known from the pioneering study of Jacobson and Gordon (1976) that the shaping of the neural plate of the California newt, *Taricha torosa*, is dependent on both extension at the midline and cell shape changes (see also Schoenwolf, 1985).

Convergence and Extension Change the Geometry of Tissue Interactions

Tissue interactions result in new cell behaviors, but these cell behaviors in turn change the geometry of tissue interactions and thus possibly their effectiveness. Convergence brings the dorsal mesoderm closer to the ventral mesoderm, perhaps aiding in dorsalization of the latter by the former (Keller and Danilchik, 1988). During planar induction of the dorsal NIMZ in the early gastrula, the inducing mesodermal tissue is initially only five to seven cells from the future hindbrain tissue, but as the dorsal NIMZ responds to this induction, it extends as much as 12-fold, progressively removing the inducing mesoderm from the brain region. The full implications of this movement for patterning of the nervous system are not yet known. Stewart and Gerhart (1990) have argued that extension of axial tissues may contribute to development of anterior structures simply by moving them farther away from the caudalizing influence expressed by the organizer at later stages.

MECHANISM OF MEDIOLATERAL INTERCALATION

We now turn our attention to the motile activities responsible for mediolateral intercalation. This discussion will focus on the IMZ, since this is the only region in which we have been able to observe deep cell behavior directly, and on mediolateral intercalation, since we still know little about radial intercalation.

Protrusive Activity During Mediolateral Intercalation

The deep cells next to the epithelium express most strongly the protrusive activity that causes mediolateral intercalation. We developed a method of observing these cells, using ‘‘shaved’’ explants, in which the deepest cell layers of the IMZ are removed until just one or two layers of deep cells remain on the inner surface of the epithelium (Fig. 2C) (Shih and Keller, in preparation). Radial intercalation is minimal, since there are few layers or only one layer of deep cells. These cells divide and oscillate about in place until the midgastrula stage, when convergence and extension begin. The cells then elongate, orient mediolaterally, and align parallel to one another in a transverse band located at the vegetal (prospective anterior) end of the dorsal IMZ (see arrowheads, Fig. 10A). During their elongation and alignment, the protrusive activity of

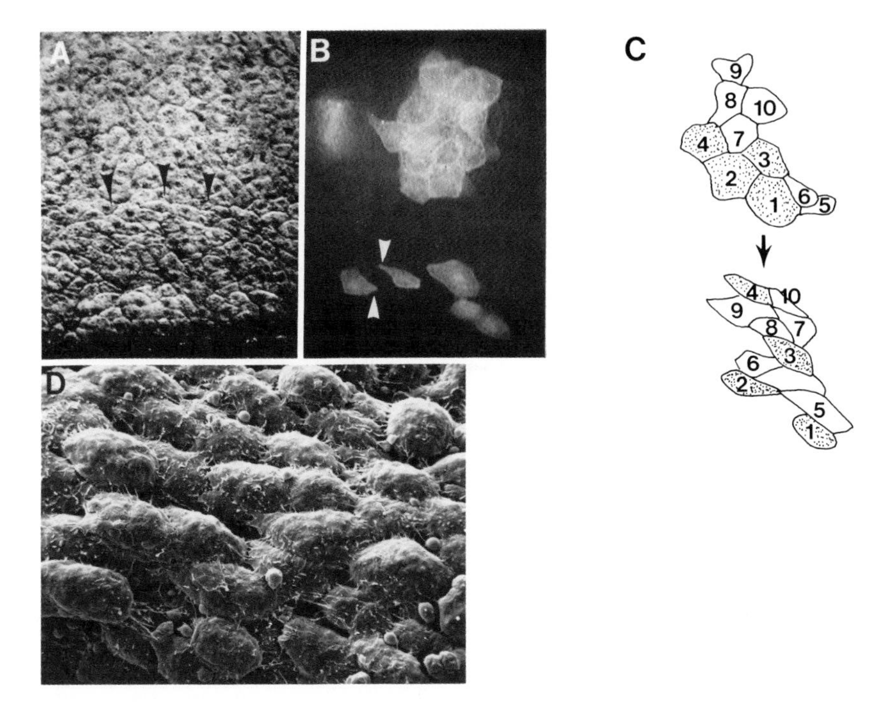

Fig. 10. The morphology and arrangement of cells undergoing mediolateral intercalation in shaved open-faced explants is illustrated (Shih and Keller, in preparation). A light micrograph taken from a time-lapse videorecording shows a band of mediolaterally elongate, aligned cells at the vegetal end of a shaved dorsal explant (arrowheads, **A**). A fluorescence videomicrograph (**B**) shows elongated cells in this band labeled with fluorescein dextran amine (FDA) as they make medial and lateral protrusions (arrowheads). Tracings from time-lapse recordings show the mediolateral intercalation of cells in this region (**C**). Scanning electron micrographs show the morphological polarization of the deep cells immediately adjacent to the endodermal epithelium in the IMZ in whole embryos (**D**). The axis of extension is approximately vertical in all cases. Based on Shih and Keller (in preparation). (D) From Keller et al., 1989b.

these cells, visualized by low-light, time-lapse fluorescence microscopy, is restricted progressively to their medial and lateral ends (Figs. 10B, 11). These protrusions are applied to the surfaces of adjacent deep cells, appear to exert traction on them, and thus move the cells between one another in the mediolateral direction, producing intercalation (Fig. 10C). By contrast, the anterior and posterior margins of the cells do not advance across adjacent cells. The morphological polarization of these cells is further characterized by scanning electron microscopy of the same cells in whole embryos (Keller et al., 1989b) (Fig. 10D). There are numerous filiform protrusions on the anterior and posterior sides of fixed cells; these probably represent focal points of contact that are turned into retraction fibers by shrinkage, since they are not seen in this

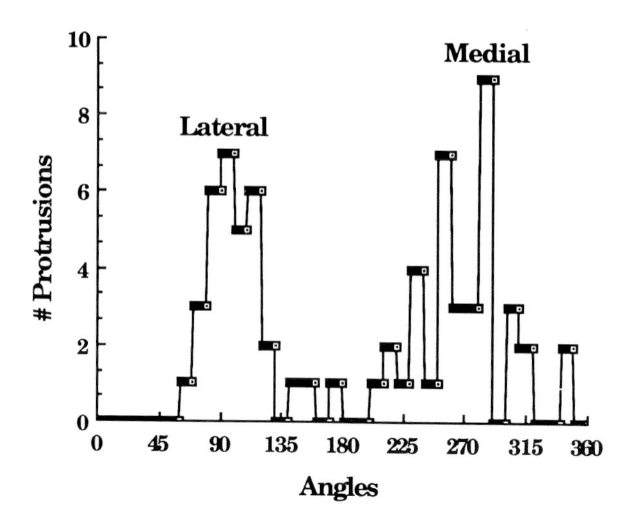

Fig. 11. The mediolateral bias of protrusive activity of the aligned deep cells during mediolateral intercalation is shown by a plot of frequency of protrusions against angle.

number and length in recordings of the closely packed, living cells. These protrusions probably hold the cells, cheek-by-jowl, in a parallel array but allow transverse shearing between cells, as they are pulled between one another by large medial and lateral protrusions.

Mechanism of Mediolateral Intercalation: A Current View

We summarize here our current view of the mechanism of mediolateral intercalation in *Xenopus*, paying particular attention to how a general, intrinsic behavior might be modified, cumulatively, by cell interactions, and how the more specific cell behavior arising out of these interactions might function in the mechanical and geometric context of the cell population.

1. Intrinsic mesodermal cell behavior: transient bipolarity. The mesodermal cells initially appear to be transiently bipolar (Fig. 12A). That is, a cell spends most of its time in a bipolar state in which two main protrusions dominate its activity until a subsidiary protrusion arises at another site to replace one of the originals. Individual *Xenopus* mesodermal cells cultured on fibronectin substrates display similar behavior (Nakatsuji and Johnson, 1983; Winklbauer, 1990). But individual cells are less polarized than those in the explant during intercalation. Cell interactions in the embryo or explant must enhance this intrinsic behavior.

2. Interactions of mesodermal cells with the endodermal epithelium. We know that the dorsal deep cells require early contact with the epithelium to produce convergence and extension, and that finally they become independent of the epithelium (Wilson et al., 1989; Shih and Keller, in preparation). We do not know why this occurs, but it appears to aid the development of the bipolar, aligned array of cells described above. The epithelium might reinforce the bipolar state by directly altering the protrusive activity of the deep cells (Fig. 12B, 1). This in turn might reinforce alignment by cell interactions discussed below (Fig. 12B, 2). It might *align* deep cell movement by providing substrate cues oriented mediolaterally (Fig. 12B, 3). Lastly, the epithelium might bias the *direction* of deep cell movement with substrate cues, perhaps by haptotaxis (Fig. 12B, 4). To decide among these possibilities, we must observe if deep cells have a specific behavior on the inner surface of the epithelium.

3. Contact-mediated alignment and reinforcement of bipolarity. Contact interactions between mesodermal cells may reinforce their bipolarity and bring them into parallel alignment (Fig. 12C). The morphology and behavior of the intercalating mesodermal cells resemble certain types of fibroblasts, which can self-organize into parallel arrays of elongate cells if allowed to interact with one another in culture (Elsdale and Wasoff, 1976; Erickson, 1978a,b). As individuals, these fibroblasts are bipolar, moreso than individual mesodermal cells, usually having an elongate morphology with two main protrusions at the narrow ends and many smaller contacts with the substratum on their long sides (see Erickson, 1978a). Numerous small, lateral protrusions inhibit underlapping by approaching cells, thus favoring parallel movements and alignment and minimizing collisions (see Erickson, 1978a; Elsdale and Wasoff, 1976). Transformed cells lacking the lateral protrusions do not align as well and crisscross readily (Erickson, 1978b). This type of behavior may align and polarize mesodermal cells (Keller et al., 1989a). We know that the large medial and lateral protrusions move along the surfaces of adjacent cell bodies without contact inhibition and with little crisscrossing (Shih and Keller, in preparation). The small contacts at the anterior and posterior surfaces may be arranged such that they channel the advance of the large protrusions. To test this idea, we must learn much more about how the large protrusions interact with one another and with the small anterior–posterior contacts.

4. Bias in the direction of intercalation. Since the main protrusive activity is directed medially and laterally, we believe that this is the primary force-generating mechanism producing mediolateral intercalation (Fig. 12D). Exactly *how* this behavior brings about intercalation is not clear. We suspect that cells are biased either in the frequency, the timing, or in the effectiveness of their protrusive activity such that they intercalate toward the midline. But we do not yet have conclusive data on these parameters in the early stages of mediolateral intercalation before notochord formation. Several investigators have devised mechanisms by which differential adhesion or cell recognition might function in intercalation (Mittenthal and Mazo, 1983; Wieschaus et al., 1991). At this point, we have no direct evidence that adhesive differences or cell recognition is involved in this system, other than in the obvious ways described above.

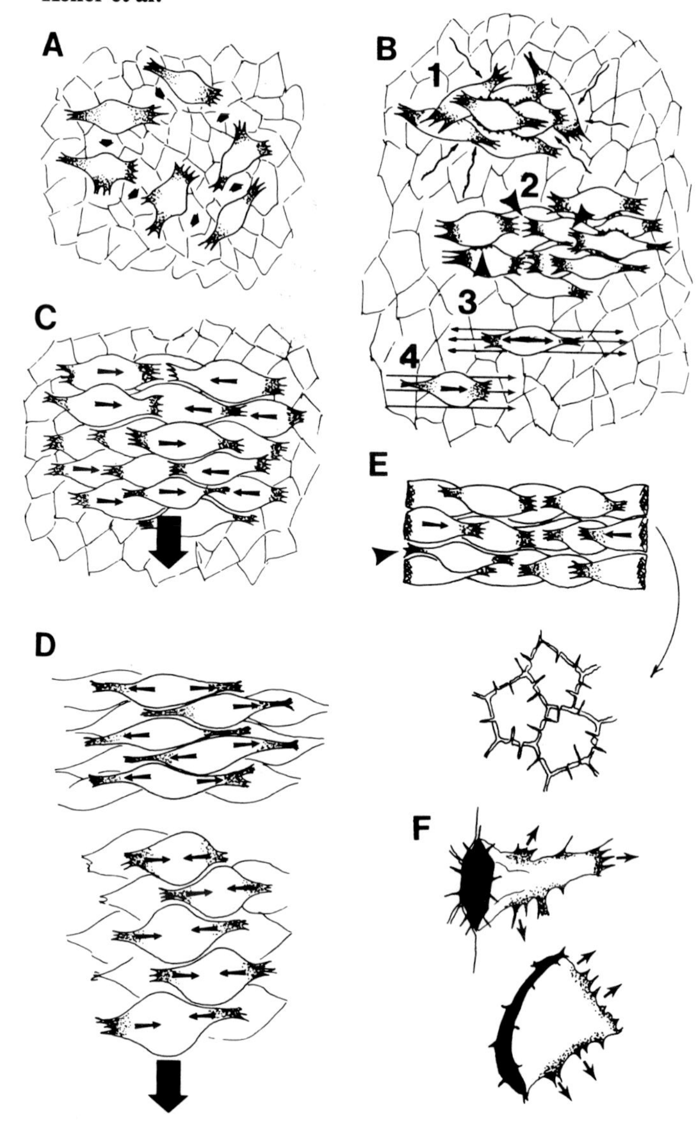

Fig. 12. A schematic diagram depicts our view of possible cellular events leading to mediolateral intercalation of deep cells, pictured here as lying on the inner surface of the endodermal epithelium. Isolated mesodermal cells display a transient, weak bipolarity (**A**). This may be reinforced by an influence from the epithelium (**B**, 1). Contact interactions between cells may produce alignment (arrowheads, B, 2). The epithelium could align cells and direct protrusive activity by mediolaterally organized stereotactic cues (B, 3), or it could produce directional protrusive activity by directional cues, perhaps by haptotaxis (B, 4). Progressive intercalation occurs as the cells gradually move between one another along the mediolateral axis to produce a longer array (**C**). Episodic intercalation involves extension of protrusions between adjacent cells followed by a contraction of the protrusions (**D**). Protrusive activity is inhibited at the boundary of the noto-

5. Relation of cell traction, tension, and compression during mediolateral intercalation. For the time being, we assume that the mediolaterally directed traction pulls the cells between one another, generating a wedging action that pushes neighboring cells apart and extends the tissue at the expense of width (Fig. 12C). This predicts that the deep cells should be under tension in the mediolateral direction and under compression in the anteroposterior direction. We are testing this notion by comparing retraction when medial/lateral connections and anterior/posterior connections are broken by micromanipulation. Tension could lend a self-reinforcing property to convergence and extension by reinforcing polarized protrusive activity. Protrusive activity is inhibited at cell margins under tension, at least in some cell types (see Kolega, 1986). If this applies to intercalating mesodermal cells, resistance to extension might increase tension in the long sides of the cells and thus increase polarization of protrusive activity.

6. Interplay of progressive intercalation and episodic intercalation. Our analysis thus far suggests that intercalation normally occurs by *progressive intercalation*, in which the cells move slowly between one another without great change in cell shape, as a function of the mediolateral traction. But under other conditions, probably involving increased resistance to extension, *episodic intercalation* occurs. As resistance rises enough to prevent intercalation, the cells appear to continue their protrusive activity and extend protrusions considerable distances between one another without changing the shape of the explant. They then undergo an episode of intercalation by contraction of these extended protrusions, pulling the cells between one another, and producing convergence and extension (Fig. 12C). Afterwards, they may return to progressive intercalation. We think episodic intercalation may be interspersed with progressive intercalation in space and time, depending on local, minute-to-minute resistance to extension. In support of this idea, the advance of the *Xenopus* blastoporal lip is sometimes punctuated by episodic contractions; it will advance at a uniform rate, slow down, and then surge forward in rapid convergence and extension before returning to the normal rate (Keller, unpublished observations). The newt neural plate normally shows episodic movements during convergence and extension (B. Burnside and A. Jacobson, personal communication). These behaviors are hard to explain by progressive intercalation alone and may reflect the existence of episodic intercalation.

7. Boundary-induced polarization of protrusive activity. The normal, invasive protrusive activity of intercalating cells is inhibited on contact with the notochord–somite boundary, and thus the previously bipolar deep cells become

---

chord (crosshatching, **E**), and the cells become monopolar in protrusive activity. The apices comprising the boundary surface become gridlocked by filiform protrusions lapping across adjacent cells (E, below). Subsequently, monopolar protrusive activity spreads to the dorsal and ventral edges of the boundary cells and the cells change shape (**F**).

monopolar and direct all protrusive activity inward, away from the boundary, when one end contacts the boundary (Keller et al., 1989a) (Fig. 12E). Inwardly directed traction would tend to pull other cells to the boundary, where they too would be "captured" by its inhibitory effect (arrowhead, Fig. 12E). Accumulation of cells at the boundary would expand its area, in this case, an increase channeled primarily into length (Keller et al., 1989a). This idea of capture of cells at a boundary has been invoked to explain why the notoplate–neural plate boundary is essential for extension of the neural anlagen of the newt (Jacobson et al., 1986). It is likely that only *invasive* protrusive activity is inhibited at the boundary, because some time after this event, the apices of the boundary cells become gridlocked by overlapping filiform protrusions, which probably stabilize the lateral surface of notochord. In the late neurula, invasive protrusive activity spreads to the top and bottom edges of the cells, which probably functions in the spreading of the cells circumferentially to form the pizza-slice shape of the cells (Fig. 12F). This shape and pattern of connections between cells sets up the biomechanical environment that channels the uniform forces generated by the subsequent vacuolation into continued extension, straightening, and stiffening of the notochord in the tailbud stages (Koehl et al., 1990).

8. Mechanical models of protrusive activity. How do we visualize and evaluate what type of tissue movement might result from a given pattern of protrusive activity and rules for cell interactions? A very powerful way is to make computer simulations of cell behavior, using realistic mechanical parameters and rules of cell behavior and interaction taken from experimental results (Weliky and Oster, 1991; Weliky et al., 1991). Using this technique, simulations show that inhibition of protrusive activity at the boundary of the notochord will produce elongation of the simulated notochord and behavior of the boundary cells but not the elongation and alignment of internal notochord cells. When intrinsic polarization is added, the internal cells elongate but do not align. Finally, addition of contact inhibition of protrusive activity to the previous rule produces all major features of notochord morphogenesis. These investigations lend credence to the idea that cumulative action of several experimentally observed intrinsic motile properties and rules of cell interaction are sufficient to account for the observed mediolateral intercalation of cells.

In summary, a large number of subtle cues or a few strong ones could bring about highly polarized protrusive activity during mediolateral intercalation. One major task before us is evaluating the contributions of those cues described above, and others as well. We also want to learn more about the mechanics of translating polarized protrusive activity into convergence and extension by studying responses of intercalation to different mechanical perturbations.

CONVERGENCE AND EXTENSION AND CELL INTERCALATION IN OTHER SYSTEMS

Convergence and extension occur in development of many vertebrates and invertebrates. These should be attacked experimentally to learn how they work, for it is very likely that there is not one mechanism of convergence and extension but several. During neurulation of the newt, extension occurs and mediolateral cell intercalation appears necessary (Jacobson and Gordon, 1976) and in fact occurs (Jacobson et al., 1986). Extension may be produced by a special condition at the boundary of the notoplate (neural plate above the notochord), and the cells lateral to it, that drives cell rearrangement (Jacobson, 1981; Jacobson et al., 1986). Mediolateral intercalation occurs in extension of the chick neural plate, and here oriented cell division and change in cell shape also play roles (Schoenwolf and Alvarez, 1989; Schoenwolf, Chapter 4, this volume). both types of intercalation, radial and mediolateral, occur in gastrulation of the teleost fish (Kimmel et al., 1990; Warga and Kimmel, 1990), in the same sequence as in the amphibian. Protrusive activity among cells converging on the embryonic shield in *Fundulus*, a teleost, is contact inhibited (Trinkaus et al., 1991), suggesting that behavior during convergence is governed by contact behavior that is different from that seen in *Xenopus*. Notochord cell intercalation occurs in the teleost fish (Thorogood and Wood, 1987) and in the ascidian (Miyamoto and Crowther, 1985) in much the same fashion as it does in *Xenopus*. Both systems are transparent and lend themselves to detailed, direct analysis of cell behavior. Archenteron elongation in the sea urchin involves cell intercalation (Ettensohn, 1985; Hardin and Cheng, 1986; Hardin, 1988), and the elongation produced is active and autonomous in the middle phase of archenteron formulation (Hardin and Cheng, 1986; Hardin, 1988). The protrusive activity during archenteron elongation in the sea urchin appears quite different from that observed in *Xenopus* (see Hardin, 1989), suggesting that this epithelial system may have a different mechanism. Germband elongation in *Drosophila* appears to involve mediolateral cell intercalation; an interesting model has been proposed to explain this rearrangement, based on adhesive preferences determined by positional values (Gergen et al., 1986; Wieschaus et al., 1991). *Drosophila* imaginal disc evagination is the system in which epithelial cell rearrangement was first described (Fristrom, 1976). Recently it has been shown that a dramatic change in cell shape also occurs in this system and drives much of the convergence and extension in part of the leg disc during evagination (Condic et al., 1991).

Since we know very little about what cells actually do in most systems, and even less about their biomechanics, we do not yet know how many significantly different fundamental mechanisms there are for convergence and extension and related mass movements. Many systems should be examined in depth to answer this question.

GENERAL CONCLUSIONS

We have only begun the analysis of the mechanism of convergence and extension at the cell and cell population level, but several general points are clear regarding cell interactions. The important behavior of the individual cell arises from cell interactions, and that behavior, in turn, takes its meaning from a specific biomechanical and geometric context in the cell population. Cells must interact to display properties that are the *specific causes* of a morphogenetic mass movement. A cell may be highly specialized to participate in complex population movements, but these specializations cannot be exercised without the essential aspect of multicellular morphogenesis—cell interactions. Functions of specialized cell surface proteins, cytoskeletal components, or signal transducers can be expressed when cells are allowed to interact in their normal configuration. The resulting behavior is not likely to be general behavior seen in culture but *morphogenetically specific behavior*. Likewise, the function of this specific behavior is not defined intrinsically but is given meaning by the geometric and biomechanical context in which it is exercised, and thus it is context-dependent or context-sensitive (see Hardin, 1990; Hardin and Keller, 1988; Koehl et al., 1990). What we know of mediolateral intercalation already makes it highly unlikely that the force-generating capacities of the cytomusculature and the mechanical properties of the cytoskeleton have significance outside the mechanics of the tissue lattice. Thus, the mechanism and function of cell behaviors and the function of the molecular aggregates underlying these behaviors, in producing the mass movements so important in embryogenesis, will be defined in terms of the cell interactions and the biomechanics of the cell population to a far greater degree than has been appreciated. For this reason we think it important to return developmental mechanics to a central position in developmental biology.

ACKNOWLEDGMENTS

We thank former members of the laboratory, particularly Jeff Hardin, Rudolf Winklbauer, and Mark Cooper, and visitors to the laboratory, Antone Jacobson and J.P. Trinkaus, who have had a hand in developing our ideas and experiments on cell intercalation, above and beyond what they have published. The same applies to our colleagues George Oster, Diane Fristrom, Beth Burnside, John Gerhart, Mike Weliky, Steve Minsuk, Connie Lane, Jessica Bolker, and Maureen Condic at Berkeley, and Ann Sutherland at University of California, San Francisco. The work described here was supported by National Institutes of Health grants HD18979 and HD25594 and National Science Foundation grant DCB89052 to Ray Keller, NSF grant DMS 8618975 to Paul Wilson and George Oster; Paul Wilson was supported in part by National Institutes of Health Training Grant HD7375. Amy Sater was supported by an NIH postdoctoral fellowship. We thank Paul Tibbetts for technical assistance.

REFERENCES

Adams D, Keller R, Koehl MAR (1990): The mechanics of notochord elongation, straightening, and stiffening in the embryo of *Xenopus laevis*. Development 110:115–130.

Aker RM, Phillips CR, Wessels NK (1986): Expression of an epidermal antigen used to study tissue induction in the early *Xenopus* embryo. Science 231:613–616.

Bautzmann H (1933): Uber Determinationgrad und Wirkungsbezeichungen der Randzonenteil anlagen (Chorda—Ursegmente, Seitenplatten und Kopfdarmanlage) bei Urodelen und Anuren. Roux Arch 128:666–765.

Burnside B (1971): Microtubules and microfilaments in newt neurulation. Dev Biol 26:416–441.

Condic M, Fristrom D, Fristrom JW (1991): Apical cell shape changes during *Drosophila* imaginal leg disc elongation: A novel morphogenetic mechanism. Development (in press).

Dale L, Slack JMW (1987): Regional specification within the mesoderm of early embryos. Development 99:527–552.

Dixon J, Kintner C (1989): Cellular contacts required for neural induction in *Xenopus* embryos: Evidence for two signals. Development 106:749–757.

Elsdale T, Wasoff F (1976): Fibroblast cultures and dermatoglyphs: The topology of two planar patterns. Roux Arch 180:121–147.

Erickson CA (1978a): Analysis of the formation of parallel arrays of BHK cells *in vitro*. Exp Cell Res 115:303–315.

Erickson CA (1978b): Contact behavior and pattern formation of BHK and polyoma virus-transformed BHK fibroblasts in culture. J Cell Sci 33:53–84.

Ettensohn C (1985): Gastrulation in the sea urchin is accompanied by the rearrangement of invaginating epithelial cells. Dev Biol 112:383–390.

Fristrom D (1976): The mechanism of evagination of imaginal discs of *Drosophila melanogaster*. III. Evidence for cell rearrangement. Dev Biol 54:163–171.

Fristrom D (1988): The cellular basis of epithelial morphogenesis. A review. Tissue Cell 20:645–690.

Gergen JP, Coulter D, Wieschaus E (1986): Segmental pattern and blastoderm cell identities. In Subtelny S (ed): "Gametogenesis and the Early Embryo." New York: Alan R. Liss, Symp Soc Dev Biol 43:195–220.

Gerhart J, Keller RK (1986): Region-specific cell activities in amphibian gastrulation. Ann Rev Cell Biol 2:201–229.

Gerhart J, Doniach T, Stewart R (1991): Organizing the organizer. In Keller R, Clark W, Griffin F (eds): "Gastrulation: Movements, Patterns, and Molecules." New York: Plenum Press (in press).

Gilbert S (1988): "Developmental Biology." Sunderland, MA: Sinauer Associates.

Gimlich R, Cooke J (1983): Cell lineage and induction of second nervous systems in amphibian development. Nature 30:471–473.

Goodale HD (1911): The early development of *Spelerpes bilineatus* (Green). Am J Anat 12:173–247.

Hamburger V (1960): "A Manual of Experimental Embryology" (rev. ed.). Chicago: University of Chicago Press.

Hardin J (1988): The role of secondary mesenchyme cells during sea urchin gastrulation studied by laser ablation. Development 103:317–324.

Hardin J (1989): Local shifts in position and polarized motility drive cell rearrangement during sea urchin gastrulation. Dev Biol 136:430–445.

Hardin J (1990): Context-sensitive cell behaviors during gastrulation. In Keller R, Fristrom D (eds): "Control of Morphogenesis by Specific Cell Behaviors." Seminars in Developmental Biology 1:335–345.

Hardin J, Cheng LY (1986): The mechanisms and mechanics of archenteron elongation during sea urchin gastrulation. Dev Biol 115:490–501.

Hardin J, Keller R (1988): The behavior and function of bottle cells during gastrulation of *Xenopus laevis*. Development 103:211–230.

Hemmati-Brivanlou A, Stewart R, Harland R (1990) Region-specific neural induction of an engrailed protein by anterior notochord in *Xenopus*. Science 250:800–802.

Holtfreter J (1933): Die totale Exogastrulation eine Selbstablosung Ektoderm von Entomesoderm. Arch Entwicklungsmech Org 129:669–793.

Holtfreter J (1939): Gewebeaffinität, ein Mittel der embryonalen Formbildung. Arch Exp Z Besonders Geweb 23:169–209.
Holtfreter J (1943a): Properties and function of the surface coat in amphibian embryos. J Exp Zool 93:251–323.
Holtfreter J (1943b): A study of the mechanics of gastrulation: Part I. J Exp Zool 94:261–318.
Holtfreter J (1944): A study of the mechanics of gastrulation: Part II. J Exp Zool 95:171–212.
Holtfreter J (1946): Structure, motility and locomotion in isolated embryonic amphibian cells. J Morphol 72:27–62.
Holtfreter J, Hamburger V (1955): Embryogenesis: Progressive differentiation, amphibians. In Willier BH, Weiss PA, Hamburger V (eds): "Analysis of Development." Philadelphia: W.B. Saunders Co., pp 230–296.
Ikushima N, Maruyama S (1971): Structure and developmental tendency of the dorsal marginal zone in the early amphibian gastrula. J Embryol Exp Morphol 25:263–276.
Jacobson A (1981): Morphogenesis of the neural plate and tube. In Connolly TG, Brinkley L, Carlson B (eds): "Morphogenesis and Pattern Formation." New York: Raven Press, pp 223–263.
Jacobson A, Gordon R (1976): Changes in the shape of the developing vertebrate nervous system analyzed experimentally, mathematically, and by computer simulation. J Exp Zool 197:191–246.
Jacobson A, Oster G, Odell G, Cheng L (1986): Neurulation and the cortical tractor model for epithelial folding. J Embryol Exp Morphol 96:19–49.
Jessell T, Bovolenta M, Placzek M, Tessier-Lavigne M, Dodd J (1989): Polarity and patterning in the neural tube: The origin and function of the floor plate. In "Cellular Basis of Morphogenesis," Ciba Foundation Symposium. New York: Wiley, pp 257–282.
Keller RE (1975): Vital dye mapping of the gastrula and neurula of *Xenopus laevis*. I. Prospective areas and morphogenetic movements of the superficial layer. Dev Biol 42:222–241.
Keller RE (1976): Vital dye mapping of the gastrula and neurula of *Xenopus laevis*. II. Prospective areas and morphogenetic movements of the deep layer. Dev Biol 51:118–137.
Keller RE (1978): Time-lapse cinemicrographic analysis of superficial cell behavior during and prior to gastrulation in *Xenopus laevis*. J Morphol 157:223–248.
Keller RE (1980): The cellular basis of epiboly: An SEM study of deep cell rearrangement during gastrulation in *Xenopus laevis*. J Embryol Exp Morphol 60:201–234.
Keller RE (1981): An experimental analysis of the role of bottle cells and the deep marginal zone in gastrulation of *Xenopus laevis*. J Exp Zool 216:81–101.
Keller RE (1984): The cellular basis of gastrulation in *Xenopus laevis*: Active post-involution convergence and extension by medio-lateral interdigitation. Am Zool 24:589–603.
Keller RE (1986): The cellular basis of amphibian gastrulation. In Browder L (ed): "Developmental Biology: A Comprehensive Synthesis. The Cellular Basis of Morphogenesis," Vol 2. New York: Plenum Press, pp 241–327.
Keller RE (1987): Cell rearrangement in morphogenesis. Zool Sci 4:763–779.
Keller RE, Schoenwolf GC (1977): An SEM study of cellular morphology, contact, and arrangement, as related to gastrulation in *Xenopus laevis*. Roux Arch 181:165–182.
Keller RE, Danilchik M, Gimlich R, Shih J (1985a): Convergent extension by cell intercalation during gastrulation of *Xenopus laevis*. In Edelman GM (ed): "Molecular Determinants of Animal Form." New York: Alan R. Liss, pp 111–141.
Keller RE, Danilchik M, Gimlich R, Shih J (1985b): The function of convergent extension during gastrulation of *Xenopus laevis*. J Embryol Exp Morphol 89 (Suppl)185–209.
Keller RE, Hardin J (1987): Cell behavior during active cell rearrangement: Evidence and speculation. J Cell Sci 8 (Suppl):369–393.
Keller RE, Trinkaus JP (1987): Rearrangement of enveloping layer cells without disruption of the epithelial permeability barrier as a factor in *Fundulus* epiboly. Dev Biol 120:12–24.

Keller RE, Danilchik M (1988): Regional expression, pattern and timing of convergence and extension during gastrulation of *Xenopus laevis*. Development 103:193–210.

Keller R, Cooper MS, Danilchik M, Tibbetts P, Wilson PA (1989a): Cell intercalation during notochord development in *Xenopus laevis*. J Exp Zool 251:134–154.

Keller RE, Shih J, Wilson PA (1989b): Morphological polarity of intercalating deep mesodermal cells in the organizer of *Xenopus laevis* gastrulae. Proceedings, 47th Annual Meeting, Electron Microscopy Society of America. San Francisco: San Francisco Press, p 840.

Keller RE, Tibbetts P (1989): Mediolateral cell intercalation is a property of the dorsal, axial mesoderm of *Xenopus laevis*. Dev Biol 131:539–549.

Keller R, Shih J, Wilson PA (1991): Cell motility, control and function of convergence and extension during gastrulation of *Xenopus*. In Keller R, Clark W, Griffin F (eds): "Gastrulation: Movements, Patterns, and Molecules." New York: Plenum Press (in press).

Kimmel CR, Warga R, Schilling T (1990): Origin and organization of the zebrafish fate map. Development 108:581–594.

Kintner C, Melton D (1987): Expression of *Xenopus* N-CAM RNA in ectoderm is an early response to neural induction. Development 99:311–325.

Koehl MAR, Adams D, Keller R (1990): Mechanical development of the notochord in *Xenopus* early tail-bud embryos. In Akkas N (ed): "Biomechanics of Active Movement and Deformation of Cells," NATO ASI series, Vol H42. Berlin: Springer-Verlag, pp 471–485.

Kolega J (1986): Effects of mechanical tension on protrusive activity and microfilament and intermediate filament organization in an epidermal epithelium moving in culture. J Cell Biol 102:1400–1411.

Lewis W (1947): Mechanics of invagination. Anat Rec 97:139–156.

London C, Akers R, Phillips C (1988): Expression of Epi-1, an epidermis-specific marker in *Xenopus laevis* embryos, is specified prior to gastrulation. Dev Biol 129:380–389.

Lundmark C (1986): Role of bilateral zones of ingressing superficial cells during gastrulation of *Ambystoma mexicanum*. J Embryol Exp Morphol 97:47–62.

McEwen R (1923): "A Textbook of Vertebrate Embryology." New York: Henry Holt.

Mittenthal J, Mazo R (1983): A model for shape generation by strain and cell-cell adhesion in the epithelium of an arthropod leg segment. J Theoret Biol 100:443–483.

Miyamoto DM, Crowther R (1985): Formation of the notochord in living ascidian embryos. J Embryol Exp Morphol 86:1–17.

Morgan TH (1897): "The Development of the Frog's Egg: An Introduction to Experimental Embryology." New York: Macmillan.

Morgan TH (1927): "Experimental Embryology." New York: Columbia University Press.

Nakatsuji N, Johnson K (1983): Cell locomotion in vitro by *Xenopus laevis* gastrula mesodermal cells. J Cell Sci 59:43–60.

Purcell S (1989): A different type of anuran gastrulation and morphogenesis as seen in *Ceratophrys ornata*. Am Zool 29:85a.

Rhumbler L (1902): Zur Mechanik des Gastrulationsvorganges, insbesondere der Invagination. Eine entwicklungsmechanishe Studie. Roux Arch Entwicklungsmech Org 14:401–476.

Sater A, Uzman JA, Steinhardt R, Keller R (1991): Neural induction in *Xenopus*: Induction of neuron differentiation by either planar or vertical signals. (Submitted for publication).

Savage R, Phillips C (1989): Signals from the dorsal blastopore lip region during gastrulation bias the ectoderm toward a nonepidermal pathway of differentiation in *Xenopus laevis*. Dev Biol 133:157–168.

Schechtman AM (1942): The mechanics of amphibian gastrulation. I. Gastrulation-producing interactions between various regions of an anuran egg (*Hyla regilla*). Univ Calif Publ Zool 51:1–39.

Schoenwolf GC (1985): Shaping and bending of the avian neuroepithelium: Morphometric analyses. Dev Biol 109:127–139.

Schoenwolf GC, Alvarez IS (1989): Roles of neuroepithelial cell rearrangement and division in shaping of the avian neural plate. Development 106:427–439.

Schroeder T (1970): Neurulation in *Xenopus laevis*. An analysis and model based upon light and electron microscopy. J Embryol Exp Morphol 23:427–462.

Shumway W (1927): ''Vertebrate Embryology.'' New York: Wiley.

Smith J, Malacinski G (1983): The origin of the mesoderm in an Anuran, *Xenopus laevis*, and a Urodele, *Ambystoma mexicanum*. Dev Biol 98:250–254.

Smith J, Schoenwolf G (1989): Notochordal induction of cell wedging in the chick neural plate and its role in neural tube formation. J Exp Zool 250:49–62.

Smith J, Dale L, Slack JMW (1985): Cell lineage labels and region-specific markers in the analysis of inductive interactions. J Embryol Exp Morphol 89 (Suppl):317–331.

Smith J, Symes K (1987): Gastrulation movements provide an early marker of mesoderm induction in *Xenopus laevis*. Development 101:339–349.

Spemann H (1938): ''Embryonic Development and Induction.'' New Haven: Yale University Press.

Stewart R (1990): ''The Active Inducing Center of the Embryonic Body Axis in *Xenopus*,'' Ph.D. dissertation, University of California, Berkeley.

Stewart R, Gerhart J (1990): The anterior extent of dorsal development of the *Xenopus* embryonic axis depends on the quantity of organizer in the late blastula. Development 109:363–372.

Stewart R, Gerhart J (1991): Induction of notochord by the organizer in *Xenopus*. (Submitted for publication).

Sudarwati S, Nieuwkoop P (1971): Mesoderm formation in the Anuran *Xenopus laevis* (Daudin). Roux Arch 166:189–204.

Thorogood P, Wood A (1987): analysis of *in vivo* cell movement using transparent tissue systems. J Cell Sci 8 (Suppl):395–413.

Townes P, Holtfreter J (1955): Directed movements and selective adhesion of embryonic amphibian cells. J Exp Zool 128:53–120.

Trinkaus JP, Fink R, Trinkaus M (1991): An in vivo analysis of convergent cell movements in the germ ring of *Fundulus*. In Keller R, Clark W, Griffin F (eds): ''Gastrulation: Movements, Patterns, and Molecules.'' New York: Plenum Press (in press).

Vogt W (1929): Gestaltanalyse am Amphibienkein mit ortlicher Vitalfarbung. II. Teil. Gastrulation und Mesodermbildung bei Urodelen und Anuren. Roux Arch Entwmech Org 120:384–706.

Waddington CH (1940): ''Organizers and Genes.'' Cambridge, UK: Cambridge University Press.

Warga R, Kimmel C (1990): Cell movements during epiboly and gastrulation in zebrafish. Development 108:569–580.

Weliky M, Oster G (1991): Dynamical models for cell rearrangement during morphogenesis. In Keller R, Clark W, Griffin F (eds): ''Gastrulation: Movements, Patterns, and Molecules.'' New York: Plenum Press (in press).

Weliky M, Minsuk S, Oster G, Keller R (1991): The mechanical basis of cell rearrangement. II. Models for cell behavior driving notochord morphogenesis in *Xenopus laevis*. (Submitted for publication).

Wieschaus E, Sweeton D, Costa M (1991): Convergence and extension during germband elongation in *Drosophila* embryos. In Keller R, Clark W, Griffin F (eds): ''Gastrulation: Movements, Patterns, and Molecules.'' New York: Plenum Press (in press).

Wilson PA (1990): ''The Development of the Axial Mesoderm in *Xenopus laevis*,'' Ph.D. dissertation, University of California, Berkeley.

Wilson PA, Oster G, Keller RE (1989): Cell rearrangement and segmentation in *Xenopus*: Direct observation of cultured explants. Development 105:155–166.

Wilson PA, Keller RE (1991): Cell rearrangement during gastrulation of *Xenopus*: Direct observation of cultured explants. Development 112:289–300.

Winklbauer R (1990): Mesodermal cell migration during *Xenopus* gastrulation. Dev Biol 142:155–168.

Cell-Cell Interactions in Early Development, pages 63–78

4. Neurepithelial Cell Behavior During Avian Neurulation

Gary C. Schoenwolf

Department of Anatomy, University of Utah, School of Medicine, Salt Lake City, Utah 84132

INTRODUCTION

Neurulation is a complex, multifactorial process involving the interplay of forces generated both intrinsically and extrinsically to the neural plate (reviewed by Schoenwolf and Smith, 1990a); as a result of these forces, the flat neural plate is transformed into the neural tube, the rudiment of the central nervous system (Fig. 1). Soon after formation of the neural plate, two processes occur—termed *shaping* and *bending*—that collectively convert the neural plate from a flat epithelial sheet of low columnar cells to an elongated trough (and, eventually, in conjunction with fusion of the neural folds flanking the trough, to an elongated tube) whose walls consist mainly of high columnar cells arranged as a pseudostratified epithelium (see especially Schoenwolf, 1982, 1983). Experiments suggest that shaping of the neural plate results chiefly from *intrinsic* forces generated by changes in the behavior of the constituent neurepithelial cells (Schoenwolf and Powers, 1987; Schoenwolf, 1988; Schoenwolf and Alvarez, 1989; Schoenwolf et al., 1989b), whereas bending of the neural plate also requires substantial extrinsic forces, presumably generated by changes in the behavior of cells constituting non-neurepithelial tissues (Schoenwolf, 1988; Schoenwolf et al., 1989b; Smith and Schoenwolf, 1991). Unfortunately, because of the traditional focus of neurulation studies on the neural plate itself, we know very little about the behaviors outside the neural plate that generate morphogenetic movements essential for neurulation. Thus my discussion is, by necessity, focused on the neural plate.

The purpose of this chapter is to discuss briefly neurepithelial cell behavior during shaping and bending of the avian neural plate. Cells display a limited repertoire of diverse behaviors during morphogenesis (Fig. 2). Neurepithelial cells are known to exhibit at least three fundamental behaviors: change in shape, number, and position. The evidence for the existence of each of these behaviors, as well as their postulated role in avian neurulation, is discussed below.

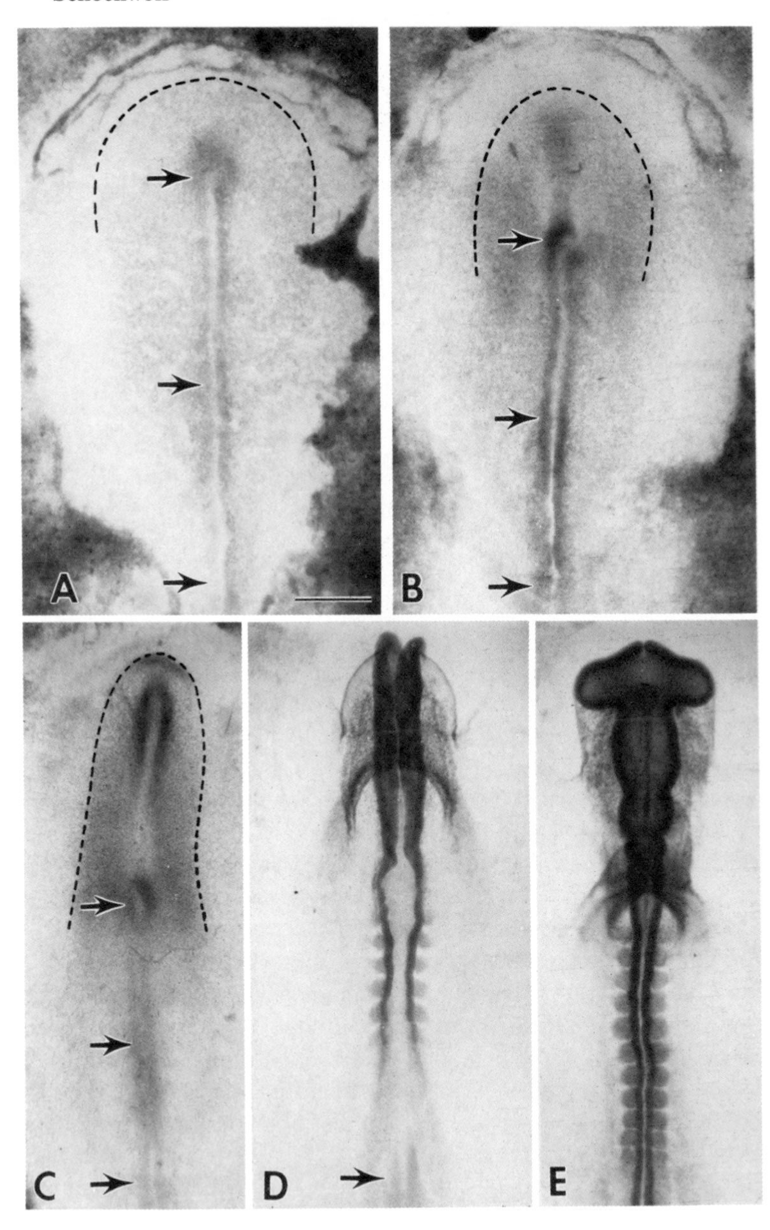

Fig. 1. Light micrographs of dorsal views of chick blastoderms at stages 4–11 (Hamburger and Hamilton, 1951). **A**, Flat neural plate (stage 4); **B**, flat neural plate with underlying head process (notochord) (stage 5); **C**, neural groove (stage 6); **D**, incipient neural tube (stage 8); **E**, definitive neural tube (stage 11). Note particularly that coordinated narrowing and lengthening of the neural plate occur prior to formation of the neural tube (at stage 8) and that elongation of the neural plate continues with bending and neural fold fusion. The boundaries of the neural plate are partially outlined. Arrows, primitive streak. Bar = 200 μm.

Fig. 2. "The agony and the ecstasy." Neurepithelial cells, like people, exhibit a diversity of behaviors, but their repertoire is limited. A single morphogenetic stimulus can generate different behaviors in adjacent cells, but the range of responses is fixed both by the genetic constitution of the organism and by the prospective potencies of its cells at the time the stimulus is applied. Here, in response to a stimulus (the camera), one organism (my daughter Jennifer Lee at age 6) smiles while the other (my son Gregory Charles at age 2.5) frowns.

CHANGE IN NEUREPITHELIAL CELL SHAPE

During neurulation, neurepithelial cells alter their shapes in two typical ways. First, during shaping of the neural plate, a process in which the neural plate undergoes true growth (i.e., an increase in cell number with an increase in tissue volume) as well as apicobasal thickening, transverse narrowing, and longitudinal lengthening, neurepithelial cells change their heights, transforming from low columnar to high columnar; subsequently, midline neurepithelial cells overlying the notochord decrease their heights, while more lateral neurepithelial cells become even taller (see especially Schoenwolf, 1985). Second, during shaping and bending of the neural plate, most midline neurepithelial cells become wedge-shaped; by contrast, most of the lateral neurepithelial cells remain spindle-shaped (spindle-shaped cells are generally designated as high columnar, but spindle-shaped is more accurate) (Schoenwolf and Franks, 1984). An additional population of neurepithelial cells becomes distinct at this time at certain neuraxial levels, namely, at future brain and caudal spinal cord levels. Like the midline neurepithelial cells, most of these cells, which are located near the dorsolateral margins of the neural plate, become wedge-shaped; moreover, these cells continue to increase their heights, becoming the tallest cells of the neural plate.

Both midline and dorsolateral wedge-shaped neurepithelial cells reside within regions termed *hinge points* (Schoenwolf, 1982; Smith and Schoenwolf, 1987). The three hinge points exhibit the following features (Fig. 3; Schoenwolf and Smith, 1990b). The median hinge point (MHP) is a single region where the

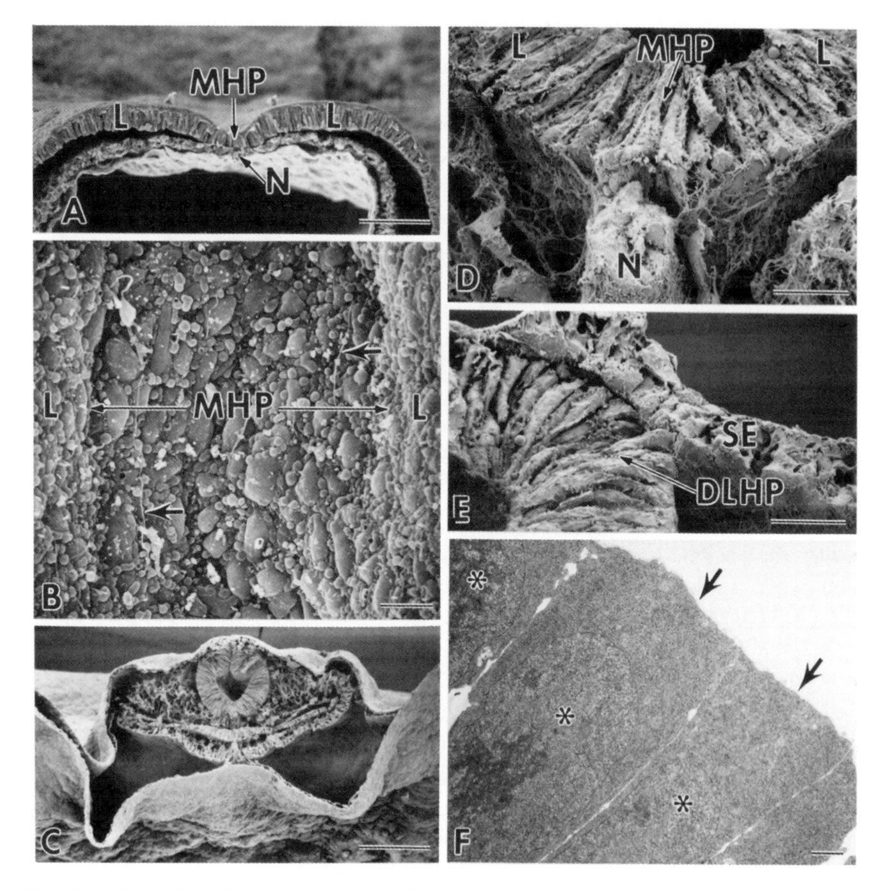

Fig. 3. Views of chick neurulation. Scanning electron micrographs (A–E); transmission electron micrograph (F). **A** and **C–E** show transverse slices at stages 5 (A) and 9 (C–E); DLHP, dorsolateral hinge point cells; L, lateral neurepithelial plate cells; MHP, median hinge point cells; N, notochord; SE, surface ectoderm of neural fold. **B** shows a dorsal view of the apices of MHP cells and flanking L cells at stage 5; arrows, telophase bridges. **F** shows the basal aspect of neurepithelial cells within one of the DLHP regions. Note the proximity of cell nuclei (asterisks) to the basal lamina (arrows). Bars = 100 μm (A,C); 5 μm (B); 10 μm (D,E); 1 μm (F).

neural plate is anchored to underlying tissue (prechordal plate mesoderm at the forebrain level and notochord at the midbrain through spinal cord levels) and change in cell shape occurs—as stated above, MHP cells (i.e., *neurepithelial* cells in the MHP) *decrease* their heights and most become wedge-shaped. The dorsolateral hinge points (DLHPs) are bilateral regions where the neural plate is anchored to adjacent tissue (the surface ectoderm of each neural fold) and change in cell shape occurs—as stated above, DLHP cells (i.e., *neurepithelial* cells in the DLHPs) *increase* their heights and most become wedge-shaped. Therefore, three morphologically distinct populations of neurepithelial cells can be identified during neurulation: MHP cells, two groups of DLHP cells, and intervening neurepithelial cells—designated as L cells (for lateral cells)—one group on each side of the MHP region.

In a series of studies, we have addressed five crucial questions about change in neurepithelial cell shape during neurulation: 1) How is change in cell shape localized within the neural plate? 2) How does cell elongation occur? 3) What is the role of cell elongation in shaping of the neural plate? 4) How does cell wedging occur? 5) What is the role of cell wedging in bending of the neural plate?

Change in Neurepithelial Cell Shape Is Localized Within the Neural Plate by Cell-Cell Interactions

To determine how the shape of neurepithelial cells in one population becomes different from that in an adjacent one, we asked whether inductive interactions occur between the notochord and overlying neurepithelial cells (i.e., prospective MHP cells). Two experiments were performed (Smith and Schoenwolf, 1989; also see van Straaten et al., 1988). In the first experiment, Hensen's node (the rudiment of the notochord) was extirpated to produce notochordless embryos. In such embryos, the morphological features characteristic of MHP cells (i.e., shortness and wedgeness) failed to develop. In the second experiment, notochords were removed from quail embryos and transplanted beneath L cells of the chick neural plate. In the presence of supernumerary notochords, ectopic MHP-like cells formed; that is, L cells decreased their heights and became wedge-shaped. Collectively, these results provide evidence that localized change in neurepithelial cell shape is mediated by inductive interactions (i.e., instructive interactions as defined by Wessells, 1977). The nature of this induction is under further study.

Neurepithelial Cell Elongation Is Driven Partially by Microtubules

Numerous studies have shown that the elongated cells of the neural plate are replete with paraxial microtubules (i.e., microtubules oriented parallel to the long axis of the cell) and that disruption of these structures ultimately results in the rounding up of neurepithelial cells at the apex of the epithelium (reviewed by Schoenwolf and Powers, 1987). In a recent experimental study, we showed that in the absence of paraxial microtubules, neurepithelial cells decrease their heights only by approximately 25% (Schoenwolf and Powers, 1987). However, they still remain highly elongated until they round up to enter metaphase where, in the absence of microtubules, they are arrested. Thus this study furnishes support for a role for paraxial microtubules in cell elongation but suggests that additional factors augment this process. The other factors involved in cell elongation are unknown; however, two candidates have been proposed: localized change in intercellular adhesion (reviewed by Ettensohn, 1985) and flow of cortical cytoplasm (cortical tractoring) (Jacobson et al., 1986).

Neurepithelial Cell Elongation Partially Drives Shaping of the Neural Plate

A causal relationship between the height of neurepithelial cells and the width of the neural plate was first recognized in the amphibian embryo (Burnside and Jacobson, 1968; Jacobson and Gordon, 1976). Similarly, in the chick embryo, the width of the neural plate decreases as the height of its cells increases (and, consequently, as their diameter decreases) (Schoenwolf, 1985). In an experimental study (Schoenwolf and Powers, 1987), we showed that when neurepithelial cell height is decreased following depolymerization of paraxial microtubules, neural plate width is correspondingly increased. Upon subsequent repolymerization, neurepithelial cell height increases, while neural plate width decreases. However, it was estimated that such cell elongation can account for only about one-third of the *total* decrease that occurs in the width of the neural plate during its normal shaping; the remaining two-thirds must be generated by other form-shaping events. Also, neurepithelial cell elongation cannot account for the lengthening of the neural plate that occurs during its shaping (in fact, a decrease in cell diameter would decrease the length of the neural plate as well as its width). These issues are dealt with below.

Neurepithelial Cell Wedging Is Driven Largely by Cell-Cycle-Regulated Interkinetic Nuclear Migration

The mechanism underlying neurepithelial cell wedging has been a topic of interest for over 100 years. Traditionally, cell wedging has been attributed to the constriction of cell apices, mediated by circumferential bands of microfilaments (reviewed by Schoenwolf et al., 1988). However, recent studies provide an alternative explanation, namely, that neurepithelial cell wedging is generated principally by basal expansion rather than by apical constriction.

In one study, we treated neurulating chick embryos with concentrations of cytochalasin D sufficient to disrupt apical microfilament bands. Neural tube defects resulted in all cases (Schoenwolf et al., 1988). However, these defects were owing to failure of the neural folds to fuse (and sometimes also to converge) at the dorsal midline—neural fold *elevation* routinely occurred. Furthermore, MHP cells (and, in about one-third of the cases, DLHP cells) still became wedge-shaped after treatment with cytochalasin D. Thus this experiment provides evidence that neurepithelial cell wedging does not require microfilament-mediated apical constriction and has led us to focus on another possible mechanism of cell wedging, namely, basal expansion.

Two observations reveal that basal expansion must occur during neurepithelial cell wedging. First, as shown by a number of investigations, cells of the avian neural plate undergo interkinetic nuclear migration as they traverse the cell cycle (e.g., Watterson, 1965; Langman et al., 1966). Second, the nucleus of a neurepithelial cell occupies the cell's widest portion (Schoenwolf and Franks, 1984). Hence, as a nucleus moves, the cell progressively changes its shape from inverted wedge-shaped (its shape just following the M phase of the cell cycle when the newly formed daughter nucleus resides at the apex) to spindle-shaped (its shape as the nucleus moves away from the apex toward the base) to wedge-shaped (its shape while the nucleus resides at the base). Prior to the next M phase, the nucleus migrates away from the base and back toward the apex again, changing the shape of the cells from wedge-shaped to spindle-shaped to inverted wedge-shaped.

Based on these observations, we proposed that neurepithelial cell wedging could result from cell-cycle-regulated interkinetic nuclear migration. We began to test this hypothesis by asking whether populations of MHP cells (most of which are wedge-shaped) have longer cycles than populations of L cells (most of which are spindle-shaped) (Smith and Schoenwolf, 1987). The answer is yes: the cell cycle length of MHP cells is about 65% greater than that of L cells. We then asked whether this difference in the length of the cell cycle was brought about by an increase in the cell cycle length of prospective MHP cells (between stages 4 and 5; Hamburger and Hamilton, 1951) or by a decrease in the cell cycle length of prospective L cells. We found that the cell cycle length of prospective MHP cells increases concomitantly with cell wedging. A subsequent analysis suggested that regulation of the length of the cell cycle involves a decrease in the length of the M phase (reducing the time mitotic figures reside at the apex of the neurepithelium) and an increase in the length of those phases during which nuclei reside at the base (the DNA synthetic phase and the gap phase—presumably G_2) (Smith and Schoenwolf, 1988). Thus these studies support the hypothesis that basal expansion, as mediated by cell-cycle-regulated interkinetic nuclear migration, plays a paramount role in neurepithelial cell wedging. How interkinetic nuclear migration occurs is unknown, although paraxial microtubules and microtubule-based motors (but not microfilaments) are possible candidates (see Schoenwolf and Smith, 1990b).

Neurepithelial Cell Wedging Facilitates Bending of the Neural Plate: The Hinge Point Model

Two experiments directly demonstrate that elevation of the neural folds during bending of the neural plate does not require wedging of MHP cells. First, when Hensen's node is extirpated to prevent formation of the notochord, prospective MHP cells remain spindle-shaped (Smith and Schoenwolf, 1989). Nevertheless, bending of the neural plate still occurs, often with complete formation of a neural tube. Second, bending of the neural plate and complete formation of the neural tube can still occur after extirpation of the entire thickness of the MHP region (Smith and Schoenwolf, 1991).

If bending of the neural plate is not driven by wedging of neurepithelial cells, what role does this process play in neurulation? We suggest that cell wedging contributes to the formation of hinge points and that these structures facilitate bending of the neural plate. Recall that hinge points are regions where the neural plate is anchored to adjacent tissues and changes in cell shape (both wedging and change in height) occur. Our model (Schoenwolf and Smith,

1990b) predicts that hinge points play a permissive rather than an obligatory role in neurulation by allowing extrinsic forces to bend the neural plate so that the neural folds rotate around a "hinge pin." This pin consists of both the area of anchorage as well as the neurepithelial cells that change their shape. Wedging, rather than driving bending, causes the formation of a longitudinal furrow (Schoenwolf, 1988), which, like a perforated line on a sheet of paper, facilitates and directs folding of the sheet. If cell wedging is prevented in the hinge point (e.g., by extirpating Hensen's node), bending can still occur but its efficacy is reduced, resulting in a greater failure rate of neurulation (Smith and Schoenwolf, 1989). Moreover, if the hinge point is extirpated, bending of the neural plate still occurs around the site of extirpation (Smith and Schoenwolf, 1991). This is likely because the hole in the blastoderm at the extirpation site does not generate resistance to bending (again acting like a perforated line) and new anchorage points form just lateral to the hole between the neural plate and subjacent endoderm and mesoderm.

CHANGE IN NEUREPITHELIAL CELL NUMBER

The previous section considered the role of regulation of the cell cycle in neurepithelial cell wedging. Here, I discuss briefly neurepithelial cell division as a component of neural plate growth, a process that provides morphogenetic forces partially responsible for shaping of the neural plate.

Positioning of Daughter Cells Within the Neurepithelium Partially Drives Shaping of the Neural Plate

During neurulation of birds and mammals, true growth of the neurepithelium occurs (Fig. 4; Schoenwolf, 1985; Jacobson and Tam, 1982). How does this growth affect shaping of the neural plate? Recall that during shaping, the neural plate on the average thickens apicobasally, narrows transversely, and lengthens longitudinally. As mentioned previously, all of the thickening and about one-third of the narrowing can be accounted for by neurepithelial cell elongation, with a corresponding decrease in cell diameter. The remaining narrowing can be accounted for by change in cell position (discussed in the next section). Lengthening of the neural plate involves principally two processes (Schoenwolf and Alvarez, 1989): cell division, with the orientation of division presumably occurring such that daughter cells are inserted into the longitudinal plane of the neural plate rather than into its transverse plane, and change in cell position (again, see next section). Additionally, at certain neuraxial levels, some daughter cells presumably are inserted into the transverse plane of the neural plate so that localized widening occurs (Fig. 5).

The mechanism by which daughter cells are positioned within the neural plate is unclear, but several studies have suggested that this process is influ-

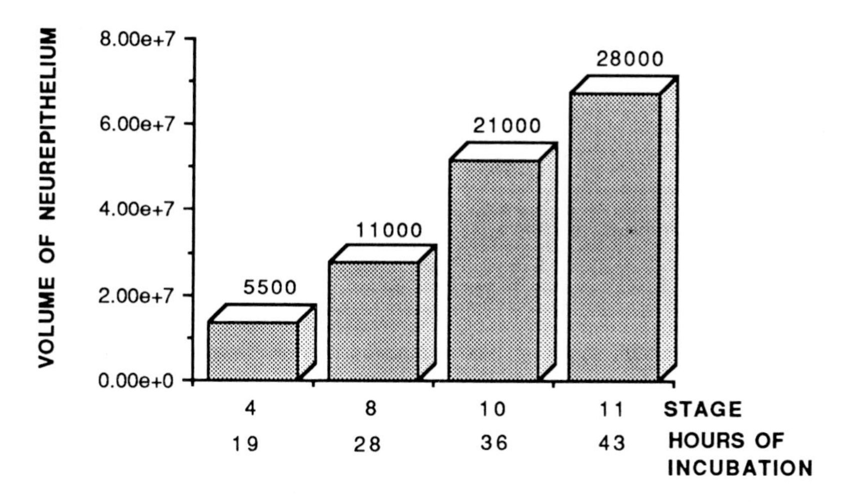

Fig. 4. Graph showing increase in the volume (from about 14×10^6 to 67×10^6 μm^3) of the neural plate during neurulation (based on data of Schoenwolf, 1985). Also shown are estimates of neurepithelial cell number (5,500 to 28,000) determined in the following way. Average neurepithelial cell size (2,425 μm^3) was calculated from the data of Schoenwolf and Alvarez (1989). Neural plate volume at each stage was then divided by the average cell size to estimate the number of cells present at each stage (it was assumed that average cell size remains constant during neurulation). The values obtained were compared to expected values based on the approximate hours of incubation between stages 4 and 11 (24 hr; data of Hamburger and Hamilton, 1951) and cell-cycle length (average cell generation time ranges from 8–12 hr during neurulation; Smith and Schoenwolf, 1987) as well as the number of cells generated from quail epiblast plugs analyzed 24 hr after they were grafted to chick blastoderms (in 24 hr, cell number slightly more than quadrupled; data of Schoenwolf and Alvarez, 1989). These comparisons revealed that estimates appearing on the graph were reasonable (i.e., the assumption that average cell size remains constant during neurulation is basically correct).

enced by the orientation of the mitotic spindle and, consequently, the orientation of the division furrow (Langman et al., 1966; Martin, 1967; Jacobson and Tam, 1982; Tuckett and Morriss-Kay, 1985; Zieba et al., 1986; Everaert et al., 1988). Although these studies provide evidence for oriented division, it is unclear whether localized differences exist that are consistent with the regional characteristics of neural plate shaping (i.e., differences at various mediolateral and craniocaudal levels of the neuraxis have not been examined). Further problems include the following: 1) It is unknown whether the orientation of the mitotic spindle determines the orientation of the division furrow in epithelial sheets (these could be independent events or the orientation of the division furrow, which could be fixed prior to division, could determine the orientation of the spindle); 2) it is unknown whether constraints of cell packing determine the ultimate position of daughter cells (either by orienting spindles or furrows or by altering cell position, owing to space availability as daughter

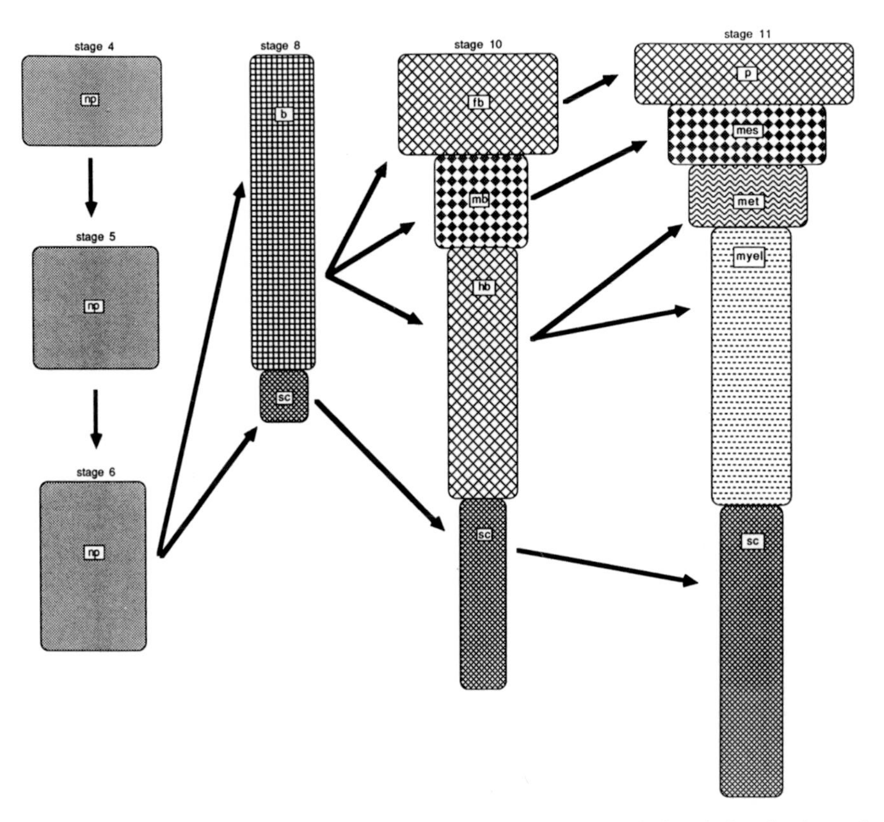

Fig. 5. Graph showing changes in the width and length of the neural plate during shaping and bending at various neuraxial levels (based on data of Schoenwolf, 1985). The neurepithelium is laid out flat at all stages (i.e., the neural tube has been ''filleted'' by ''cutting'' its roof plate). np, flat neural plate; b, brain; sc, spinal cord; fb, forebrain; mb, midbrain; hb, hindbrain; p, prosencephalon; mes, mesencephalon; met, metencephalon; myel, myelencephalon.

cell nuclei begin interkinetic nuclear migration by moving toward the base of the epithelium); and 3) it is unknown whether the orientation of division planes is even meaningful; that is, cell rearrangement alone might determine the cell's ultimate position regardless of the orientation of the division plane. These possibilities deserve further investigation.

CHANGE IN NEUREPITHELIAL CELL POSITION

Change in cell position is a common behavior underlying morphogenesis. For example, cell rearrangement occurs during both gastrulation (sea urchins: Ettensohn, 1985; Hardin, 1989; fish: Keller and Trinkaus, 1987; Warga and Kimmel, 1990; and amphibians: Keller, 1980; Keller et al., 1985) and neuru-

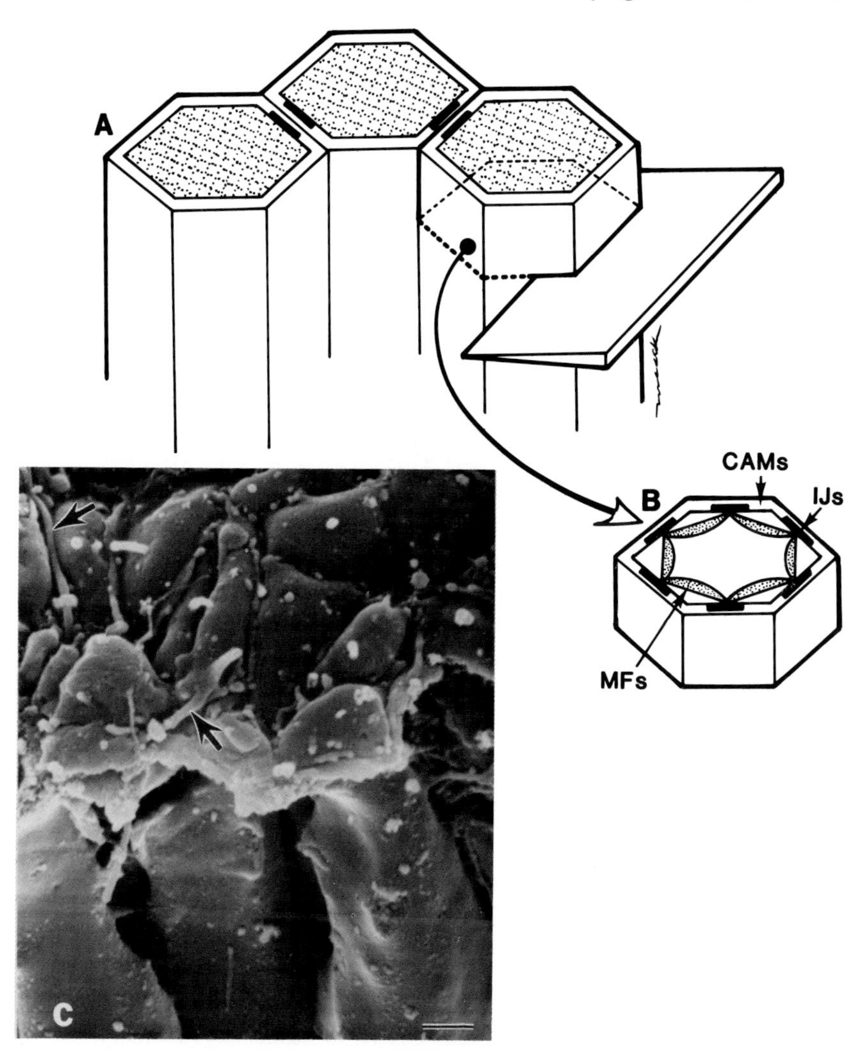

Fig. 6. Views of neurepithelial cell apices shown diagrammatically (**A,B**) and by scanning electron microscopy (**C**). CAMs, cell adhesion molecules; IJs, intercellular junctional complexes; MFs, microfilament bands; arrows (C), telophase bridges. Bar = 1 μm (C).

lation (amphibians: Jacobson and Gordon, 1976; and birds: Schoenwolf et al., 1989a; Schoenwolf and Alvarez, 1989). Intuitively, based on structural grounds, one would not expect extensive cell rearrangement to occur in a tight epithelial sheet like the neural plate (Fig. 6). However, experiments, not intuition, reveal to us how the embryo develops.

The Rearrangement Behavior of Neurepithelial Cells Differs Among Populations

The initial sites of prospective MHP, L, and DLHP cells in the flat neural plate stage have been partially mapped by constructing quail/chick transplantation chimeras (Schoenwolf and Alvarez, 1989; Schoenwolf et al., 1989a; Alvarez and Schoenwolf, 1991) and by microinjecting blastoderms with a cell marker (rhodamine-conjugated horseradish peroxidase) (Schoenwolf and Sheard, 1989, 1990). Prospective MHP cells arise from a midline area cranial to and overlapping Hensen's node, whereas prospective L cells arise from bilateral areas flanking the cranial half of the primitive streak. The origin of prospective DLHP cells is less certain: The most lateral cells of the postnodal neural plate extend down the length of the axis, remaining at the surface ectoderm–neurepithelium interface. This suggests that DLHP cells might originate from the most lateral neurepithelial regions. However, because the most lateral cells have been mapped principally at the spinal cord level (throughout most of which DLHPs do not form), further studies are necessary to clarify this issue.

Reconstructions of chimeric embryos containing grafts of epiblast plugs have revealed clearly the patterns of neurepithelial cell rearrangement during neurulation (Schoenwolf and Alvarez, 1989). Prospective MHP cells originate from the nodal and prenodal midline and extend caudally in the wake of the regressing Hensen's node; the MHP cell group always spans the midline. In contrast, prospective L cells originate from paired rudiments flanking the cranial part of the primitive streak and extend caudally, with most cells remaining lateral to the MHP region. Thus prospective MHP cells extend caudally in a single group, between paired groups of prospective L cells, and each of the latter groups extends caudally between prospective MHP cells (residing medially) and prospective surface ectodermal cells (residing further laterally).

Scanning electron microscopy of the dorsal surface of the epiblast reveals the shape of the forming MHP region during the regression of Hensen's node (Fig. 7). This region undergoes a change in shape similar to that occurring throughout the entire neural plate; namely, it narrows transversely and lengthens longitudinally. The forming L regions undergo similar movements, as known from fate mapping studies, but these regions are more difficult to track by scanning electron microscopy.

Two questions remain for our consideration: 1) What role does change in neurepithelial cell position play in shaping and bending of the neural plate? 2) How are patterns of neurepithelial cell rearrangement established?

Change in Neurepithelial Cell Position Partially Drives Shaping of the Neural Plate and Facilitates Bending of the Neural Plate

Modeling studies, based on estimates of the amount of cell rearrangement that occurs during neurulation, have shown that cell rearrangement plays a

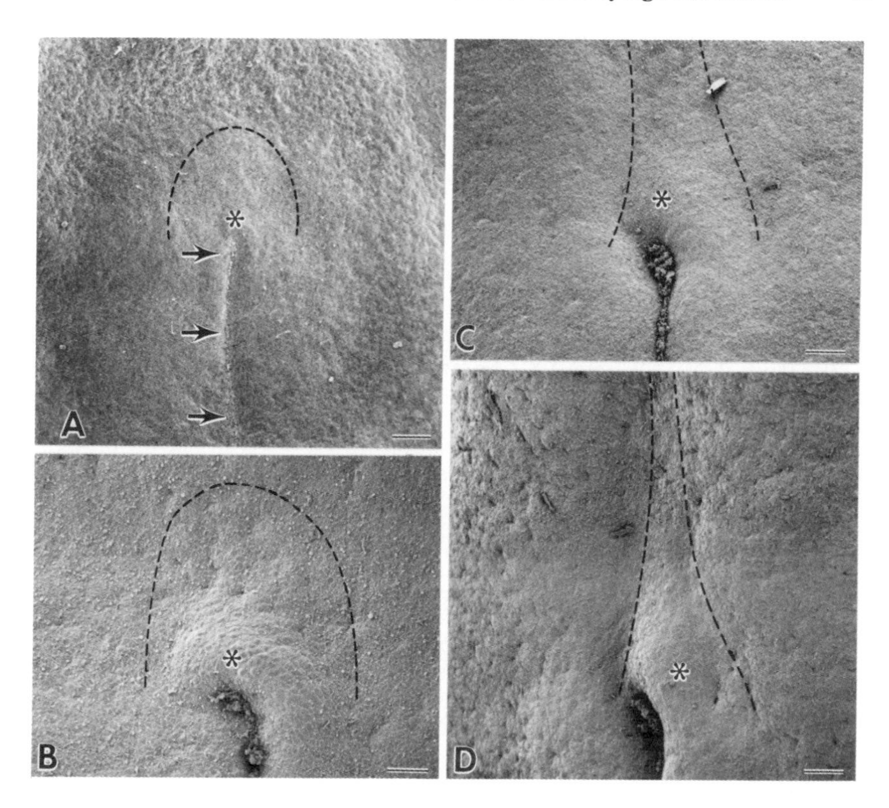

Fig. 7. Scanning electron micrographs of dorsal views of chick blastoderms showing the deformation of cells of the prospective MHP region during neural plate shaping and bending. The boundaries of the prospective MHP rudiment, which overlaps Hensen's node (asterisks; the cranial end of the primitive streak), are partially outlined. Arrows, primitive streak. Bars = 100 μm (**A**); 20 μm (**B–D**).

major role in neural plate narrowing and lengthening (Schoenwolf and Alvarez, 1989). That is, cells rearrange such that they move from the transverse plane of the plate (contributing to its narrowing) to the longitudinal plane of the plate (contributing to its lengthening). Approximately two rounds of cell rearrangement occur during neurulation (each round of rearrangement is defined as cell-cell interdigitation resulting in a halving of the width of the neural plate with a corresponding doubling in its length). Coordinately, two to three rounds of cell division occur, with about half contributing to increase in the length of the neural plate. These two behaviors—change in cell number and change in cell position—in conjunction with change in cell shape (i.e., neurepithelial cell elongation) are sufficient to generate a *flat* neurepithelium having the proper size and shape. Conversion of this flat sheet into a neural tube (i.e., bringing about bending of the neural plate and fusion of the neural folds) requires extrinsic forces generated by tissues outside the neural plate (e.g., see Schoenwolf, 1988). Narrowing of the neural plate facilitates bending by decreasing the distance between the neural folds and, thereby, allowing extrinsic forces to elevate and converge the neural folds toward the dorsal midline, where fusion occurs. In the absence of sufficient narrowing, the broad neural plate becomes an impediment, and extrinsic forces are unable to bring the neural folds together; this results in neural tube defects such as anencephaly (Schoenwolf, unpublished observations).

Patterns of Neurepithelial Cell Rearrangement Are Generated by Restriction of Cell Intermingling

In a recent study, we asked whether neurepithelial cell rearrangement behavior was determined prior to formation of the notochord. Recall that MHP cells can be distinguished from L cells because the notochord induces them to modify their behavior; that is, MHP cells change their shape and alter their cell cycles. As just discussed, the pattern of cell rearrangement differs for MHP and L cells. Therefore, we wondered if this difference was established by an inductive interaction.

To examine this question, we made quail/chick transplantation chimeras by heterotopically grafting prospective L cells to prospective MHP cell territory and vice versa (Alvarez and Schoenwolf, 1991). Under these conditions, cells ''remember'' their original position: L cells move off the midline and extend laterally to intermix with host L cells, whereas MHP cells move into the midline and extend caudally to intermix with host MHP cells. This suggests that the unique rearrangement behavior of MHP and L cells is established by restriction of cell intermingling between heterologous cell populations. That is, property lines seem to exist between different populations of neurepithelial cells, and such populations respect these boundaries rather than trespass. How restriction of cell intermingling is established is unknown, but cell surface differences, substrate differences, and gradients are all possibilities.

CONCLUDING REMARKS

The study of avian neurulation over the past decade has yielded a number of exciting surprises. Traditionally, neurulation has been viewed as a relatively simple process that is driven by intrinsic forces generated exclusively by cytoskeleton-mediated changes in neurepithelial cell shape. Recent studies have emphasized the fact that embryos are iconoclastic, being in no way limited by our tradition. These studies have shown that both intrinsic and extrinsic forces are required for neurulation and that multiple cell behaviors, not just change in cell shape, play important roles. Furthermore, these studies have suggested that a conceptionally simple process such as change in cell shape can have complex mechanisms underlying it. Embryos are secretive, but with persistent prodding they can be coerced to tell us amazing things. Let the prodding continue!

ACKNOWLEDGMENTS

I wish to thank Drs. Ignacio S. Alvarez and Jodi L. Smith for their constructive criticisms of the manuscript as well as for many stimulating discussions of neurulation. I also thank Fahima Rahman for technical assistance and Jennifer Parsons for secretarial assistance. The original research described herein was supported by grants principally from the NIH.

REFERENCES

Alvarez IS, Schoenwolf GC (1991): Patterns of neurepithelial cell rearrangement during avian neurulation are determined prior to notochordal inductive interactions. Dev Biol 143:78–92.

Burnside MB, Jacobson AG (1968): Analysis of morphogenetic movements in the neural plate of the newt *Taricha torosa*. Dev Biol 18:537–552.

Ettensohn CA (1985): Mechanisms of epithelial invagination. Q Rev Biol 60:289–307.

Everaert S, Espeel M, Bortier H, Vakaet L (1988): Connecting cords and morphogenetic movements in the quail blastoderm. Anat Embryol 177:311–316.

Hamburger V, Hamilton HL (1951): A series of normal stages in the development of the chick embryo. J Morphol 88:49–92.

Hardin J (1989): Local shifts in position and polarized motility drive cell rearrangement during sea urchin gastrulation. Dev Biol 136:430–445.

Jacobson AG, Gordon R (1976): Changes in the shape of the developing vertebrate nervous system analyzed experimentally, mathematically and by computer simulation. J Exp Zool 197:191–246.

Jacobson AG, Oster GF, Odell GM, Cheng LY (1986): Neurulation and the cortical tractor model for epithelial folding. J Embryol Exp Morphol 96:19–49.

Jacobson AG, Tam PPL (1982): Cephalic neurulation in the mouse embryo analyzed by SEM and morphometry. Anat Rec 203:375–396.

Keller RE (1980): The cellular basis of epiboly: An SEM study of deep cell rearrangment during gastrulation in *Xenopus laevis*. J Embryol Exp Morphol 60:201–234.

Keller RE, Danilchik M, Gimlich R, Shih J (1985): The function and mechanism of convergent extension during gastrulation of *Xenopus laevis*. J Embryol Exp Morphol 89 (Suppl):185–209.

Keller RE, Trinkaus JP (1987): Rearrangement of enveloping layer cells without disruption of the epithelial permeability barrier as a factor in *Fundulus* epiboly. Dev Biol 120:12–24.

Langman J, Guerrant RL, Freeman BG (1966): Behavior of neuro-epithelial cells during closure of the neural tube. J Comp Neurol 127:399–412.

Martin AH (1967): Significance of mitotic spindle fibre orientation in the neural tube. Nature 216:1133–1134.

Schoenwolf GC (1982): On the morphogenesis of the early rudiments of the developing central nervous system. Scanning Electron Microsc 1982/I:289–308.

Schoenwolf GC (1983): The chick epiblast: A model for examining epithelial morphogenesis. Scanning Electron Microsc 1983/III:1371–1385.

Schoenwolf GC (1985): Shaping and bending of the avian neuroepithelium: Morphometric analyses. Dev Biol 109:127–139.

Schoenwolf GC (1988): Microsurgical analyses of avian neurulation: Separation of medial and lateral tissues. J Comp Neurol 276:498–507.
Schoenwolf GC, Alvarez IS (1989): Roles of neuroepithelial cell rearrangement and division in shaping of the avian neural plate. Development 106:427–439.
Schoenwolf GC, Bortier H, Vakaet L (1989a): Fate mapping the avian neural plate with quail/chick chimeras: Origin of prospective median wedge cells. J Exp Zool 249:271–278.
Schoenwolf GC, Everaert S, Bortier H, Vakaet L (1989b): Neural plate- and neural tube-forming potential of isolated epiblast areas in avian embryos. Anat Embryol 179:541–549.
Schoenwolf GC, Folsom D, Moe A (1988): A reexamination of the role of microfilaments in neurulation in the chick embryo. Anat Rec 220:87–102.
Schoenwolf GC, Franks MV (1984): Quantitative analyses of changes in cell shapes during bending of the avian neural plate. Dev Biol 105:257–272.
Schoenwolf GC, Powers ML (1987): Shaping of the chick neuroepithelium during primary and secondary neurulation: Role of cell elongation. Anat Rec 218:182–195.
Schoenwolf GC, Sheard P (1989): Shaping and bending of the avian neural plate as analysed with a fluorescent-histochemical marker. Development 105:17–25.
Schoenwolf GC, Sheard P (1990): Fate mapping the avian epiblast with focal injections of a fluorescent-histochemical marker: Ectodermal derivatives. J Exp Zool 255:323–339.
Schoenwolf GC, Smith JL (1990a): Mechanisms of neurulation: Traditional viewpoint and recent advances. Development 109:243–270.
Schoenwolf GC, Smith JL (1990b): Epithelial cell wedging: A fundamental cell behavior contributing to hinge point formation during epithelial morphogenesis. In Keller RE, Fristrom D (eds): ''Control of Morphogenesis by Specific Cell Behaviors. Seminars in Developmental Biology.'' London: W.B. Saunders, pp 325–334.
Smith JL, Schoenwolf GC (1987): Cell cycle and neuroepithelial cell shape during bending of the chick neural plate. Anat Rec 218:196–206.
Smith JL, Schoenwolf GC (1988): Role of cell-cycle in regulating neuroepithelial cell shape during bending of the chick neural plate. Cell Tissue Res 252:491–500.
Smith JL, Schoenwolf GC (1989): Notochordal induction of cell wedging in the chick neural plate and its role in neural tube formation. J Exp Zool 250:49–62.
Smith JL, Schoenwolf GC (1991): Further evidence of extrinsic forces in bending of the neural plate. J Comp Neurol 307:225–236.
Tuckett F, Morriss-Kay GM (1985): The kinetic behaviour of the cranial neural epithelium during neurulation in the rat. J Embryol Exp Morphol 85:111–119.
van Straaten HWM, Hekking JWM, Wiertz-Hoessels EJLM, Thors F, Drukker J (1988): Effect of the notochord on the differentiation of a floor plate area in the neural tube of the chick embryo. Anat Embryol 177:317–324.
Warga RM, Kimmel CB (1990): Cell movements during epiboly and gastrulation in zebrafish. Development 108:569–580.
Watterson RL (1965): Structure and mitotic behavior of the early neural tube. In DeHaan RL, Ursprung H (eds): ''Organogenesis.'' New York: Holt, Rinehart and Winston.
Wessells NK (1977): ''Tissue Interactions and Development.'' Menlo Park, CA: Benjamin/Cummings.
Zieba P, Strojny P, Lamprecht J (1986): Positioning and stability of mitotic spindle orientation in the neuroepithelial cell. Cell Biol Intl Rep 10:91–100.

Cell-Cell Interactions in Early Development, pages 79–91

5. Activins and the Induction of Axial Mesoderm and Dorsoanterior Structures in *Xenopus*

Gerald Thomsen, Tod Woolf, Malcolm Whitman, Sergei Sokol, and Douglas A. Melton

Department of Biochemistry and Molecular Biology, Harvard University, Cambridge, Massachusetts 02138

INTRODUCTION

In this chapter we summarize our work showing that mammalian activin A and *Xenopus* activin B have the capacity to induce dorsal (axial) mesoderm and anterior structures in explants of competent *Xenopus* blastula cells. We have obtained clones for *Xenopus* activin β_A and β_B genes and have used these to study the expression of activins during development. Activin β_B mRNA is first detectable in the late blastula (stage 9), during the latter period of mesoderm induction, and around the time that the Spemann organizer is established. Activin β_A, on the other hand, is first detected in the late gastrula (stage 13). Maternal transcripts of either gene were not detected. Furthermore, recombinant *Xenopus* activin B induces dorsoanterior and mesodermal structures (such as notochord, eyes, and brain), just like activin A. Ectopic expression of activin B in ventral marginal zone blastomeres results in the formation of a second body axis, further suggesting that activin B may contribute to the formation or function of the Spemann organizer. We emphasize a role for activin B in axial mesoderm induction principally because its mRNA is first detected in the late blastula, whereas activin β_A mRNA is detected much later than β_B mRNA. Thus our findings on the frog, together with those of the chick (see Mitrani et al., 1990), suggest that activin B is important for the induction and patterning of axial mesoderm in vertebrates.

The vertebrate body plan arises through a series of inductive interactions between neighboring cells of the developing embryo. Induction in vertebrates was first demonstrated by Spemann and Mangold (1924), who showed that a dorsal blastopore lip of an early amphibian gastrula induced a second body axis when implanted in the ventral side of a host embryo. The majority of induced structures were host-derived. The dorsal lip region responsible for such inductions is often referred to as the Spemann organizer. Since the time of these early experiments, many other examples of induction in vertebrates have been described (see Nieuwkoop et al., 1985; Slack, 1983; Hamburger, 1989 for reviews), but it is only within the past several years that the molecular nature of inducing substances has become more clear. Indeed, the theme emerging from a variety of recent studies is that peptide growth factors are likely candidates for endogenous inducing substances.

The first inductive interaction in the amphibian embryo occurs between cells of the animal and vegetal hemispheres, and it gives rise to mesodermal derivatives such as blood, mesenchyme, muscle, and notochord. In *Xenopus*, this occurs within the blastula'a equatorial margin through the action of a maternal signal(s) from cells of the vegetal hemisphere (Nieuwkoop, 1969). This signal(s) acts upon overlying competent cells of the animal hemisphere. In fact, all regions of the animal hemisphere appear to be competent to respond to the inducing signal(s), but the physical separation of animal and vegetal hemispheres by the blastocoel limits the extent of inductive interactions to the equatorial band. In a Spemann organizer graft, many of the structures induced in the host, such as the secondary nervous system, reflect the inductive capacities of the dorsal blastopore lip, which itself is a piece of presumptive dorso-anterior mesoderm induced earlier by the vegetal signal(s).

The inductive ability of vegetal cells was clearly demonstrated by Nieuwkoop and colleagues (1969), who showed that animal hemisphere explants (''caps'') went on to form mesodermal structures after limited contact with vegetal hemisphere explants. More recent experiments employing molecular markers for mesoderm induction have confirmed the earlier findings (Gurdon et al., 1985). Induction by the vegetal cells was further shown to be mediated by a soluble factor(s) that could traverse a nucleopore filter placed between animal and vegetal halves (Saxen et al., 1976). When cultured alone, animal caps form ectodermal tissue such as ciliated epidermis, and vegetal clumps form endoderm-like tissue. An assay system employing isolated animal caps has proven invaluable in the hunt for mesoderm-inducing substances. Tiedeman and colleagues early on showed that a yet-unidentified substance from chick embryo extracts could induce mesoderm in animal cap sandwiches (reviewed by Tiedeman, 1978). More recent work has now shown that peptide growth factors of the fibroblast growth factor (FGF) (Kimmelman and Kirschner, 1987; Slack et al., 1987) and transforming growth factor beta (TGF-β) families induce mesoderm in animal caps (Smith, 1987; Smith et al., 1988; Rosa et al., 1988; Asashima et al., 1990; Sokol et al., 1990; Thomsen et al., 1990). Members of these families, namely FGF and Vg1, are also present as maternal mRNA or proteins in the *Xenopus* egg (Weeks and Melton, 1987; Kimmelman et al., 1988; Slack and Isaacs, 1989).

FGF induces mesoderm of a more ventral sort, such as mesenchyme and mesothelia (Godsave et al., 1988), whereas TGF-β-related peptides like XTC-MIF (Smith, 1987; Smith et al., 1988), TGF-β2 (Rosa et al., 1988), PIF (Sokol et al., 1990), and activin (Asashima et al., 1990; Smith et al., 1990;

Thomsen et al., 1990; van den Eijnden-Van Raaij et al., 1990; this report) induce a wide spectrum of mesodermal tissues, including dorsal types such as somitic muscle and notochord. Recently, XTC-MIF has been identified as a *Xenopus* homologue of mammalian activin A (Smith et al., 1990). Unlike other studies (Smith, 1987; Smith et al., 1988; Asashima et al., 1990), however, our own studies with PIF and purified activin A have revealed that these factors not only induce mesoderm, but they also can induce axially patterned "embryoids" from animal caps. These embryoids often contain notochord, brain rudiments, and eyes, all arranged in polarized fashion (Sokol et al., 1990; Thomsen et al., 1990; and see results below).

Activins were first described as factors that stimulate the release of follicle-stimulating hormone (FSH) from the anterior pituitary (Mason et al., 1985; Ling et al., 1986). Activins are members of the TGF-β family, and they have been purified from natural sources as a homodimer of two β_A chains (activin A) or a heterodimer of β_A and β_B chains (activin AB) (Vale et al., 1986). Activin A also stimulates erythroid differentiation (Yu et al., 1987). Recombinant activin β_B chain homodimer (activin B) is also biologically active (Mason et al., 1989), though it has not been reported to occur naturally. A related molecule, inhibin, blocks the FSH-releasing capacity of pituitary cells, and it consists of an inhibin-specific α chain paired with either of the β chains (Mason et al., 1985).

We find that both mammalian activin A and recombinant *Xenopus* activin B induce a variety of dorsoanterior mesodermal and neural structures, including eyes, which are often arranged in a rudimentary axis. The timing of activin gene expression suggests that these molecules play a key role in mesoderm induction and axial patterning in *Xenopus* embryos.

MATERIALS AND METHODS

See Thomsen et al. (1990) for details of experimental procedures.

RESULTS AND DISCUSSION

Our studies of activin arose while we attempted to identify new candidates for mesoderm inducing factors, since we had not been able to demonstrate any direct inducing activity for Vg1 or TGF-β5 (D.A. Melton et al., unpublished observations). In one approach, conditioned media from a variety of cell lines were screened for inducing activity, and this yielded PIF, a factor produced by the mouse macrophage cell line P388D1 (Sokol et al., 1990). Purified PIF was shown to be antigenically related to activin A (Thomsen et al., 1990). In another approach we used PCR to pull out new members of the TGF-β family, with the expectation that one of these might correspond to the maternal dorsal-type mesoderm inducer. A number of new TGF-β-related genes have been cloned in this fashion (G. Thomsen and D. Melton, in preparation), including *Xenopus* activin genes, as we discuss below.

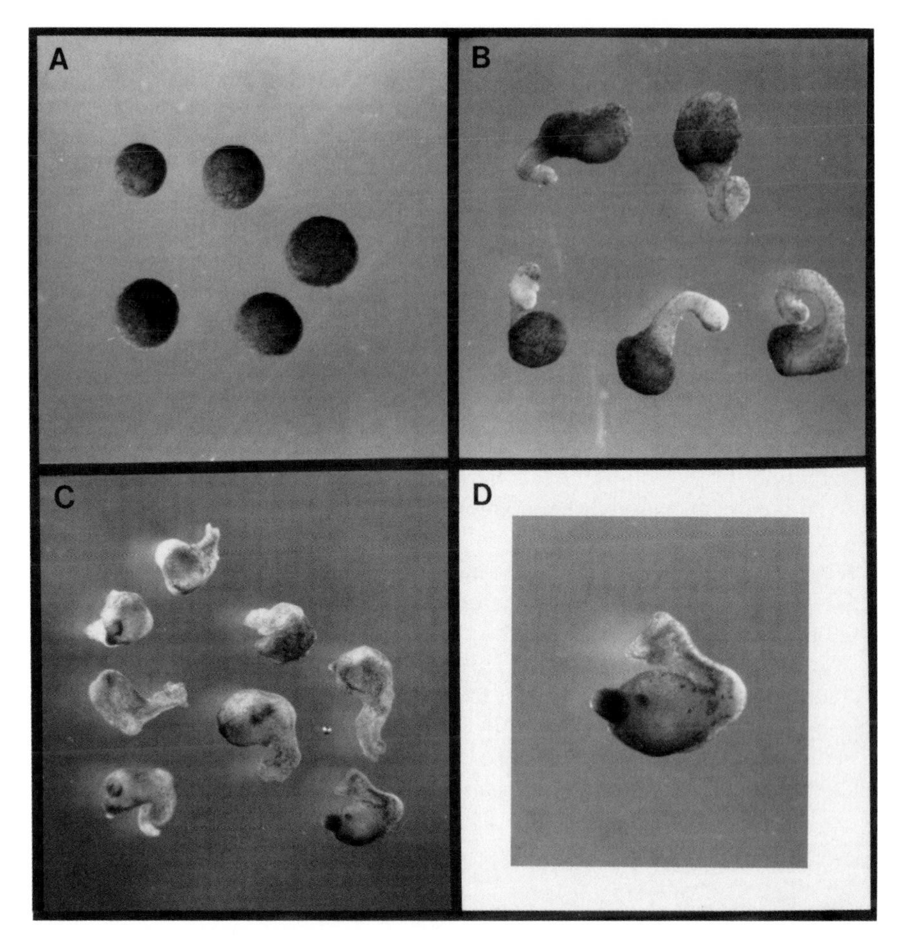

Fig. 1. Induction of animal caps with purified activin. Animal caps from stage 7–10 blastula were cultured in the absence (**A**) or presence (**B–D**) of purified porcine activin A. The untreated caps (A) went on to form ciliated epidermis after one day. Caps treated with activin elongate after 8–12 hr (B), and later develop into embryoids with anterior–posterior polarity. C shows a field of induced caps after two days and D is a closeup view of an embryoid with pronounced axial polarity.

Induction of Dorsoanterior Structures by Activin

As part of the studies with PIF, we wished to directly compare the inducing capacities of PIF to purified activin, which had been reported to induce mesoderm in *Xenopus* animal caps (Asashima et al., 1990). The kinds of inductions we observed with purified porcine activin A are shown in Figures 1 and 2. Figure 1 shows the external morphology of uninduced and induced animal caps at different times after activin application. Approximately 8–12 hr after

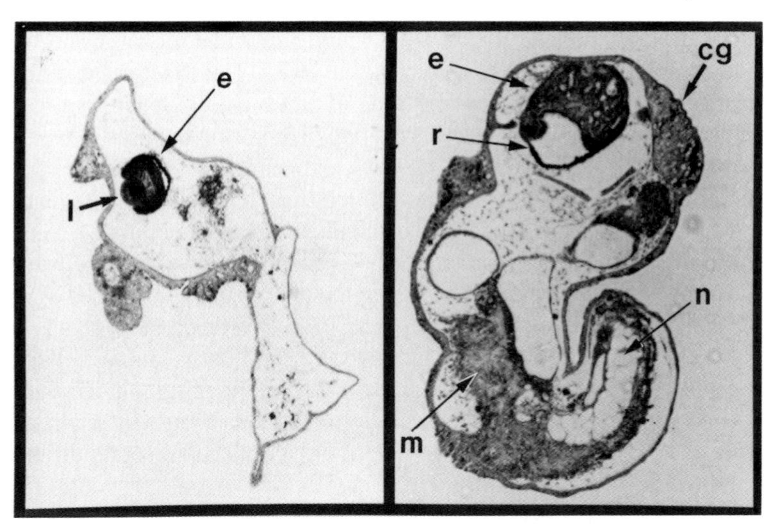

Fig. 2. Histological sections of activin-induced animal caps. Note the presence of muscle (m), notochord (n), cement gland (cg), and eyes (e) with lens (l) and retina (r).

treatment, the animal caps elongate in a manner characteristic of dorsal mesoderm induction, as seen with TGF-β2 and XTC-MIF (Rosa et al., 1988; Smith, 1987; Smith et al., 1988). After two days, the activin-treated caps develop into "embryoids" possessing significant anteroposterior polarity, characterized by the presence of anterior structures such as sucker glands and eyes at one end, and small tails at the other. These tails often move, indicating the presence of muscle. Histological examination of the induced embryoids (Fig. 2) further reveals the presence of axial mesoderm such as notochordal tissue and blocks of muscle, together with well-formed eyes (containing a pigmented retina and often a lens), and brain ventricles (also see Sokol et al., 1990; Thomsen et al., 1990).

Activin is a very potent substance that induces muscle actin (a molecular marker for mesoderm induction) at concentrations as low as 5 pM. The induction response plateaus between about 80–240 nM activin A (Thomsen et al., 1990). We have shown that activin is indeed the functional inducer in our preparation since its ability to induce is abolished by preincubating it with antiactivin A antisera. This antiserum also blocks the inducing activity of purified PIF and recognizes PIF in a Western blot (Thomsen et al., 1990). Thus, PIF appears to correspond to mouse activin A.

The induction of dorsal mesodermal tissues such as muscle and notochord, along with some neural tissue, has been observed by others with purified activin A (Asashima et al., 1990) or the *Xenopus* activin A homologue XTC-MIF (Smith et al., 1988, 1990). Those studies, however, did not reveal the capacity of activin to induce a rudimentary axial pattern in animal caps, nor has activin been shown to induce extensive neural structures such as brain ventricles, or complex anterior structures such as eyes. The exact reasons for the differences between our studies and those of others are unknown, but perhaps subtle variations in the way the animal caps are cut or cultured, or perhaps differences in the source of the factors, could explain the discrepancies. Whether activin induces neural tissue directly also requires resolution, since it is possible that tissues induced by activin in turn induce neural structures.

An important question raised by our observations pertains to the origin of polarity in the induced animal cap. Since the animal caps are placed in a uniform solution of activin, polarity might arise stochastically, or perhaps some differences in the ability to respond to activin already exist among cells of the cap. In other words, some inherent polarity or prepatterning might be present at the time of activin treatment. In fact, recent experiments in this lab have shown that the dorsal and ventral halves of an animal cap respond differently to activin, resulting in the formation of dorsal and ventral mesoderm (Sokol and Melton, submitted) from those respective regions of the cap.

Cloning and Developmental Expression of *Xenopus* Activin Genes

The remarkable inductive properties of purified activin A beg the question of whether activin is present in the embryo at the appropriate time and place consistent with its potential role as an endogenous mesoderm inducer and axial patterning substance. Dorsal vegetal cells of the blastula can induce dorsal mesoderm when recombined with animal caps (Boterenbrood and Nieuwkoop, 1973). Therefore, one prediction is that activin might correspond to the endogenous, maternally encoded dorsal mesoderm inducer. Another is that activin might be an axial patterning substance of the Spemann organizer. In any case, eventual resolution of questions related to activin gene expression and potential embryonic function required cloning of the *Xenopus* activin genes.

Several *Xenopus* activin genes were amplified and cloned from the genomic DNA of a single animal using degenerate oligonucleotide primers in conjunction with the PCR technique (Fig. 3) (methods in Innis et al., 1990). The primers were targeted to regions corresponding to conserved peptides at the N and C termini of the mature portion of activin β_A and β_B chains. Three activin homologues were obtained, one β_A chain and two β_B chains. The protein encoded by the *Xenopus* activin β_A (Xactβ_A) gene is about 85% homologous to mammalian activin A. The sequence of our activin β_A clone differs at position 9 of the 10 amino acids reported for XTC-MIF (Smith et al., 1990). The Xactβ_A gene encodes a serine, whereas XTC-MIF contains an asparagine at that position. The two *Xenopus* activin β_B chains, Xactβ_B1 and β_B2, encode proteins that are respectively 94% and 96% homologous to mammalian activin β_B. Additionally, when these degenerate primers were applied to the cDNA of the PIF-producing P388D1 cells, we were able to

Beta A chains:

```
mammal        G L E C D G V N I C C K K Q F F V S F K D I G W N D W I I A P S G Y H A N Y
XactßA                    * * S * * * * * H * Y * * * * * * * * S * * * * * * P * * * * *

mammal        C E G E C P S H I A G T S G S S L S F H S T V I N H Y R M R G H S P F A N L
XactßA        * * * D * * * * * * * * T * * * * * * * * * * * * Q * * L * * Q * * * T S I

mammal        K S C C V P T K L R P M S M L Y Y D D G Q N I I K K D I Q N M I V E E C G C
XactßA        * * * * * * S * * * A * * * * * * * * * * * * * * * * * * * *
```

Beta B chains:

```
mammal        G L E C D G R T N L C C R Q Q F F I D F R L I G W S D W I I A P T G Y Y G N Y
XactßB1                   * * * * * * * * * * Y * * * * * * * * N * * * * * * A * * * * * *
XactßB2                   H * * * * * * * * * Y V * * * * * * * N * * * * * * A * * * * * *
XactßBcDNA    * * * * * * H * * * * * * * * * Y * * * * * * * * N * * * * * * A * * * * * *

mammal        C E G S C P A Y L A G V P G S A S S F H T A V V N Q Y R M R G L N P G T V N
XactßB1       * * * * * * * * * * * * * * * * * * * * * * * * * * * * * * * * * * * * * *
XactßB2       * * * * * * * * * * * * * * * * * * * * * * * * * * * * * * * * * * * * * *

mammal        S C C I P T K L S T M S M L Y F D D E Y N I V K R D V P N M I V E E C G C
XactßB1       * * * * * * * * * * * * * * * * * * * * * * * * * * * * * *
XactßB2       * * * * * * * * * * * * * * * * * * * * * * * * * * * * * *
XactßBcDNA    * * * * * * * * * * * * * * * * * * * * * * * * * * * * * * * * * * * * *
```

Fig. 3. Amino acid sequences of the mature region of *Xenopus* and mammalian activins. The predicted protein sequences of the *Xenopus* PCR clones of β_A (Xactβ_A) and β_B genes (Xactβ_B1 and Xactβ_B2) are shown along with the mature portion of the full-length β_B cDNA (Xactβ_{Bc}). The peptides for which degenerate PCR primers were designed are underlined. The sequence of the primers are given in Thomsen et al. (1990). Amino acids conserved between mammalian and *Xenopus* activins are marked with asterisks. The β_B cDNA encodes amino acids identical to those of the primers, as indicated by the asterisks at those positions.

amplify and clone an activin β_A chain gene whose predicted protein sequence matched that of other mammalian activins. We did not obtain an activin β_B gene from the cDNA of these cells. These findings are consistent with PIF's apparent identity as mouse activin A.

The expression of the activin genes during *Xenopus* embryogenesis was assayed by RNAse protection with antisense probes synthesized from the Xactβ_A and Xactβ_B2 clones. The Xactβ_B1 clone was not tested, since its DNA sequence is about 93%, identical to the Xactβ_B2 clone, and would likely result in a protection pattern nearly indistinguishable from that of the Xactβ_B2 gene. The expression patterns of the two genes are summarized in Figure 4. As indicated in the figure, activin β_B transcripts are first detectable in the late blastula (stage 9), whereas activin β_A transcripts don't appear until the late gastrula (stage 13). At their time of appearance, the activin β_A and β_B

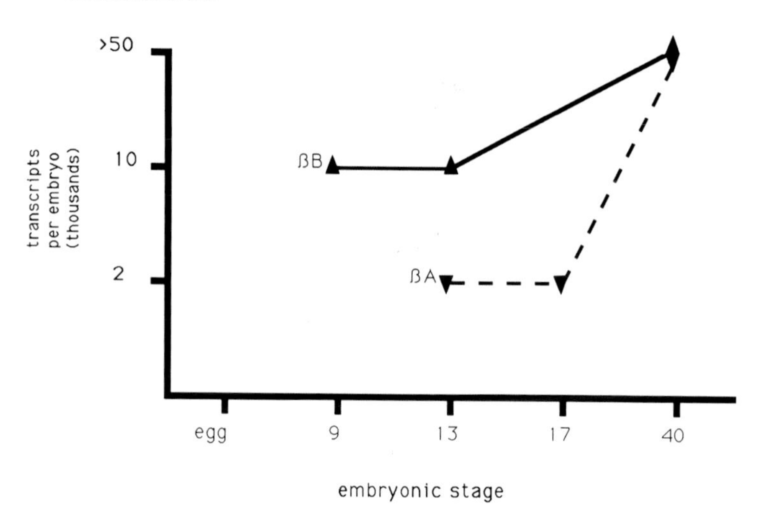

Fig. 4. Summary of activin gene expression during *Xenopus* embryogenesis. The relative levels of activin β_A and β_B mRNA are depicted over the course of early development. Activin β_B mRNA is first detectable in late blastula (stage 9), at about 10,000 copies per embryo. Activin β_A transcripts are first detectable in late gastrula (stage 13) at about 2,000 copies per embryo. Both transcripts are highly abundant (greater than approximately 50,000 copies per embryo) in tailbud tadpoles (stage 40). See Thomsen et al. (1990) for original RNAse protection gels.

transcripts are at about 2,000 and 10,000 copies per embryo, respectively, and they remain at these levels into the neurula stage (Fig. 4; Thomsen et al., 1990). Transcript levels from both genes are greatly elevated late in development (stage 40, tailbud tadpole stage), and both are expressed in XTC cells. We have not detected activin β_A or β_B mRNA in the maternal RNA of transcriptionally quiescent, pre-midblastula transition (MBT) (Newport and Kirschner, 1982) embryonic stages. We are currently investigating whether activin mRNA is localized in the embryo.

The developmental expression profile of the *Xenopus* activin genes leads us to suggest that activin B, rather than activin A, may play a crucial role in the induction of mesoderm and other dorsoanterior structures in the developing embryo. Activin β_B transcripts appear in the late blastula, stage 9, which is too late for it to function as a maternal inducer, but its expression is within the period of competence for mesoderm induction and patterning. Activin β_B expression also precedes the formation of the Spemann organizer (stage 10). Presumably, expression of the activin β_B gene on its own leads to the production of the homodimer activin B.

It is tempting to speculate that activin B might establish the Spemann organizer or perhaps be the agent responsible for its remarkable inductive and patterning properties. Resolution of this question will in part require information about the localization of activin β_B mRNA and protein in the embryo. Activin β_A mRNA, on the other hand, first appears well after the period of competence for mesoderm induction, around the time (stage 13) that the involuting chordamesoderm has extended to its most anterior position. Thus it seems unlikely that activin β_A participates in the induction of mesoderm. Perhaps instead it participates in later induction and patterning events such as development of the nervous system. The recent identification of activin A as a neuronal cell survival factor (Schubert et al., 1990) is consistent with this possibility. We cannot be sure if our findings on activin β_A expression will hold for the gene encoding XTC-MIF, since the protein sequence of this factor differs by one amino acid of 10 from that predicted for our clone.

Inductive Capacity of *Xenopus* Activin β_B

To determine whether *Xenopus* activin B is capable of inducing mesoderm, we retrieved a full-length activin β_B cDNA by screening a stage 28 (tailbud tadpole) cDNA library with activin β_A and β_B probes. The DNA sequence of this clone that encodes the mature portion of the β_B protein matches the PCR clone Xactβ_B2 at all but one nucleotide. This difference results in a single amino acid alteration between genomic clone Xactβ_B2 (a Val) and the full-length cDNA (an Ile). This difference could be the result of allelic polymorphism or an alteration introduced in the genomic clone during PCR amplification.

The full-length β_B cDNA was subcloned into the mammalian expression vector pcDNA1 (Invitrogen Inc.) and this was transfected into COS cells. After two days, conditioned medium was collected and tested for inducing activity. The inducing capacity of the recombinant *Xenopus* activin B conditioned medium was indistinguishable from purified porcine activin A and PIF (data not shown, but refer to Figs. 1 and 2, and to Sokol et al., 1990; Thomsen et al., 1990). It is noteworthy that activin B has qualitatively the same inductive characteristics as mammalian activin A, even though the mature chain of the *Xenopus* activin β_B cDNA encodes a protein with only 65% homology to mammalian activin β_A.

To test whether activin B might mimic the action of the Spemann organizer, we injected synthetic (Krieg and Melton, 1984), full-length activin β_B mRNA into a ventral, subequatorial blastomere of 32-cell blastula hosts. Injection of the mRNA into this area of a normal embryo was intended to mimic the placement of a donor organizer (Spemann and Mangold, 1924; Hamburger, 1988), and to examine the effect of activin on a region normally fated to produce ventral mesoderm. Figure 5 shows an example of an injected embryo that has developed a second axis resembling a dorsal trunk and head. The presence of pigmented melanophores in the second body axis is indicative of neural crest cell derivatives from induced neural tissue. Similar results are

Fig. 5. Synthetic activin β_B mRNA injected into a ventral marginal zone blastomere induces a second body axis. The **top** embryo is an uninjected control. The **bottom** embryo received about 100 pg of synthetic *Xenopus* activin β_B mRNA. Note the trunk and head features of the ectopically induced axis.

obtained with injected mammalian activin β_A (Thomsen et al., 1990). Histological sections of these secondary axes (not shown) indicate the presence of dorsal mesoderm (blocks of muscle and notochord) and melanocytes (which are derived from neural crest cells). These findings are consistent with the suggestion that activin B might be an agent of the Spemann organizer. Knowledge of the local expression pattern of activin mRNA and protein is required to assess this possibility.

CONCLUSION

The studies by ourselves and others cited above have revealed that activins are likely to play important roles in mesoderm induction and axial patterning in vertebrate embryos. Our results on the developmental expression of cloned *Xenopus* activin genes points to activin B as a critical player in these processes. We emphasize the role of activin B principally because its mRNA is detected in the late blastula at a time when cells are still competent to respond to mesoderm inducers. The onset of its expression also nearly coincides with the time of the appearance of the Spemann organizer. Activin β_A, on the other hand, is first detected in the late gastrula. The fact that recombinant activin B induces mesodermal and dorsoanterior structures, coupled with the ability of synthetic β_B mRNA to induce a secondary axis when injected into a host, argues in favor of the importance of this molecule in embryogenesis.

Among the many questions to pursue on the heels of these studies are those addressing the nature of prepatterning in the animal hemisphere cells, and the molecular identity of the maternal inducer(s) of the mesoderm. It seems unlikely that activin is involved in the early, maternal (pre-MBT) period of mesoderm induction since this is underway before the onset of activin β_B expression (the caveat being that maternal activin mRNA might be present at levels below our assay sensitivity). Perhaps instead other TGF-β genes, or members of other growth factor families, function in this capacity.

It is further noteworthy that activin A and B can also induce axial structures in the hypoblast of the pregastrula (stage XIII) chick embryo (Mitrani et al., 1990). During normal development, axial pattern is induced in the epiblast by the underlying hypoblast, and the hypoblast has been shown to express activin β_B at the time of axial induction. The results with the frog and chick systems thus suggest an important role for activin B in axial patterning in vertebrates.

ACKNOWLEDGMENTS

We thank our colleagues Joan Vaughan and Wylie Vale for their gift of purified activin A. This work was supported by an NIH grant to D.A.M. G.T. was supported by a postdoctoral fellowship from the American Cancer Society. M.W. was supported by a postdoctoral fellowship from the Lucille P. Markey Charitable Trust.

REFERENCES

Asashima M, Nakano H, Shimada K, Kinoshita K, Ishii K, Shibai H, Ueno N (1990): Mesodermal induction in early amphibian embryos by activin A (erythroid differentiation factor). Roux Arch Dev Biol 198:330–335.

Boterenbrood EC, Nieuwkoop PD (1973): The formation of the mesoderm in urodelean amphibians. V. Its regional induction by the endoderm. Roux Arch Dev Biol 173:319–332.

Godsave SF, Isaacs HV, Slack JMW (1988): Mesoderm inducing factors: A small class of molecules. Development 102:555–566.

Gurdon JB, Fairman S, Mohun TJ, Brennan S (1985): Activation of muscle-specific actin genes by an induction between animal and vegetal cells of a blastula. Cell 41:913–922.

Hamburger V (1988): ''The Heritage of Experimental Embryology.'' New York: Oxford University Press.

Innis MA, Gelfand DH, Sninsky JJ, White TJ (1990): ''PCR Protocols. A Guide to Methods and Applications. New York: Academic Press.

Kimmelman D, Abraham JA, Haaparanta T, Palisi TM, Kirschner MW (1988): The presence of fibroblast growth factor in the frog egg: Its role as a natural mesoderm inducer. Science 242:1053–1056.

Kimmelman D, Kirschner M (1987): Synergistic induction of mesoderm by FGF and TGF-β and the identification of FGF in the early *Xenopus* embryo. Cell 51:869–877.
Krieg PA, Melton DA (1984): Functional messenger RNAs are produced by SP6 *in vitro* transcription of cloned DNAs. Nucleic Acids Res 12:7057–7070.
Ling N, Ying S-Y, Ueno N, Shimasaki S, Esch F, Hotta M, Guillemin R (1986): Pituitary FSH is released by a heterodimer of the β-subunits from the two forms of inhibin. Nature 321:779–782.
Mason A, Berkenmeier L, Schmelzer C, Schwall R (1989): Activin B: Precursor sequences, genomic structure and in vitro activities. Mol Endocrinol 3:1352–1358.
Mason AJ, Hayflick JS, Ling N, Esch F, Ueno N, Ying SY, Guilleman R, Niall H, Seeburg PH (1985): Complementary DNA sequences of ovarian follicular fluid inhibin show precursor structure homology with transforming growth factor-β. Nature 318:659–663.
Mitrani E, Ziv T, Thomsen J, Shimoni Y, Melton DA, Bril A (1990): Activin can induce the formation of axial structures and is expressed in the hypoblast of the chick. Cell 63:495–501.
Newport J, Kirschner M (1982): A major developmental transition in early *Xenopus* embryos: II. Control of the onset of transcription. Cell 30:687–696.
Nieuwkoop PD (1969): The formation of mesoderm in early urodelean amphibians. I. Induction by the endoderm. Roux Arch Dev Biol 162:341–373.
Nieuwkoop PD, Johnen AG, Albers B (1985): "The Epigenetic Nature of Early Chordate Development. Inductive Interaction and Competence." New York: Cambridge University Press.
Rosa F, Roberts AB, Danielpour D, Dart LL, Sporn MB, Dawid IB (1988): Mesoderm induction in amphibians: The role of TGF-β2-like factors. Science 239:783–785.
Saxen L, Lehtonen E, Karkinen-Jaaskeleinen M, Nordling S, Wartiovarra J (1976): Are morphogenetic tissue interactions mediated by transmissible signal substances or through cell contacts? Nature 259:622–623.
Schubert D, Kimura H, LaCorbiere M, Vaughan J, Karr D, Fischer WH (1990): Activin is a nerve cell survival molecule. Nature 344:868–870.
Slack J (1983): "From Egg to Embryo." New York: Cambridge University Press.
Slack JMW, Darlington BG, Heath JK, Godsave SF (1987): Mesoderm induction in early *Xenopus* embryos by heparin-binding growth factors. Nature 326:197–200.
Slack JMW, Isaacs HV (1989): Presence of fibroblast growth factor in the early *Xenopus* embryos by heparin-binding growth factors. Nature 326:197–200.
Slack JMW, Isaacs HV (1989): Presence of fibroblast growth factor in the early *Xenopus* embryo. Development 105:147–153.
Smith JC (1987): A mesoderm inducing factor is produced by a *Xenopus* cell line. Development 99:3–14.
Smith JC, Price BMJ, Van Nimmen K, Huylebroeck D (1990): Identification of a potent *Xenopus* mesoderm-inducing factor as a homolog of activin A. Nature 345:729–731.
Smith JC, Yaqoob M, Symes K (1988): Purification, partial characterization and biological effects of the XTC mesoderm-inducing factor. Development 103:591–600.
Sokol S, Wong G, Melton DA (1990): A mouse macrophage factor induces head structures and organizes a body axis in *Xenopus*. Science 249:561–564.
Spemann H, Mangold H (1924): Induction of embryonic primordia by implantation of organizers from different species. Roux Arch Dev Biol 100:555–638.
Thomsen G, Woolf T, Whitman M, Sokol S, Vaughan J, Vale W, Melton DA (1990): Activins are expressed early in *Xenopus* embryogenesis and can induce axial mesoderm and anterior structures. Cell 63:485–493.
Tiedeman H (1978): Chemical approach to inducing agents. In Nakamura O, Toivonen S (eds): "Organizer: A milestone of a Half Century from Spemann," pp 91–117. Amsterdam: Elsevier/North-Holland Biomedical Press.
Vale W, Rivier J, Vaughan J, McClintock R, Corrigan A, Wilson W, Karr D, Spiess J (1986):

Purification and characterization of an FSH releasing protein from porcine ovarian follicular fluid. Nature 321:776–779.

van den Eijnden-Van Raaij AJM, van Zoelent EJJ, van Nimmen K, Koster CH, Snoek GT, Durston AJ, Huylebroeck D (1990): Activin-like factor from a *Xenopus laevis* cell line responsible for mesoderm induction. Nature 345:732–734.

Weeks DL, Melton DA (1987): A maternal messenger RNA localized to the vegetal hemisphere in *Xenopus* eggs codes for a growth factor related to TGF-β. Cell 51:861–867.

Yu J, Shao L, Lemas V, Yu AL, Vaughan J, Rivier J, Vale W (1987): Importance of FSH-releasing protein and inhibin in erythrodifferentiation. Nature 330:765–767.

Cell-Cell Interactions in Early Development, pages 93–107

6. Effects of the Dorsal Blastopore Lip and the Involuted Dorsal Mesoderm on Neural Induction in *Xenopus laevis*

Carey R. Phillips
Department of Biology, Bowdoin College, Brunswick, Maine 04011

NEURAL INDUCTION

Transplanting a blastopore lip to the ventral side of a host embryo results in the induction of a second axis (Spemann and Mangold, 1924). Many of the dorsal structures found in the second axis are derived from the ventral mesoderm and ventral ectoderm of the host (Gimlich and Cooke, 1983; Smith and Slack, 1983). Spemann (1938) proposed two possible mechanisms to explain the inductive actions of the grafted dorsal blastopore lip region. By one mechanism, the grafted lip induces the adjacent marginal zone cells to become functional dorsal mesoderm. The induced dorsal mesoderm, in concert with lip-derived tissue, would then undergo the convergent extension movements of gastrulation and eventually come to underlie and induce ectoderm to follow a neural pathway. We refer to inductions that occur across two tissue layers as "vertical inductions." By the second mechanism, the grafted lip sends a set of signals that induce both the adjacent ectoderm and the adjacent mesoderm to form dorsal structures; the mesoderm forms notochord and somites while the adjacent ectoderm forms neural plate structures (Spemann, 1938). Spemann suggested that signals directing the induction of neural plate might originate from the dorsal blastopore lip region and travel through the plane of the ectoderm (Spemann, 1938; see Hamburger, 1988 for review). Inductions within the plane of the tissue may be called "planar inductions." However, experiments by two of Spemann's students provided support for the model that the dorsal blastopore lip region induces the adjacent mesoderm, which, in turn, induces the overlying ectoderm by vertical signals.

J. Holtfreter placed late blastula embryos in a hypertonic solution, forcing them to exogastrulate. In these embryos, the mesoderm fails to invaginate during gastrulation and therefore does not underlie the ectoderm, leaving the ectoderm as an empty sac forming an epidermal-like tissue. The ectoderm, connected by only a thin isthmus of tissue to mesoderm of the blastopore lip region, does not undergo any morphologically observable neural differentiation (Holtfreter, 1933). Holtfreter concluded that neural induction normally occurs exclusively by information passing from involuted dorsal mesoderm to overlying ectoderm, that is, by "vertical induction." In the second experiment, Otto Mangold implanted pieces of the dorsal mesoderm from neural plate stage Urodeles into the ventral blastocoel cavity of a host Urodele embryo. The host embryos developed a secondary axis appropriate to the anterior–posterior type of mesoderm implanted into the blastocoel; the anterior dorsal mesoderm induced head-like structures and the posterior dorsal mesoderm induced trunk-like structures (Mangold, 1933). He concluded that the involuted dorsal mesoderm is responsible for both the induction of neural development and the establishment of neural pattern in overlying ectoderm.

Although these results have long been taken to favor vertical neural induction, recent advances in molecular technology have provided experimental evidence that neural induction may be much more complex than previously believed. Molecular probes allow the investigator to study early events involved in the induction process and free the experimenter from relying on a recognizable morphology as the basis for determining whether an early induction event has taken place. In this chapter, we address some of the major issues involved in the neural induction process and discuss the experimental evidence, both recent and historical, that relates to these issues.

Study of Neural Competence

The process of neural induction can be divided into two very broad areas of study, the production of a signal capable of eliciting a neural response and the response of a tissue to this signal. The ability of a tissue to respond to inductive signals is termed "competence." Assays designed to determine if or when a neural induction event has occurred require that the state of neural competence of the potentially responding tissue, the ectoderm, be known. Originally, neural competence of ectoderm was studied using a battery of microsurgical transplantation techniques. In one such experiment, a portion of the presumptive neural plate and adjacent presumptive epidermis were removed and rotated so that the presumptive epidermal region came to lie within the region that is normally induced to form neural structures, while the presumptive neural region then resided outside the region fated to become neural tissue (Lehmann, 1929). Lehmann found that the ectodermal tissues of *Triton* embryos had lost their neural competence by midgastrula stages. In a variation of this experiment, Mangold placed neural plate tissue, using this as a neural inducer, into the ventral blastocoel cavity and determined at what stage the adjacent host ventral ectoderm can no longer respond to neural induction signals of the implanted neural plate. Mangold found that the ability of the ventral ectoderm to respond to neural induction signals is lost by midgastrula (Mangold, 1929). Blastopore lip was also used as a source of neural induction signals when implanted into the ventral portion of a host embryo. Again, it was found that neural competence in the ectoderm is lost by midgastrula. In addition, it was found that neural competence was lost first in the posteriormost regions of the ventral ectoderm and subsequently lost in the more anterior regions (Machemer, 1932; Schechtman, 1938). A progressive loss of competence to respond to neural induction signals from midventral to dorsal lateral was also observed (Machemer, 1932).

Holtfreter found that the progressive loss of neural competence is an autonomous function of the ectoderm. He removed portions of the ectoderm, cultured them for varying lengths of time in vitro, and implanted them into the presumptive neural plate region of either *Triton* or *Bombinator*. Ectoderm incubated in vitro to the equivalent of midgastrula resulted in the loss of neural competence in the ectoderm (Holtfreter, 1938). Recently, neural competence has been reexamined using fluorescein-labeled dextran (FLDx) cell lineage markers so that the identity of the responding cell could be verified, while neural-specific antibodies (neural cell adhesion molecule, NCAM) were used as indicators of early steps in the neural induction process. Early gastrula ventral ectoderm grafted into the presumptive lens-forming region, lateral and not contiguous to the anterior neural plate of an early neurula, will express NCAM and exhibit a neural-plate-like morphology (Servetnick and Grainger, 1991). When ectoderm older than midgastrula is used, it develops into lens instead of neural plate (Servetnick and Grainger, 1991). In addition, ectoderm incubated in vitro and then grafted into the presumptive lens-forming region will change its developmental competence from forming a neural plate to forming a lens. These experiments indicate that the ability of ectoderm to respond to neural induction signals is normally lost by midgastrula and that the loss of neural competence is autonomously programmed within the ectoderm.

Bias of Dorsal Ectoderm

This observation on the time of loss of neural competence begs the question of when the dorsal ectoderm is normally induced to become neural plate. Animal cap ectoderm removed prior to midgastrula and grown in culture will develop into atypical epidermis. However, if animal cap ectoderm is removed at midgastrula, or later, and cultured in a buffered saline solution, it will form neural plate structures that express NCAM intensely (Jacobson and Rutishauser, 1986; Fig. 1). Therefore, neural induction has occurred, in at least a portion of the presumptive neural plate region, by the middle of gastrulation. At present, we do not have the necessary molecular probes to determine if the entire neural plate is specified to become neural tissue by midgastrula, or whether a portion of the presumptive neural plate, in *Xenopus*, remains responsive to neural signals after midgastrula.

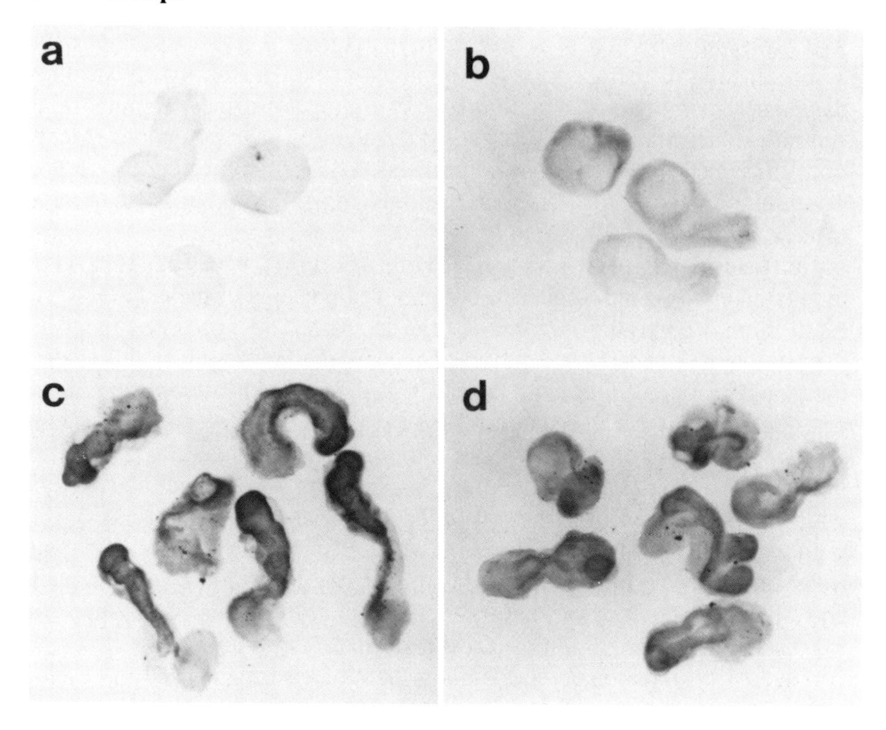

Fig. 1. The developmental time when the presumptive neural plate becomes autonomous in its ability to express NCAM. Animal caps were dissected from above the point where cells will enter the blastopore lip and cultured to the equivalent of stage 24. Animal caps removed at early gastrula, stage 10 (**a**) and 10.5 (**b**) do not express NCAM. Animal caps removed at midgastrula, stage 11 (**c**) or late gastrula, stage 12 (**d**) express NCAM very well.

By midgastrula, at least some neural induction has taken place. However, the dorsal mesoderm has not yet extended into place under the presumptive anterior neural plate region by the time ectoderm loses neural competence. There are several possible explanations for how the anterior neural plate might be induced without the presence of underlying dorsal mesoderm. First, the dorsal ectoderm (presumptive neural ectoderm) may not be totally naive, at least relative to the neural pathway, in that it may maintain neural competence longer than does other ectoderm. It has been suggested that dorsal ectoderm has a different set of developmental potentials than does the ventral ectoderm. For example, dorsal ectoderm (presumptive neural) and ventral ectoderm (presumptive epidermal) already differ at early cleavage stages of *Xenopus* development. Epi-1, a cell surface molecule expressed only on ventral ectoderm (presumptive epidermal ectoderm), has been used as a molecular marker to indicate when early ectodermal decisions have been initiated within the animal cap ectoderm (Akers et al., 1986). Animal cap ectoderm was isolated at successive stages of early cleavage and blastula. The animal caps were divided into dorsal and ventral halves, cultured, and stained for the presence of Epi-1. Epi-1 is poorly expressed on isolated and cultured dorsal animal blastomeres, while Epi-1 is intensely expressed in isolated and cultured ventral animal blastomeres (London et al., 1988). This suggests that dorsal animal blastomeres acquire a predisposition toward certain dorsoventral fates, and away from others, long before the onset of gastrulation. This early dorsoventral bias may differentially affect the loss of neural competence within the dorsal ectoderm during gastrulation.

It should be pointed out that these early differences are not strictly required for neural induction in some experimental situations. For example, a dorsal blastopore lip transplanted into the ventral marginal zone of a host embryo can induce ventral ectoderm to form anterior neural structures (Spemann and Mangold, 1924; Gimlich and Cooke, 1983; Smith and Slack, 1983). Therefore, the blastopore lip is capable of organizing neural development without the benefit of the early dorsoventral ectodermal bias. Holtfreter has also shown that animal cap ectoderm will form neural tissue when wrapped around blastopore lip tissue (Holtfreter, 1936). However, it has also been observed that the early dorsal blastopore lip can inhibit the subsequent expression of Epi-1 in adjacent ventral ectoderm and thereby recreate a dorsal ectoderm bias during early gastrulation. Thus, the bias may have an important role even in these experimental situations.

Inhibition of Epi-1 Expression

Since the blastopore lip and the adjacent ectoderm normally reside within the same plane of tissue within the embryo, we were interested in knowing whether the blastopore lip tissue could inhibit the expression of the Epi-1 molecule if a planar configuration were maintained between the two tissues. This was tested by holding the blastopore lip/ectoderm explant flat under a coverslip during the culture period. The ectoderm adjacent to the blastopore lip region did not express the Epi-1 antigen (Savage and Phillips, 1989). Therefore, inhibitory signal(s) from the blastopore lip region must be moving in a planar fashion from the blastopore lip through the adjacent ectoderm. It was also observed that the cells residing at the dorsal lip at early gastrula (stage 10) are more potent at signaling the inhibition of Epi-1 than cells residing at the lip during mid- to late gastrula (Savage and Phillips, 1989). These results indicate that the early blastopore lip can provide enough information, which appears to travel through the plane of the ectoderm, to inhibit expression of Epi-1 in adjacent ectoderm, thus mimicking the original dorsoventral pattern of Epi-1 expression. It is expected that signals from the dorsal blastopore lip region reinforce and extend the earlier dorsoventral bias during normal development.

The dorsoventral ectodermal bias, either the early bias or the bias initiated at the beginning of gastrulation, may also play a role in subsequent induction events associated with other neural-specific molecules or neural anatomical differentiations. For example, dorsal ectoderm cocultured with involuted dorsal mesoderm is much more easily induced to express either NCAM or X1hbox6 than is ventral ectoderm cocultured with involuted dorsal mesoderm (Sharpe et al., 1987; Phillips and Doniach, 1991). However, both the dorsal and the ventral ectoderm from an early gastrula will readily express NCAM if cocultured with a dorsal blastopore lip in a planar configuration (Phillips and Doniach, 1991). Therefore, both the dorsal and the ventral ectoderm are competent to respond to neural inducing signals by expressing a neural-specific marker molecule.

Planar Signals

It appears then that the signal from the blastopore lip, prior to involution, may present a stronger signal than does the involuted dorsal mesoderm (Phillips and Doniach, 1991; Sater et al., 1991). Alternatively, the signals of the blastopore lip cells and the involuted dorsal mesoderm may be the same, both in quality and quantity, but the delivery of the signal is more efficient, at least in vitro, in the planar configuration than it is between the involuted dorsal mesoderm and the overlying ectoderm. Recently, a difference in the ability of dorsal or ventral ectoderm to respond to activin B has also been observed. Dorsal ectoderm incubated in activin B will form embryoids containing dorsal structures, complete with neural tissue, while the ventral ectoderm incubated in activin B will form only atypical epidermis (Thomsen et al., 1990). These results also indicate that, in some experimental situations, the dorsal ectoderm is more competent to respond to signals that induce mesoderm and neural structures than is the ventral ectoderm. Therefore, the observation discussed above concerning the loss of ectodermal competence for neural induction prior to the full extension of invaginated dorsal mesoderm may not be applicable to the presumptive anterior neural plate ectoderm, because this tissue is already predisposed toward the neural pathway. Anterior neural ectoderm may maintain neural competence longer than has been shown for ectoderm in general.

A planar path for neural induction might also explain how the anterior neural plate becomes induced at a time before the involuted dorsal mesoderm has reached the anterior regions of the neural plate and before neural competence is lost. The generally accepted mechanism for neural induction is that the invaginated dorsal mesoderm induces the overlying ectoderm and provides the necessary information to form the different regions and structures of the brain and spinal cord. However, the involuted dorsal mesoderm may not be the only source of neural induction signals. In fact, embryos dissociated during early cleavage and reassociated during midgastrula stages will not make differentiated mesoderm of any type, but will still synthesize NCAM (Sato and Sargent, 1989). These results suggest that invaginated dorsal mesoderm is not needed for early neural events. Therefore, signals that induce and pattern the neurectoderm might also be supplied by sources other than the invaginated dorsal mesoderm and by different paths of transmission, as well. A likely alternative is that of planar signals from the blastopore lip.

Planar signals have already been shown to affect the expression of Epi-1. Experiments using exogastrulated embryos have provided other examples of induction by planar signals. Exogastrulated embryos fail to invaginate the mesoderm, and the ectoderm is never underlain by dorsal mesoderm in the ''vertical'' configuration found in normal development. Instead, the ectoderm is connected with the mesoderm of the blastopore lip in a planar configuration. Even though the resultant sac of ectoderm does not show any neural morphology, it will express some neural-specific molecules (Kintner and Melton, 1987; Dixon and Kintner, 1989; Ruiz i Altaba, 1990). Therefore, it appears that enough information to initiate the synthesis of neural-specific molecules like NCAM, as well as the inhibition of some epidermal-specific molecules discussed above, can occur when the blastopore lip region maintains contact with ectoderm within a single plane of tissue.

A second method often used to study the inductive potentials of planar signals between two tisues is the ''Keller sandwich'' (see Keller et al., Chapter 3, this volume). In the Keller sandwich, the dorsal blastopore lip region is dissected from the embryo, along with the tissues fated to involute into the blastocoel (the ''involuting marginal zone'' or IMZ; Gerhart and Keller, 1986), and those cells fated to form the neural plate region (the ''non-involuting marginal zone''). Two of the dorsal strips are placed face to face and cultured under a glass coverslip so that the culture maintains a flat configuration (Gerhart and Keller, 1986). Specific groups of cells within the explanted tissues undergo two periods of cellular rearrangements often referred to as convergent extension movements. Convergent extension movements occur in the Keller sandwich at approximately the correct distances from the limit of involution and at approximately the same times as in vivo (Keller and Danilchik, 1988). This limit of involution represents the boundary between mesoderm cells fated to involute into the blastocoel by way of the blastopore lip, and ectodermal cells fated to become posterior neural plate. The mesodermal cells undergo convergent extension movements first and extend away from the presumptive neural plate ectoderm while maintaining only an edgewise contact with the ectoderm at the original boundary. The presumptive posterior neural plate ectoderm then undergoes convergent extension movements. This movement also maintains a planar configuration between the two tissue types. After a day in culture, the resulting explant has a dorsal mesoderm piece, complete with notochord and somites, connected in a planar fashion to tissues normally fated to become neural plate.

The second set of convergent extension movements, normally associated with the posterior neural plate, will not occur in the Keller sandwich unless this tissue has had edgewise, or planar, contact with the mesoderm. The mesoderm does not need to underlie the presumptive neural plate region to induce the second set of convergent extension movements; a planar association will suffice (Keller and Danilchik, 1988). In addition, an experimentally devised planar association between dorsal mesoderm cells and ventral ectoderm cells, not normally fated to become neural plate, is sufficient to induce cells of the ventral ectoderm to undertake convergent and extension behavior (Keller, personal communications). Therefore, it appears that some form of information travels in a planar path from the IMZ region (dorsal mesoderm) into the adjacent ectoderm, which induces the convergent extension movements normally associated with dorsal neurectoderm.

In addition, the Keller sandwich explant also expresses NCAM intensely within the ectodermal portion, and at a considerable distance from the boundary between he mesoderm of the IMZ and the posterior neural plate (Dixon and Kintner, 1989; Phillips and Doniach, 1991; Sater et al., 1991). The only communication permitted between the mesoderm and the ectoderm is edge-to-edge between the two tissue types. Therefore, a signal that induces the synthesis of NCAM must travel through the plane of the tissue in a manner similar to that which initiates the convergent extension movements within the ectoderm portion of the explant.

METHODS AND RESULTS

Verification of the inductive sufficiency of planar signals requires a demonstration that cells from the IMZ portion of the explant (potential inducing mesoderm) do not migrate into the ectoderm portion of the explant and induce neural tissue vertically. The migratory behaviors of cells within the two tissue types were studied using IMZ material from embryos prelabeled with the cell lineage tracer fluorescein dextran (FLDx). The IMZ region of an FLDx-labeled embryo is removed at early gastrula (stage 10.25) and divided into three regions from anterior to posterior. The anteriormost region represents the portion of the IMZ that involutes over the blastopore lip during early gastrulation, while the posteriormost region represents that portion involuting over the blastopore lip at the end of gastrulation. The FLDx prelabeled portion of the IMZ was grafted onto one edge of a ventral ectoderm sheet and the entire explant was held flat with a coverslip during culture. IMZ-ventral ectoderm cultures were incubated until the tissues reached the equivalent of early tailbud (stage 24). Cultured explants were observed with a Wild fluorescent dissecting microscope equipped with a Video International light intensifier. In some cases, the explant was dissected open to better expose cells in the interior positions of the explant. We could observe no fluorescently labeled cells from any of the IMZ regions within the ventral ectoderm tissue more than two to three cell diameters from the original IMZ–ectoderm boundary. Each explant was stained for the expression of NCAM and we observed that a subset of the cells within the ventral ectoderm portion of the explant expresses NCAM intensely and at some distance from the boundary of the IMZ cells (Phillips and Doniach, 1991). We also observed that the posterior portion of the IMZ, the late blastopore lip tissue, induced the expression of NCAM in adjacent ectoderm much better than did the anterior portions of the IMZ.

Migration of cells across the boundary between the IMZ region and presumptive neural plate region was also analyzed by time-lapse photomicroscopy. Cells can be observed directly in an open-face Keller sandwich, which contains only one strip of tissue consisting of the blastopore lip, IMZ, and presumptive neural plate region. The superficial cells are removed and the tissue cultured under a coverslip in a modified Danilchik's medium. Filming the boundary between the two cell types revealed no migration of the IMZ cells into the presumptive neural plate region for more than two to three cell diameters. However, the ectodermal portion of the explant does express NCAM intensely under these conditions, indicating that the ectoderm was induced to express neural-specific molecules even though it has only a planar association with dorsal mesoderm and there is no apparent migration of IMZ cells into the ectoderm (Sater et al., 1991).

Experiments in which the animal cap ectoderm is isolated at different developmental stages, cultured, and assayed for the presence of NCAM indicate that neurectoderm is specified by midgastrula. The period between the beginning of gastrulation, when isolated animal cap ectoderm cultures make only atypical epidermis, and midgastrula, when ectoderm is specified to be neural, is approximately 2 hr. Therefore, the dorsal ectoderm obtains enough information to initiate development of neural tissue sometime within the 2 hr between the initiation of gastrulation and the midpoint of gastrulation, when the dorsal ectoderm has become specified and most ectoderm has lost neural competence. Blastopore lip tissue was grafted onto the edge of ventral ectoderm, incubated for 2 hr, and removed to determine if 2 hr' contact is sufficient to induce the synthesis of NCAM when the only means of induction is through planar signals. We found that 2 hr' contact is sufficient to induce a significant amount of NCAM synthesis in cultured ectoderm (Phillips and Doniach, 1991; see Fig. 2).

DISCUSSION

As yet, it is not known how many different signals, either planar or vertical, are required to induce the full complement of neural structures. It is also not known how many different signals travel through the plane of the ectoderm or how far they travel. Recently, a few molecular probes for neural,

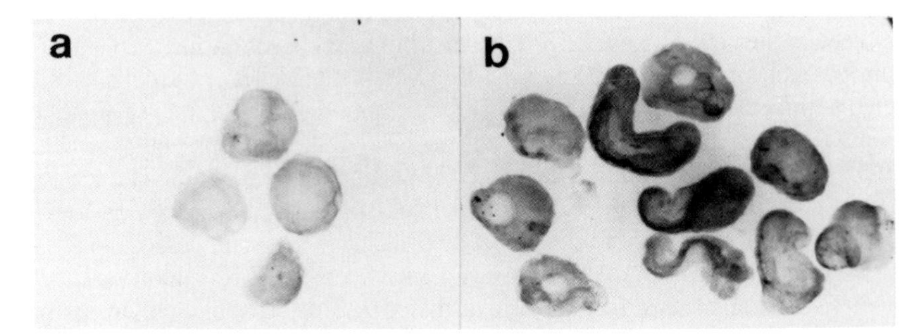

Fig. 2. The effects of a transient, planar association of the blastopore lip on the induction of NCAM synthesis. Animal caps isolated from UV-irradiated embryos do not express NCAM when grown in culture to the equivalent of stage 24 (**a**). Midgastrula blastopore lips were grafted to one edge of the animal cap ectoderm and removed after 2 hr of culture. The blastopore lip tissues were prelabeled with fluorescent dextran so that every cell of the blastopore lip could be monitored and removed from the recombinate. Six of nine blastopore lip/animal cap recombinants synthesized a significant amount of NCAM (**b**).

regionally specific molecules have become available. These probes have been used to determine if planar signals can induce the synthesis of the region-specific molecules at a distance from the inductor and if planar signals can establish localized groups of cellular expression instead of only inducing a general expression pattern. For example, signals that induce the formation of cement gland appear to originate in the posterior mesoderm during midgastrula, move into the overlying ectoderm, and then travel through the plane of the ectoderm to their final position (Sive et al., 1989). Engrailed, another neural-specific molecule, is expressed in a narrow band of cells at the early neural plate stage in an area fated to reside between the midbrain and the hindbrain regions (Hemmati-Brivanlou and Harland, 1989). Different regions of the involuted dorsal mesoderm were cultured with ventral ectoderm to determine the source of inducing signals for engrailed. Notochord, but not somites, will induce a small patch of engrailed-expressing cells. However, the cells expressing the engrailed molecule reside some distance from the notochord tissue (Hemmati-Brivanlou et al., 1990). Blastopore lip incubated within a sheet of ventral ectoderm will also induce the synthesis of engrailed protein within the ventral ectoderm, again at some distance from the implanted mesoderm (Doniach, personal communications). These results indicate that signals capable of inducing regionally specific patterns of neural expression can travel within the ectoderm. It also appears that both the involuted dorsal mesoderm and the blastopore lip are capable of eliciting expression of at least some region-specific molecular synthesis events and that, regardless of the inducing source, the ectodermal cells can respond at some distance from the source of induction.

It remains, however, a formal possibility that ectoderm contains a prepattern relative to the anterior–posterior axis and that a single neural induction event might elicit many different regional responses. It is also possible that the responding ectoderm can "self-organize" into region-specific tissue types after a more general neural induction event. In fact, Chaung (1938, 1939, 1940) and Holtfreter (1934) found that boiled hampster liver and kidney will induce well-organized anterior neural structures in competent ectoderm of Urodeles. This would argue that the presence of organized dorsal mesoderm is not required for neural pattern and that ectoderm has some inherent self-organizing ability. However the regionalization of the neurectoderm is accomplished, it must occur quickly, since engrailed is first expressed immediately after the end of gastrulation. It is likely that the neural plate is already fairly well patterned by early neural plate stage.

Servetnick and Grainger (1991) have also shown that signals traveling through the plane of the ectoderm can induce neural tissue and have provided insights into one possible mechanism for regional specificity. Early gastrula stage ectoderm was grafted into the presumptive lens region of a host embryo. This region is not in contact with dorsal mesoderm and is outside of the region normally fated to become neural plate. However, the early gastrula stage ventral ectoderm is induced to form neural plate-like structures. This would indicate that signals capable of inducing neural plate tissues travel much further within the ectoderm than is normally fated to become neural plate. It was also observed that the type of neural tissue induced appears to be dependent upon the age of the ectoderm. For example, ectodermal grafts of midgastrula tissues will form NCAM-expressing neural plate tissue, while older ectoderm (late gastrula) grafted into the same region will form lens instead (Servetnick and Grainger, 1991). The ectoderm appears to lose neural competence and acquire lens competence with age. If early gastrula ectoderm is isolated and cultured until the equivalent of late gastrula, then it will form lens instead of neural plate when it is grafted into the presumptive lens-forming region. Therefore, the loss of neural competence and the acquisition of lens competence occurs even when the ectoderm is cultured in vitro (Servetnick and Grainger, 1991).

SUMMARY

It appears that neural induction is a multiple-step process. It is likely that this process begins during the first cell cycle when the dorsoventral axis is established by cytoplasmic rearrangements. The dorsal animal blastomers appear to inherit a somewhat different developmental program than do the ventral animal blastomeres. One such documented difference is a bias in the ability of dorsal animal blastomeres to express an epidermal-specific molecule. During early gastrulation, this bias is reinforced and extended by signals from the blastopore lip region. This bias in the dorsal animal cells, toward a neural pathway and away from an epidermal pathway, is also reflected in the ability of the dorsal ectoderm to more easily respond to induction for the synthesis of NCAM than does the ventral ectoderm. The dorsal ectoderm is also much more responsive to induction by activin B than the ventral ectoderm.

Neural competence appears to be lost by midgastrula in *Xenopus* ectoderm. This is a curious observation, since the involuted dorsal mesoderm has not yet come to underlie the anterior portion of the neural plate by midgastrula. In fact, the head mesenchyme, which is closest to the anterior portion of the neural plate, has been found to be a much less potent inducer of neural tissue than the posterior chordamesoderm (Dixon and Kintner, 1989; Phillips and Doniach, 1991). Two possible explanations for the induction of the anterior portion of the neural plate have been offered. One involves an alteration in the rate of loss of neural competence within the anterior neural plate region due to the observed neural-like bias in dorsal ectoderm. There is also evidence from studies in other amphibian species that the loss of neural competence does not occur simultaneously throughout the ectoderm. In fact, some experimental data suggests that the dorsal anterior ectoderm is the last of the ectoderm to lose neural competence. Therefore, it is possible that the two early events that bias the dorsal ectoderm toward a neural pathway also enable the dorsal ectoderm to maintain neural competence until stages older than midgastrula.

Another possibility for inducing anterior neural plate without the presence of underlying dorsal mesoderm involves the method of delivering the neural induction signals. Several investigators have recently provided evidence suggesting that signals capable of inducing the expression of neural-specific molecules can travel through the plane of the ectoderm. It appears that signals can travel an appreciable distance and that expression of even regionally specific neural markers can be induced by planar signals. In fact, Suzuki and collaborators (1984) suggested that anterior neural induction does not occur by interaction with underlying dorsal mesoderm in *Cynops*, but that induction occurs through the plane of the ectoderm from the blastopore lip. Therefore, at this time there is ample evidence to suggest that both an alteration in the loss of neural competence and additional mechanisms for delivering neural induction signals may be operating within the embryo.

Several laboratories have shown that both planar signals, from the blastopore lip region through the ectoderm, and vertical signals, from involuted dorsal mesoderm into overlying ectoderm, operate to induce neural structures. It would therefore appear that a somewhat redundant system for neural induction exists within the amphibians. However, it also appears that some contact with underlying mesoderm is necessary for development of neural morphology, but this contact is not necessary for the synthesis of neural-specific molecules. It is certainly possible that the dorsal mesoderm is needed for mechanical support as much as any neural inductive signals it may offer.

Regional specification within the neural plate occurs during late gastrula, early neural plate stages. Engrailed is already expressed within a very narrow band of cells by early neural plate stage. it has also been observed that presumptive forebrain (neural plate stage) transplanted into the presumptive midbrain regions will still develop as forebrain (Jacobson, 1959). Therefore, the neural ectoderm is at least partially determined by early neural plate stage, and changing the underlying dorsal mesodermal environment does not respecify the neurectoderm. Thus, the time between when the neurectoderm is induced (i.e., midgastrula) and the time when at least a portion of the neurectoderm is regionally specified (i.e., early neural plate stage) is less than 3 hr. This would leave a very short time for the involuted dorsal mesoderm to ''imprint'' a spatial pattern of neural development onto the overlying neurectoderm. This is especially true since the cells of the dorsal mesoderm are actively moving with respect to one another, even during late gastrula, such that daughter cells may come to reside many cell diameters apart in the final mesodermal structures (Keller, personal communication). In addition, it should also be pointed out that the information necessary for the spatial patterns of the dorsal mesoderm are far less complex than those of the neurectoderm.

One alternative to the ''imprinting'' hypothesis is that the ectoderm, once induced to become neural, can sort itself into regionally distinct areas expressing regionally specific neural molecules. The involuted dorsal mesoderm may only provide ''imprinting'' information at the level of whether the neural induced ectoderm will be of a general spinal-type or a general cephalic-type pattern. As discussed previously, signals traveling through the plane of the ectoderm can both induce general neural molecules (NCAM) and induce the synthesis of regionally specific neural molecules within a well-defined group of cells. Induction of region-specific neural molecules may be a reflection of gradients of inducing molecules or possibly a reflection of an anterior–posterior prepattern within the ectoderm. It is not yet clear how much prepattern, relative to the anterior–posterior axis, might exist within ectoderm prior to neural induction or how much information such a prepattern might impart to the process of regional specification.

ACKNOWLEDGMENTS

The author wishes to thank Dr. John Gerhart for inviting this chapter and for editing it. Many thanks to the following who shared unpublished work: Marc Servetnick, Rob Grainger, Amy Sater, Ray Keller, Ron Stewart, and Tabitha Doniach. This work was supported by a grant from NASA, NIH grant HD23112, and an American Heart Association grant.

REFERENCES

Akers RM, Phillips CR, Wessells NK (1986): Expression of an epidermal antigen used to study tissue induction in the early *Xenopus laevis* embryo. Science 231:613–616.

Chaung HH (1938): Spezifische Induktionsleistungen von Leber und Niere im Explantatversuch. Biol Zentralbl, 58:472–480.
Chaung HH (1939): Induktionsleistungen von frischen und gekochten Organteilen (Niere, Leber) nach ihrer Verpflanzung in explantate und verschiedene Wirtsregionen von Tritonkeimen. Roux Arch Entw Mech 139:556–638.
Chaung HH (1940): Weitere Versuche über die Veränderung der Induktionsleistungen von gekochten Organteilen. Roux Arch Entw Mech, 140:25–38.
Dixon JC, Kintner CR (1989): Cellular contacts required for neural induction in *Xenopus* embryos: Evidence for two signals. Development 106:749–757.
Gerhart JC, Keller R (1986): Region-specific cell activities in amphibian gastrulation. Ann Rev Cell Biol 2:201–229.
Gimlich RL, Cooke J (1983): Cell lineage and the induction of second nervous systems in amphibian development. Nature 306:471–473.
Hamburger V (1988): "The Heritage of Experimental Embryology. Hans Spemann and the Organizer." Oxford: Oxford University Press.
Hemmati-Brivanlou A, Harland RM (1989): Expression of an engrailed-related protein is induced in the anterior ectoderm of early *Xenopus* embryos. Development 106:611–617.
Hemmati-Brivanlou A, Stewart R, Harland R (1990): Region-specific neural induction of an engrailed protein by anterior notochord in *Xenopus*. Science 250:800–802.
Holtfreter J (1933): Die totale Exogastrulation, eine Selbstablösung des Ektoderms von Entomesoderm. Entwicklung and funktionelles Verhalten nervenloser Organe. Roux Arch Entw Mech Org 129:669–793.
Holtfreter J (1934): Über die Verbreitung induzierender Substanzen und ihre Leistungen im *Triton*-keim. Roux Arch Entw Mech Org, 132:307–383.
Holtfreter J (1936): Regional Induktionen in xenoplastisch zusammengesetzten Explantaten. Roux Arch Entw Mech Org 134:466–561.
Holtfreter J (1938): Veränderungen der Reaktionsweise im alternden isolierten Gastrulaektoderm. Roux Arch Entw Mech Org 138:163–196.
Jacobson CO (1959): The localization of the presumptive cerebral regions in the neural plate of the Axolotl larva. J Embryol Exp Morphol 7:1–21.
Jacobson M, Rutishauser U (1986): Induction of neural cell adhesion molecule (N-CAM) in *Xenopus*. Dev Biol 116:524–531.
Keller R, Danilchik M (1988): Regional expression, pattern and timing of convergence and extension during gastrulation of *Xenopus laevis*. Development 103:193–209.
Kintner CR, Melton DA (1987): Expression of the *Xenopus* N-CAM RNA in ectoderm is an early response to neural induction. Development 99:311–325.
Lehmann FE (1929): Entwicklungsstorungen an der Medullaranlage von Triton, erzeugt durch Unterlagerungsdefekte. Roux Arch Entw Mech Org 108:243–283.
London C, Akers R, Phillips CR (1988): Expression of Epi 1, an epidermal specific marker, in *Xenopus laevis* embryos is specified before gastrulation. Dev Biol 129:380–389.
Machemer H (1932): Experimentelle Untersuchung über die Induktionsleistungen der oberen Urmundlippe in alteren Urodelenkeim. Roux Arch Entw Mech Org 126:391–456.
Mangold O (1929): Experimente zur Analyse der Determination und Induktion der Medullarplatte. Roux Arch Entw Mech Org 117:586–696.
Mangold O (1933): Über die Induktionsfähigkeit der verschiedenen Bezirke der Neurula von Urodelen. Naturwissenschaften 21:761–766.
Phillips CR, Doniach T (1991): Effects of the dorsal blastopore lip and the involuted dorsal mesoderm on neural induction in *Xenopus laevis*. (In preparation.)
Ruiz i Altaba A (1990): Neural expression of the *Xenopus* homeobox gene Xhox3: Evidence for a patterning neural signal that spreads through the ectoderm. Development 108:595–604.
Sater AK, Uzman JA, Steinhardt RA, Keller R (1991): Neural induction in *Xenopus* embryos: Induction of neuronal differentiation by either planar or vertical signals. (In preparation.)

Sato SM, Sargent TD (1989): Development of neural inducing capacity in dissociated *Xenopus* embryos. Dev Biol 134:263–266.

Savage R, Phillips CR (1989): Signals from the dorsal blastopore lip region during gastrulation bias the ectoderm toward a nonepidermal pathway of differentiation in *Xenopus laevis*. Dev Biol 133:157–168.

Schechtman AM (1938): Competence for neural plate formation in *Hyla* and the so-called nervous layer of the ectoderm. Proc Soc Exp Biol Med 38:430–433.

Servetnick M, Grainger R (1991): Changes in neural competence occur autonomously in *Xenopus* ectoderm. (Submitted.)

Sharpe CR, Fritz A, DeRobertis EM, Gurdon JB (1987): A homeobox-containing marker of posterior neural differentiation shows the importance of predetermination in neural induction. Cell 50:749–758.

Sive HL, Hattori K, Weintraub H (1989): Progressive determination during formation of the anteroposterior axis in *Xenopus laevis*. Cell 58:171–180.

Smith JC, Slack JMW (1983): Dorsalization and neural induction: Properties of the organizer in *Xenopus laevis*. J Embryol Exp Morphol 78:299–317.

Spemann H (1938): ''Embryonic Development and Induction.'' New Haven: Yale University Press.

Spemann H, Mangold H (1924): Über Induktion von Embryonanlagen durch Implantation artfremder Organisatoren. Roux Arch Entw Mech, 100:599–638.

Suzuki AS, Mifune Y, Kaneda T (1984): Germlayer interactions in pattern formation of amphibian mesoderm during primary embryonic induction. Dev Growth Diff, 26(1):81–94.

Thomsen G, Woolf T, Whitman M, Sokol S, Vaughan J, Vale W, Melton D (1990): Activins are expressed early in *Xenopus* embryogenesis and can induce axial mesoderm and anterior structures. Cell 63:485–493.

Cell-Cell Interactions in Early Development, pages 109–127

7. A Hierarchy of Signals Mediates Neural Induction in *Xenopus laevis*

A.J. Durston and A.P. Otte
Hubrecht Laboratorium, Uppsalalaan 8, 3584 CT Utrecht, The Netherlands

INTRODUCTION

A milestone was reached in experimental embryology when Spemann and Mangold (1924) demonstrated that the dorsal lip of the blastopore of a gastrula stage amphibian embryo has special properties. If this piece of dorsal mesoderm (which later becomes notochord and muscle) was taken from one gastrula and transplanted into the ventral side of another, it induced a secondary neural plate, leading later to the development of a twinned tadpole. This experiment (Fig. 1) was the first indication for neural induction, namely, that neural differentiation is triggered in vivo in embryonic ectoderm by inducing signals from dorsal mesoderm.

Neural induction was already evident in this original experiment, because the implanted organizer tissue was taken from an unpigmented species (*Triturus cristatus*), while the host embryo was pigmented (*Triturus teaniatus*), so that it could be seen clearly that the secondary neural plate came from the pigmented host. This fact has since been confirmed repeatedly, in recent experiments using lineage markers (Smith and Slack, 1983; Gimlich and Cooke, 1983; Jacobson, 1984). Despite more than 60 years of research since the Nobel Prize-winning discovery of the organizing capacities of the dorsal mesoderm and of neural induction, many of the most important details of this process and, notably, the nature of the inducing signals, have remained obscure. We are concerned, here, with the possible involvement of three known signaling pathways in neural induction. Three important aspects of the induction process are as follows.

The first aspect concerns a controversy, which was initiated between Spemann and Holtfreter in the 1930s (Spemann, 1931, 1938; Holtfreter, 1933; see also Hamburger, 1988) and still continues today (cf. Kintner and Melton, 1987; Dixon and Kintner, 1989; Ruiz i Altaba, 1990; Savage and Phillips, 1989; with Jones and Woodland, 1989; Sharpe and Gurdon, 1990). The controversy is over whether the neural inducing signals emitted by dorsal mesoderm mainly travel "homeogenetically," from cell to cell through the preneural ectoderm, or whether they mainly travel transversely, across from invaginated dorsal meso-

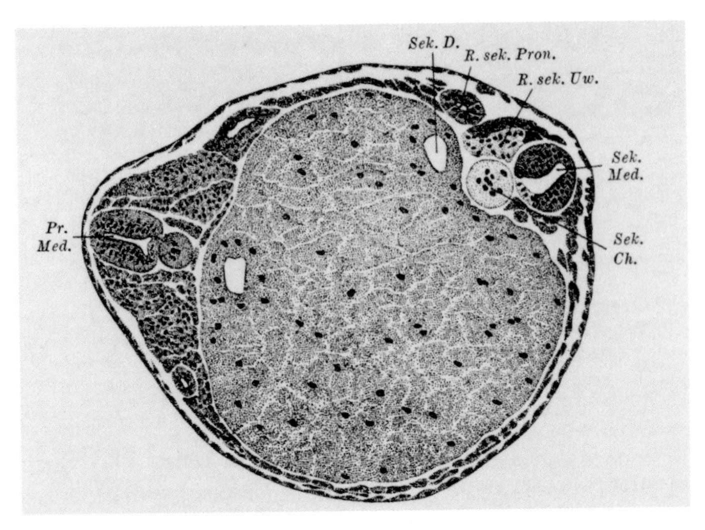

Fig. 1. A drawing of a cross section of a tadpole that has developed from a *Triturus* embryo into which an implanted secondary blastopore lip was implanted, taken from the original Spemann and Mangold paper (1924). It shows the host's neural tube (Pr. Med.), as well as a secondary neural tube (Sek. Med.), which was induced by the implanted dorsal lip.

derm, when this contacts cells in the future neural plate, after the dorsal mesoderm has invaginated during gastrulation. There is strong reason to suspect, however, that both types of signals are important, and, for example, that homeogenetic signals probably mediate the spreading of an inductive signal from the midline to the edges of the future neural plate (Nieuwkoop, 1958; Albers, 1987; see also Nieuwkoop et al., 1985). Homeogenetic signals may also be important for regulation of localized neural induction competence (see below).

The second important aspect of the induction process is that the dorsal mesoderm induces the regional differentiation of the nervous system during gastrulation. Spemann (1927, 1931) showed that the dorsal blastopore lip from an early gastrula tends to induce head structures (and forebrain) if implanted in a host embryo, while advanced blastopore lips tend to induce tail structures (and spinal cord) (Fig. 2), and Mangold (1933) showed that anterior invaginated mesoderm induced head structures, while posterior invaginated mesoderm induced tail structures. These early findings have been elaborated and built on by many authors since [see, e.g., Nieuwkoop et al. (1952); Saxen and Toivonen (1962); Ruiz i Altaba and Melton (1989); Cooke and Smith (1989)], and many questions have arisen. An important one is whether or not the neural inducing signals from anterior (or early) mesoderm and from posterior (or late) mesoderm are actually qualitatively different or only quantitatively

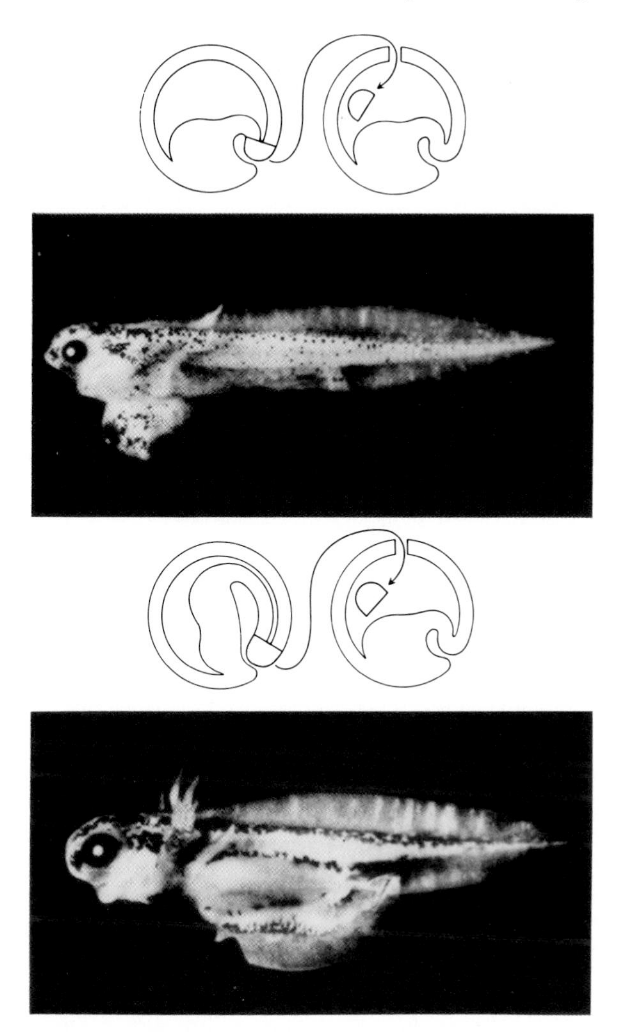

Fig. 2. The effect of implanting an early blastopore lip (which leads to development of a secondary head: **above**), and of implanting a late blastopore lip (which leads to development of a secondary tail: **below**). Reproduced from Saxen and Toivonen (1962).

different (such, for example, that the signal intensity from posterior mesoderm is greater than that from anterior mesoderm). An important detailed analysis of the inducing properties of different regions of the mesoderm by Nieuwkoop and his coworkers (1952) provided evidence that there must be more than one type of inducing signal, and Nieuwkoop (1952; see also Nieuwkoop et al., 1985) explained his findings in terms of two classes of signals: "activation signals," which induce ectoderm to differentiate to forebrain, and "transformation signals," which convert presumptive forebrain to presumptive hindbrain and spinal cord. Toivonen and Saxen (1968) have also proposed a substantially similar model. The Nieuwkoop model has not been contradicted by later work, but neither has it been substantiated biochemically. There are many agents known, including endogenous but unidentified factors from *Xenopus* embryo extracts (Born et al., 1989), which induce only forebrain differentiation and thus mimic activation signals. Early reports of transforming (or mesodermalizing or spinocaudal) activity (Lehman, 1950; Toivonen and Saxen, 1955; Saxen and Toivonen, 1961; Tiedemann, 1959) were confused by the fact that the extracts concerned contained mesoderm inducers. These initially induced differentiation of mesoderm cells, which then secondarily induced differentiation of posterior neural tissue (Toivonen and Saxen, 1968; Tiedemann et al., 1963). We note also that a cascade of mesoderm inducing and neural inducing signals cannot yet be ruled out as an explanation for the very intriguing recent finding that an activin-like factor can apparently induce neural tissue in ectoderm explants (Sokol et al., 1990). It was only very recently that a factor was identified (retinoic acid: see below) that has at least some of the properties expected of a transformation signal, in that it is directly able to transform presumptive forebrain tissue to more posterior neural tissue. In this review, we are concerned with the possible roles of the protein kinase C (PKC) and cyclic AMP (cAMP) signal transduction pathways and of retinoic acid in neural activation and in neural transformation, respectively.

The third aspect of induction that concerns us is the competence of ectoderm cells to respond to neural inducing signals. It has long been known that Amphibian gastrula stage ectoderm can be induced to differentiate to neural tissue (as above), whereas Amphibian blastula stage ectoderm cannot, but can be induced to differentiate to mesoderm instead. It has recently also been shown that the competence of ectoderm to differentiate to neural tissue is not only regulated in time but is localized as well. The dorsal ectoderm, which normally comes in contact with the inducing mesoderm during gastrulation, is much more easily induced to differentiate to neural tissue than is ventral ectoderm (Sharpe et al., 1987). This timing and localization of the competence of ectoderm to respond to neural inducing signals is likely to be of key importance in regulating pattern formation in the embryo (see, e.g., Nieuwkoop, 1958; Albers, 1987).

CELLULAR SIGNALING PATHWAYS IN EMBRYONIC INDUCTION

In 1932, Spemann, together with Holtfreter, Otto Mangold, and Bautzmann, published the puzzling observation that dorsal mesoderm which was killed either by freezing, crushing, or even cooking was still able to induce neural tissue to some extent (Bautzmann et al., 1932). This observation suggested that the neural differentiation inducing substance (or substances) was chemical in its nature. This led to an extensive search to clarify the chemical nature of the signal, a search that has been unsuccessful to this day.

Although the neural inducing signals were not identified, there were, however, important clues indicating the existence of diffusible substances involved in mediating neural induction. Niu and Twitty (1953) were among the first to perform experiments that indicated the involvement of diffusible factors in neural induction. They cultured a small piece of mesoderm in a drop of culture medium. Upon taking away the mesoderm, they incubated a piece of early gastrula ectoderm in the ''conditioned'' medium. The ectoderm became neuralized. These results strongly suggest the existence of diffusible factors, which were able to induce neural structures.

In another series of experiments, which also indicated the existence of diffusible factors, mesoderm and ectoderm were cultured as recombinates, but with a Nucleopore filter separating the ectoderm and mesoderm. Nucleopore filters with pore diameters down to 0.05 μm failed to prevent neural induction. No cellular structures were found in the pores, as investigated by using electron microscopy, again indicating that diffusible factors were responsible for the observed neural differentiation (Toivonen et al., 1975, 1976; Toivonen, 1979). Yet other experiments have also shown that (unidentified) neural inducing factors from Amphibian embryo extracts still remain active when these are bound to Sephadex particles (Born et al., 1989). This indicates that these inducing factors work via plasma membrane receptors, and that they can be expected to activate intracellular signal transduction pathways.

These indications that diffusible, membrane receptor binding factors and, in retrospect, growth factors are involved in mediating embryonic inductions were substantiated only recently, and these crucial advances were made not for neural induction but for the induction of mesoderm, which has been shown to differentiate due to interactions between ectoderm and endoderm, in the blastula stage embryo (Nieuwkoop, 1969; Nieuwkoop and Ubbels, 1972; Ogi, 1967, 1969; see also Nieuwkoop et al., 1985). A breakthrough came when Smith (1987) discovered that the culture medium from the *Xenopus* XTC cell line was able to induce mesoderm. With larger quantities of inducing material available, it was easier to characterize the mesoderm inducing factor, and very recent evidence shows that the mesoderm inducing factor in XTC medium is activin A, a member of the TGF-β family (Smith et al., 1990; Van den Eijnden-Van Raaij et al., 1990). Basic fibroblast growth factor (bFGF) and several members of the TGF-β family, including activins and bone morphogenetic proteins, have also recently been shown to be active mesoderm inducers (Slack et al., 1987; Knöchel et al., 1987; Kimelman and Kirschner, 1987; Rosa et al., 1988; Asashima et al., 1990; Sokol et al., 1990). It was also found that *Xenopus* embryos contain a mRNA-encoding bFGF (Kimelman and Kirschner,

1987), bFGF itself (Kimelman et al., 1988; Slack and Isaacs, 1989) and FGF receptors (Gillespie et al., 1989), as well as mRNAs for various TGF-β-family factors, including activins (Weeks and Melton, 1987; Tannahill and Melton, 1989; Kondiah et al., 1990; Thomsen et al., 1990). The inducing abilities of FGF and TGF-β family factors and the presence of both types of factors in the early embryo suggest that FGF and TGF-β family factors could actually mediate mesoderm induction in vivo (for review see Whitman and Melton, 1989; Smith, 1990). Microinjection of murine int-1 mRNA into the *Xenopus* zygote causes double axis formation in the *Xenopus* embryo (McMahon and Moon, 1989), suggesting that members of the int-1 family of growth factors may be also important for axis formation.

It has further been shown during recent years that several *Drosophila* genes code for growth factors or important regulatory proteins in signal transduction pathways. The *Drosophila decapentaplegic* gene (DPP-C) (Padgett et al., 1987) codes a TGF-β-like growth factor. This gene was found to be important for the formation of dorsal structures in *Drosophila*. *Sevenless*, a gene that is involved in the development of the *Drosophila* eye, was found to code for a tyrosine kinase (Hafen et al., 1987). *Notch* and *slit*, genes that play an important role in *Drosophila* neurogenesis, were found to contain epidermal growth factor (EGF)-like repeats (Wharton et al., 1985; Rothberg et al., 1988; Hartley et al., 1988). The *Drosophila* gene *torpedo*, which plays an important role in establishing the dorsoventral polarity of the embryo, is homologous to the mammalian EGF receptor (Price et al., 1989). *Torso*, another gene involved in *Drosophila* pattern formation, codes a tyrosine kinase (Sprenger et al., 1989), and the segment polarity gene *fused* and the *shaggy* gene each code for serine/threonine kinases (Preat et al., 1990; Bourouis et al., 1990).

So, within a short period, it has become clear that growth factors and signal transduction processes play important roles in early embryonic development. And now, in retrospect, some earlier observations that were puzzling at the time that they were made are more understandable. Raising the pH of the culture medium can lead to neural differentiation (Holtfreter, 1945, 1948) or cement gland formation (Picard, 1975; Sive et al., 1989). While this treatment appeared to be aspecific, it may well be that raising the pH mimics part of the PKC signal transduction pathway. It is now known that activating PKC activates the Na^+/H^+ exchanger, a plasma membrane localized ion exchanger (Moolenaar et al., 1984). Upon activation, H^+ ions are extruded from the cells, leading to a higher intracellular pH, and Na^+ ions are taken up. Bypassing the activation of PKC by raising the intracellular pH directly might lead to neural differentiation (see also below). In agreement with this hypothesis is the earlier finding that Na^+ is taken up by ectodermal cells during neural induction (Barth and Barth, 1972). At that time, this observation was also puzzling, but seems more logical now.

THE PKC AND cAMP SIGNAL TRANSDUCTION PATHWAYS IN NEURAL INDUCTION

Knowledge of signal transduction pathways that might be involved in mediating embryonic inductions has been scarce until recently. One approach is to identify activated signal transduction pathways in induced tissue by activity measurements of components of the signal transduction pathways. Artificial activation of these pathways using pharmacological agents can then also be tested for effects on differentiation. By following this approach, evidence has accumulated recently that both the PKC and the cAMP signal transduction pathways are involved in mediating neural induction.

PKC was discovered by Nishizuka and coworkers in 1977 (Inoue et al., 1977) and is defined as a kinase that is dependent on Ca^{2+} and phospholipids for its activation. It has become clear that activation of PKC is linked to receptor-mediated inositol phospholipid breakdown, which leads to the formation of two second messengers. The enzyme phospholipase C hydrolyzes phosphatidyl inositol bisphosphate (PIP_2) to diacylglycerol and inositol triphosphate $(IP_3)_{(1,4,5)}$ (Nishizuka, 1984; Berridge, 1987; Berridge and Irvine, 1989). Diacylglycerol stays in the plasma membrane and activates PKC, which in turn phosphorylates a range of cellular proteins. When diacylglycerol is formed after receptor stimulation, a membrane-bound complex between PKC, diacylglycerol, Ca^{2+}, and phosphatidylserine is formed (Bell, 1986). In agreement with this model, it has been found that several growth factors and hormones that stimulate inositol phospholipid turnover induce translocation of PKC from the cytosol to the membrane (Niedel and Blackshear, 1986). Phorbol esters such as 12-O-tetradecanoyl phorbol-13-acetate (TPA) also cause PKC to translocate from the cytosol to the membrane, thereby activating PKC (Castagna et al., 1982).

Recently it was shown that PKC translocates from the cytosol to the plasma membrane, specifically in neural induced ectoderm, during neural induction (Otte et al., 1988), and also that PKC is already almost maximally translocated at an early phase of neural induction, namely from the midgastrula (stage 11) on (Otte et al., 1990a). That this translocation is correlated with PKC activation via phospholipase C activation is suggested by the observation that there is also an increase in inositol phosphate levels in induced stage 11 ectoderm, concomitantly with PKC translocation. Several proteins that become phosphorylated in neural induced ectoderm due to PKC activation were also identified (Otte et al., 1990a). That induced neuroectoderm contains higher levels of protein kinase activities was found independently by Davids (1988). TPA, a well-known activator of mammalian PKC, not only caused activation of *Xenopus* PKC in early gastrula ectoderm but also induced this to differentiate to neural tissue (Otte et al., 1988). The neural inducing effect of TPA was also found in *Triturus alpestris* (Davids et al., 1987). Taken together, these data indicate that PKC is involved in the response to the endogenous neural inducing signals.

Many studies have shown that the PKC pathway interacts with another well-known signal transduction pathway, the cAMP pathway. This second messenger is produced by the enzyme adenylate cyclase (AC), which is activated when certain ligands activate their receptors. cAMP activates the cAMP-dependent protein kinase. Interactions or cross-talk between the PKC and cAMP pathways occur extensively (Nishizuka, 1984; Sugden et al., 1985; Sibley et al., 1986). Each of these activated pathways can influence the activity of the other. The interactions between signal transduction pathways leave the cell with many possibilities for modulating its response to external, activating signals. The mode of interaction can be positive, leading to signal enhancement, or negative, leading to attenuation of the signal. Since multiple inducing factors that modulate one another's activity are implicated in embryonic inductions, cross-talk between signal transduction pathways may play a crucial role in embryonic inductions. Both positive and negative interactions may play a role. In the case of FGF and TGF-β (or activin A) in mesoderm induction, a positive interaction can be envisioned between the corresponding signal transduction pathways. Negative feedback at the level of signal transduction may be equally important for restricting the response to inducing signals to a limited population of cells.

It was suggested previously that the cAMP analogues 8-Br-cAMP or dibutyryl cAMP, which enter the living cell, were able to evoke neural differentiation in amphibians (Wahn et al., 1975), but these results could not be confirmed (Grunz, 1985). Recent measurements of components of the cAMP pathway, however, showed that both AC activity and cAMP concentration increase during neural induction, in neural induced ectoderm (Otte et al., 1989). The enhancement of AC activity requires PKC activation, indicating cross-talk between these two signal transduction pathways (Rozengurt et al., 1987). This cross-talk appeared to be essential for neural induction. Whereas cAMP analogues alone were unable to induce neural differentiation, they had a synergistic inducing effect, if the ectoderm waas first incubated with TPA. Based on these data, it was postulated that there are at least two signals that mediate neural induction. The first signal activates PKC and the second signal then and only then activates the cAMP pathway effectively (Otte et al., 1989). This model was also indicated by the timing of activation of the two signal transduction pathways. Whereas the PKC pathway becomes activated early on during neural induction (stage 11), the cAMP pathway becomes activated from stage 12–12.5 on.

That the involvement of both the PKC and cAMP pathways in neural induction is very specific is indicated by the negative results found with another signal transduction pathway. It was shown that the cGMP pathway is not activated during neural induction. Further, manipulations of this pathway neither induced nor inhibited neural differentiation (Otte et al., 1990b).

In conclusion, whereas there are as yet no clues about the nature of the neural inducing substances, recent years have seen some progress in the identification of known signal transduction pathways that become activated during the earliest phases of neural induction in response to the neural inducing action of dorsal mesoderm. Considering the parallel with the increasing level of knowledge about the growth factors that are implicated in mesoderm induction, it is striking that highly conserved phenomena like growth factors, signal transduction pathways, and the widely occurring interactions between them are involved in the complicated process of embryonic induction.

PKC AND NEURAL INDUCTION COMPETENCE

Whereas it has long been known that amphibian dorsal mesoderm is able to induce ventral ectoderm to differentiate to neural tissue (Spemann and Mangold, 1924; Smith and Slack, 1983; Gimlich and Cooke, 1983; Jacobson, 1984; see above), Sharpe et al. (1987) found in *Xenopus* that dorsal mesoderm induces the expression of neural markers strongly in dorsal ectoderm but hardly at all in ventral ectoderm. The same difference between dorsal and ventral ectoderm was found when an epidermal marker called Epi-1 was used. When dorsal and ventral early gastrula ectoderm is excised and cultured, this marker is expressed strongly in ventral ectoderm but not in dorsal ectoderm (London et al., 1988). This result also suggests a predetermination in *Xenopus* preneural early gastrula ectoderm. The difference between these recent results and the older results, which clearly showed the ability of ventral ectoderm to become induced by dorsal mesoderm, has not been resolved and the molecular mechanisms that underlie this predetermination are unknown. Since the strength of the inducing signal (emitted from dorsal mesoderm) is kept constant, the lower capacity of ventral ectoderm to express neural markers must depend on a difference in its capacity to receive, transduce, or respond to the inducing signals. The recent evidence that embryonic inductions are mediated by signal transduction pathways and growth factors open the way to investigating the molecular mechanism of embryonic competence. It was found that contact with dorsal mesoderm caused substantially more PKC activation in dorsal ectoderm than in ventral ectoderm. TPA mainly activated PKC in dorsal ectoderm, thereby inducing neural differentiation in dorsal ectoderm only. It was further shown that PKC preparations, isolated from dorsal and ventral ectoderm, have different biochemical characteristics. The dorsal PKC is much more susceptible to activation by TPA and diacylglycerol than ventral PKC. Monoclonal antibodies against the bovine PKC α plus β or γ isozymes (Young et al., 1988; Sheu et al., 1990) also immunostained dorsal and ventral ectoderm respectively, suggesting different localizations of PKC isozymes (Nishizuka, 1988; Huang et al., 1987; Brandt et al., 1987; Coussens et al., 1986). These results raise the possibility that PKC participates in the establishment of localized embryonic competence (Otte et al., 1991).

Nothing is known about the mechanisms involved in regulating the timing of neural competence. It was shown recently that the competence of ectoderm to differentiate to neural tissue is lost rather abruptly between stage 12 and 13 (Nieuwkoop, 1958; Sharpe and Gurdon, 1990). It is possible, for example, that the timing of competence is regulated by the abundance of receptors for the neural inducing signals. In the case of mesoderm competence, it has been found that the number of FGF receptors on ectodermal cells increases from the early blastula stage on and declines during the late blastula stages, precisely matching the timing of mesoderm competence (Gillespie et al., 1989). The indications that the PKC pathway is activated first, during neural induction, thereby making the adenylate cyclase pathway susceptible to activation (Otte et al., 1989, 1990a) may also indicate that the timing of competence also involves multiple signals, the first being needed to make the ectoderm competent to respond to the next.

PKC AND ANTEROPOSTERIOR SPECIFICATION

The classical work by Nieuwkoop and coworkers (Nieuwkoop et al., 1952) and by Toivonen and Saxen (1968) provided evidence for two types of neural inducing signal: an early "activation" signal (to use Nieuwkoop's terminology), which induces anterior neural tissue (presumptive forebrain), and a later "transformation" signal, which converts anterior neural tissue to posterior neural tissue (presumptive hindbrain and spinal cord). It is therefore interesting to know what type of neural tissue is induced via the PKC pathway (TPA); i.e., whether this is anterior or posterior neural tissue.

Ectoderm explants induced to differentiate to neural tissue via TPA treatment prove to develop neuroanatomical structures associated with the forebrain (Davids et al., 1987; Durston et al., 1991). They also express antigens that are associated with the forebrain and they fail to develop neuroanatomical structures and to express mRNAs associated with the hindbrain and spinal cord. These results thus suggest strongly that the neural tissue induced via the PKC pathway is presumptive forebrain. In fact, the set of anterior neural markers induced via PKC activation is the same as that induced via natural neural activation signals in anterior neurectoderm excised from an early gastrula (stage 11), while the same posterior neural markers fail to be induced by PKC activation and these neural activation signals (Durston et al., 1991). Induction via the cAMP pathway has not been examined in detail, but the available data indicate that this also induces forebrain. It is thus probable that the PKC and cAMP pathways are associated with the "activation" step in Amphibian neural induction. It is of interest to look for pathways that are implicated in the "transformation" step.

RETINOIC ACID AND ANTEROPOSTERIOR SPECIFICATION

All trans retinoic acid (RA) has recently become very well known as a biologically active form of vitamin A, which is suspected of being a morphogen during embryonic development. The identification of nuclear receptors for this ligand suggests strongly that it is an endogenous signal molecule (Petkovitch et al., 1987; Giguere et al., 1987) and measurements of RA and teratogenic manipulations have suggested further that RA may be a morphogen that specifies the anteroposterior axis during limb development. It has, namely, been shown that RA, applied locally to the anterior side of a developing chicken limb, causes duplication of the posterior side of the limb, thus mimicking a posteriorly located organizer region: the zone of plarizing activity (ZPA) (Tickle et al., 1982). It has also been shown that the developing chicken limb contains an endogenous anteroposterior gradient of RA in an effective concentration range that could specify anteroposterior differences (Thaller and Eichele, 1987).

Recent results show that RA also has an interesting effect on anteroposterior specification in the developing *Xenopus* nervous system. If early *Xenopus* gastrulae are treated with RA ($\geq 10^{-7}$ M), they develop axis deficiencies, and notably, become microcephalic (Fig. 3). RA treatment at this stage gives no drastic reduction in neural induction (since neither the volume of induced neural tissue nor numbers of induced explants nor levels of expression of pan-neural mRNAs are noticeably reduced). However, it does strongly inhibit development of forebrain structures and expression of anterior neural (forebrain- and midbrain-associated), as well as anterior ectodermal mRNAs and antigens. It also actually increases the volume of the hindbrain and enhances the expression of posterior neural (hindbrain- and spinal cord-associated) mRNAs (Durston et al., 1989, 1991; Sive et al., 1990). RA acts directly, both on the axial mesoderm and on the neuroectoderm, and results from experiments using RA treatment of isolated neuroectoderm and of either ectoderm or mesoderm in ectoderm/mesoderm recombinates indicate that the effects described above depend at least partly on a direct effect of RA on neuroectoderm (Durston et al., 1989). These characteristics of the RA effect (direct conversion of anterior neural to posterior neural tissue, without inhibition of neurogenesis) are strongly reminiscent of the characteristics proposed for neural transformation by Nieuwkoop and coworkers (1952), and the possibility that RA actually is a natural morphogen, responsible for neural transformation, is also supported by the finding that *Xenopus* gastrulae contain endogenous RA (Durston et al., 1989). It will be interesting to determine the spatial distribution of RA in the embryo during axial specification of the nervous system to determine, for example, whether there is an axial gradient of endogenous RA. It should not be forgot ten that other forms of vitamin A may be active in vivo during embryogenesis (Thaller and Eichele, 1990; Wagner et al., 1990). The effects of RA on devel oping mesoderm and neural tissue are also complex and time-dependent (Sive et al., 1990; Sharpe, 1991; Durston et al., 1991), suggesting multiple steps in anteroposterior specification.

Fig. 3. The effect of RA on axis formation. Shown are a normal *Xenopus* tadpole (stage 45, **above**) and a microcephalic tadpole (**below**), which developed from an embryo that was treated at the early gastrula stage with a pulse of RA (10^{-5} M, 30 min).

CONCLUDING REMARKS

We have summarized above evidence concerning the possible roles of three different signaling pathways in neural induction as depicted in Figure 4. We suggest that neural induction requires sequential activation of the PKC and cAMP signal transduction pathways, in neural competent ectoderm. These pathways are presumably activated by two different ligands, which bind to different membrane receptors. Activation of the PKC pathway by the first ligand

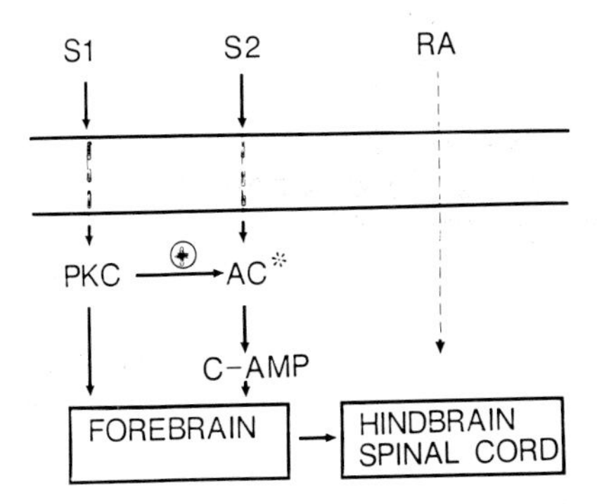

Fig. 4. A working hypothesis for neural induction. We propose that two extracellular signal molecules (S_1 and S_2) act sequentially on ectoderm to induce neural differentiation. The first (S_1) works via PKC, giving a signal for anterior neural (forebrain) differentiation, and also enhancing the level of activatable AC. It thus presets the cAMP pathway to be turned on by a second signal (S_2) and makes the ectoderm competent to be neural induced via the cAMP pathway. S_2 then activates the cAMP pathway, providing a second signal for anterior neural (forebrain) differentiation. These two signals may mediate Nieuwkoop's activation step, during neural induction, and they make ectoderm competent to respond to RA. We propose that endogenous RA then acts on the neurectoderm and converts some of the presumptive forebrain to presumptive posterior neural tissue (hindbrain and spinal cord). RA may thus mediate Nieuwkoop's transformation step. In addition to the S_1 and S_2 signals and RA, there may be a fourth signal (S_0), which acts in the pregastrula embryo and regulates competence for a response to S_1 by determining the localization of the dorsal PKC isozyme.

(S_1) induces neural differentiation and also enhances the level of activatable AC. This makes it possible for a second ligand (S_2) to activate the cAMP pathway, thereby enhancing the level of neural differentiation.

The competence of ectoderm to be neural induced via natural induction signals, or via artificial activation of the PKC pathway, is localized. Dorsal ectoderm from the early gastrula (which later becomes neural tissue in vivo) is very easily neural induced. Ventral ectoderm is not. This localization of competence is matched by the localization of in vivo PKC activation and of a PKC isozyme. Dorsal PKC is better activated by dorsal mesoderm and TPA than is ventral PKC, both in vivo and in vitro, and the dorsal ectoderm contains a PKC isozyme that cross-reacts with an antibody against the mammalian α and β isozymes, while the ventral ectoderm contains a different PKC isozyme, which cross-reacts with antibodies against the mammalian γ isozyme. This striking parallel between the in vivo competence of dorsal and ventral ectoderm and the in vitro properties of the dorsal and ventral PKC isozymes suggests that the localization of the dorsal PKC isozyme may be at least part of the basis of dorsal neural induction competence. Considering also that the availability of the dorsal and ventral PKC isozymes can be manipulated by manipulating the dorsoventrality of the embryo prior to gastrulation, we suggest that dorsal-localized neural induction competence could also be set up by an early dorsal homeogenetic signal (S_0), in the pregastrula embryo.

The neural tissue induced via the PKC and cAMP pathways expresses anterior neural (forebrain) markers, but not posterior neural (hindbrain and spinal cord) markers. In fact, it is exactly the type of neural tissue one would expect to be induced by the anterior-specific "activation" signals proposed by Nieuwkoop et al. (1952), and not the type that is expected to be induced by the posterior-specific "transformation signals." Evidence summarized above, on the other hand, shows that RA acts as expected for a transformation signal. If added early during neural induction, it inhibits the expression of forebrain markers while enhancing the expression of many posterior neural markers. The RA is present in vivo, is effective if applied directly to the (neur)ectoderm, and does not strongly enhance or inhibit neural induction. it will also only induce the expression of posterior neural markers in tissue that has previously been neural induced by activation signals, not in non-neural induced ectoderm. The activation signals thus appear to make ectoderm competent to respond to RA. These properties of RA strongly parallel those deduced from embryological experiments for the neural transformation signal.

The results above thus already suggest that at least three, and probably four, sequential signals may be involved in mediating neurogenesis. Each signal seems to make the neurectoderm competent to respond to the next. These signals presumably constitute part of an in vivo cascade of cell interactions that mediate neural induction; it is by no means excluded that there are other, as-yet unsuspected, signals. Each of the S_1 and S_2 signals and RA could either mediate an intertissue interaction (a signal from mesoderm to ectoderm), or a homeogenetic signal (between the neurectoderm cells), or both. The putative S_0 signal, on the other hand, which is presumably active before the ectoderm on which it acts is in face-to-face contact with mesoderm, is likely to be a homeogenetic signal.

These findings thus already indicate that neural induction is mediated by a complex hierarchy of interactions between different signaling pathways, some details of which are set out in Figure 4. It is likely that work in the coming years will reveal that the ideas set out here concern only the tip of the iceberg.

REFERENCES

Albers B (1987): Competence as the main factor determining the size of the neural plate. Dev Growth Differ 29:535–545.

Asashima M, Nakano H, Shimada K, Kinoshita K, Shibai H, Ueno N (1990): Mesodermal induction in early amphibian embryos by activin A (erythroid differentiation factor). Roux Arch Dev Biol 198(6):330–335.

Barth LG, Barth LJ (1972): ^{22}Sodium and ^{45}Calcium uptake during embryonic induction in *Rana pipiens*. Dev Biol 26:18–34.

Bautzmann H, Holtfreter J, Spemann H, Mangold O (1932): Versuche zur Analyse der Induktionsmittel in der Embryonal-Enwicklung. Naturwissenschaften 20:971–974.

Bell RM (1986): Protein kinase C activation by diacylglycerol second messengers. Cell 45:631–632.

Berridge MJ (1987): Inositol triphosphate and diacylglycerol: Two interacting second messengers. Ann Rev Biochem 56:159–193.

Berridge MJ, Irvine RF (1989): Inositol phosphates and cell signalling. Nature 346:197–205.

Born J, Janeczek J, Schwarz W, Tiedemann H, Tiedemann Hi (1989): Activation of masked neural determinants in amphibian eggs and embryos, and their release from the inducing tissue. Cell Differ Dev 27:1–7.

Bourouis M, Moore P, Ruel L, Grail Y, Heitzler P, Simpson P (1990): An early embryonic product of the gene *shaggy* encodes a serine threonine kinase related to the cDc28/cDc2 subfamily. EMBO J 9(9):2877.

Brandt SJ, Niedel JE, Bell RM, Scott Young W (1987): Different patterns of expression of different protein kinase C mRNAs in rat tissues. Cell 49:57–63.

Castagna M, Takai Y, Kaibuchi K, Kikkawa V, Nishizuka Y (1982): Direct activation of calcium-activated, phospholipid-dependent protein kinase by tumour promoting phorbol esters. J Biol Chem 257:7847–7851.

Cooke J, Smith J (1989): Gastrulation and larval pattern in *Xenopus* after blastocoelic injection of a *Xenopus*-derived inducing factor. Dev Biol 131:383–400.

Coussens L, Parker PJ, Rhee L, Yang-Feng TL, Chen E, Waterfield MD, Francke U, Ullrich A (1986): Multiple, distinct forms of bovine and human protein kinase C suggest diversity in cellular signalling pathways. Science 233:859–866.

Davids M, Loppnow B, Tiedemann H, Tiedemann Hi (1987): Neural differentiation of amphibian gastrula ectoderm exposed to phorbol ester. Roux Arch Dev Biol 196:137–140.

Davids M (1988): Protein kinases in amphibian ectoderm induced for neural differentiation. Roux Arch Dev Biol 197:339–344.

Dixon JE, Kintner CR (1989): Cellular contacts required for neural induction in *Xenopus* embryos: Evidence for two signals. Development 106:749–757.

Durston A, Timmermans A, Hage WJ, Hendriks HFJ, de Vries NJ, Heideveld M, Nieuwkoop PD (1989): Retinoic acid causes an anteroposterior transformation in the developing central nervous system. Nature 340:6229, 140–144.

Durston A, Heideveld M, Otte A, Koster K, Timmermans A (1991): Personal observations.

Giguere V, Ong ES, Segui P, Evans RM (1987): Identification of a receptor for the morphogen retinoic acid. Nature 330:624–629.

Gillespie L, Paterno GP, Slack JMW (1989): Analysis of competence: Receptors for fibroblast growth factor in early *Xenopus* embryos. Development 106:203–208.

Gimlich RL, Cooke J (1983): Cell lineage and the induction of second nervous systems in amphibian development. Nature 306:471–473.

Grunz H (1985): Information transfer during embryonic induction in Amphibians. J. Exp Embryol 89(Suppl):349–363.

Hafen E, Basler K, Edstroem JE, Rubin G (1987): *Sevenless*, a cell-specific homeotic gene of *Drosophila*, encodes a putative transmembrane receptor with a tyrosine kinase domain. Science 236:55–63.

Hamburger V (1988): ''The Heritage of Experimental Embryology. Hans Spemann and the Organizer.'' Oxford: Oxford University Press.

Hartley DA, Preiss A, Artavanis-Tsakonas S (1988): A deduced gene product from the *Drosophila* neurogenic locus, *Enhancer of split*, shows homology to mammalian G protein β subunit. Cell 55:785–795.

Holtfreter J (1933): Die totale Exogastrulation, eine Selbstäblosung des Ektoderms vom Entomesoderm, Entwicklung und funktionelles verhalten nervenloser Organe. Roux Arch Entw Mech Org 129:669–793.

Holtfreter J (1945): Neuralisation and epidermalisation of gastrula ectoderm. J Exp Zool 98:161–209.

Holtfreter J (1948): Concepts on the mechanism of embryonic induction and its relation to parthenogenesis and malignancy. Symp Soc Exp Biol 2:17–48.

Huang FL, Yoshida T, Nakabayashi H, Huang K-P (1987): Differential distribution of protein kinase C isozymes in the various regions of the brain. J Biol Chem 262:15714–15720.

Inoue M, Kishimoto A, Takai Y, Nishizuka Y (1977): Studies on a cyclic nucleotide-independent protein kinase and its proenzyme in mammalian tissues. II. Proenzyme and its activation by calcium-dependent protease from rat brain. J Biol Chem 252:2610–2616.

Jacobson M (1984): Cell lineage analysis of neuronal induction: Origins of cells forming the induced nervous system. Dev Biol 102:122–129.

Jones EA, Woodland HR (1989): Spatial aspects of neural induction in *Xenopus laevis*. Development 107:785–792.

Kimelman D, Kirschner MW (1987): Synergistic induction of mesoderm by TGF and TGFβ, and the identification of a mRNA coding for FGF in the early *Xenopus* embryo. Cell 51:869–877.

Kimelman D, Abraham JA, Haaparanta T, Palisi TM, Kirschner MW (1988): The presence of fibroblast growth factor in the frog egg: Its role as a natural mesoderm inducer. Science 242:1053–1056.

Kintner CR, Melton DA (1987): Expression of *Xenopus* N-CARA mRNA in ectoderm is an early response to neural induction. Development 99:311–325.

Knöchell W, Born J, Hoppe P, Loppnow-Blinde B, Tiedemann H, Tiedemann Hi, McKeehan WL, Grunz H (1987): Mesoderm inducing factors. Their possible relationship to heparin-binding growth factors and transforming growth factor β. Naturwissenschaften 74:604–606.

Kondiah P, Sands MJ, Smith JM, Fields A, Roberts AB, Sporn MB, Melton DA (1990): Identification of a novel transforming growth factor β (TGF-β5) mRNA in *Xenopus laevis*. J Biol Chem 265:1089–1093.

Lehman FE (1950): Die Morphogenese in ihrer Abhängigkeit von elementaren biologischen Konstituenten des Plasmas. Rev Suisse Zool (Suppl) 57:1, 141–151.

London C, Akers R, Phillips C (1988): Expression of epi 1: An epidermis-specific marker in *Xenopus* embryos is specified prior to gastrulation. Dev Biol 129:380–389.

Mangold O (1933): Über die Induktionsfähigkeit der verscheidenen Bezirke der Neurula von Urodelen. Naturwiss. 21, 43:761–766.

Mangold O, Spemann H (1927): Über Induktion von Medullarplatte durch Medullarplatte in jungren Kiem, ein Beispiel homoögentischer oder assimilatorischer Induktion. Arch Entw Mech Org 111:341–422.

McMahon A, Moon R (1989): Ectopic expression of the protooncogene int-1 in *Xenopus* embryos leads to duplication of the embryonic axis. Cell 58,6:1075–1084.

Moolenaar WH, Tertoolen LGJ, De Laat SW (1984): Phorbol ester and diacylglycerol mimic growth factors in raising cytoplasmic pH. Nature 312:371–374.

Niedel JE, Blackshear PJ (1986): Protein kinase C. In: Putney JW (ed): ''Phosphoinositides and Receptor Mechanisms. New York: Liss, pp 47–88.

Nieuwkoop PD (1958): Neural competence of the gastrula ectoderm in *Ambystoma mexicanum*. Acta Embryol Morphol 2:13–52.

Nieuwkoop PD (1969): The formation of the mesoderm in Urodelan amphibians. I: Induction by the endoderm. Roux Arch 162,4:341–373.

Nieuwkoop PD, Boterenbrood EC, Kremer A, Bloemsma F, Hoessels E, Verheyen F (1952): Activation and organisation of the central nervous system in Amphibians. J Exp Zool 120:1–108.

Nieuwkoop PD, Ubbels GA (1972): The formation of the mesoderm in Urodelan amphibians. IV: Quantitative evidence for the purely ectodermal origin of the entire mesoderm and of the pharyngeal endoderm. Roux Arch Entw Mech Org 169,3:185–199.

Nieuwkoop PD, Johnen AG, Albers B (1985): "The Epigenetic Nature of Early Chordate Development. Inductive Interaction and Competence." Cambridge, UK: Cambridge University Press, 373 pp.
Nishizuka Y (1984): The role of protein kinase C in cell surface signal transduction and tumour promotion. Nature 308:693–698.
Nishizuka Y (1988): The molecular heterogeneity of protein kinase C and its implications for cellular regulation. Nature 334:661–665.
Niu MC, Twitty VC (1953): The differentiation of gastrula ectoderm in medium conditioned by axial mesoderm. Proc Natl Acad Sci USA 39:985–989.
Ogi KI (1967): Determination in the development of the amphibian embryo. Science Report Tohoku University, Ser IV (Biology) 33:239–247.
Ogi KI (1969): Regulative capacity in the early amphibian embryo. Res Bull 13:31–40.
Otte AP, Koster CH, Snoek GT, Durston AJ (1988): Protein kinase C mediates neural induction in *Xenopus laevis*. Nature 334:618–620.
Otte AP, van Run P, Heideveld M, van Driel R, Durston AJ (1989): Neural induction is mediated by crosstalk between the protein kinase C and cyclic AMP pathways. Cell 58:641–648.
Otte AP, Bruinooge E, van Driel R, de Vente J, Durston AJ (1990a): Cyclic GMP is not involved in neural induction in *Xenopus laevis*. Roux Arch Dev Biol (in press).
Otte AP, Kramer YM, Manesse M, Lambrechts C, Durston A (1990b): Characterisation of protein kinase C in early *Xenopus* embryogenesis. Development (in press).
Otte AP, Kramer IM, Durston AJ (1991): Protein kinase C is involved in regulating the local competence of *Xenopus* ectoderm to differentiate to neural tissue. Science (in press).
Padgett RW, St. Johnston D, Gelbart WM (1987): A transcript from a *Drosophila* pattern gene predicts a protein homologous to the transforming growth factor β family. Nature 325:81–84.
Petkovitch M, Brand NJ, Krust A, Chambon P (1987): A human retinoic acid receptor which belongs to the family of nuclear receptors. Nature 330:444–450.
Picard JJ (1975): *Xenopus laevis* cement gland as an experimental model for embryonic differentiation. J Embryol Exp Morphol 33:957–967.
Preat T, Thérond P, Samour-Isnard C, Limbourg-Bouchon B, Tricoire H, Erk I, Mariol MC, Busson D (1990): A putative serine-threonine kinase, encoded by the segment polarity fused gene of *Drosophila*. Nature 341:87–89.
Price JV, Clifford RJ, Schupbach J (1989): The maternal ventralising locus *torpedo* is allelic to *faint little ball*, an embryonic lethal, and encodes the *Drosophila* EGF receptor homologue. Cell 56:1085–1092.
Rosa F, Roberts AB, Danielpour D, Dart LL, Sporn MB, Dawid IB (1988): Mesoderm induction in amphibians: The role of TGFβ2-like factors. Science 239:783–785.
Rothberg JM, Hartley DA, Walther Z, Artavanis-Tsakonas S (1988): *slit*: An EGF-homologous locus of *D. melanogaster* involved in the development of the embryonic central nervous system. Cell 55:1047–1059.
Rozengurt E, Murray M, Zachary I, Collins M (1987): Protein kinase C activation enhances c-AMP accumulation in Swiss 3T3 cells: Inhibition by pertussis toxin. Proc Natl Acad Sci USA 84:2282–2286.
Ruiz i Altaba A, Melton D (1989): Interaction between peptide growth factors and homeobox genes in the establishment of anteroposterior polarity in frog embryos. Nature 341:33–38.
Ruiz i Altaba A (1990): Neural expression of the *Xenopus* homeobox gene Xhox3: Evidence for a patterning signal that spreads through the ectoderm. Development 108:595–604.
Savage R, Phillips CR (1989): Signals from the dorsal blastopore lip region during gastrulation direct the ectoderm toward a nonepidermal pathway of differentiation in *Xenopus laevis*. Dev Biol 133:157–168.
Saxen L, Toivonen S (1961): The two gradient hypothesis in primary induction. The combined effects of two types of inductors mixed in different ratios. J Embryol Exp Morphol 9:514–533.

Saxen L, Toivonen S (1962): "Primary Embryonic Induction." New York: Logos Press/Academic Press.

Sharpe CR, Fritz A, de Robertis EM, Gurdon JB (1987): A homeobox containing gene shows the importance of predetermination in neural induction. Cell 50:749–758.

Sharpe CR, Gurdon JB (1990): The induction of anterior and posterior neural genes in *Xenopus laevis*. Development 109:765–774.

Sharpe C (1991): Personal communication.

Sheu FS, Marais RM, Parker PJ, Bazan WG, Ronttenberg A (1990): Neuron-specific protein Fl/GAP-43 shows substrate specificity for the subtype of protein kinase C. BBRC 171, 3:1236–1243.

Sibley DR, Jeffs RA, Daniel K, Nambi P, Lefkowitz RJ (1986): Phorbol ester treatment promotes enhanced adenylate cyclase activity in frog erythrocytes. Arch Biochem Biophys 244:373–381.

Sive HL, Hattori K, Weintraub H (1989): Progressive determination during formation of the anteroposterior axis in *Xenopus laevis*. Cell 58:171–180.

Sive HL, Draper BW, Harland RM, Weintraub H (1990): Identification of a retinoic acid sensitive period during primary axis formation in *Xenopus laevis*. Genes and Development 4:932–942.

Slack JMW, Darlington BG, Heath JK, Godsave SF (1987): Mesoderm induction in early *Xenopus* embryos by heparin-binding growth factors. Nature 326:197–200.

Slack JMW, Isaacs HV (1989): Presence of basic fibroblast growth factor in the early *Xenopus* embryo. Development 105:147–154.

Smith JC (1987): Mesoderm induction and mesoderm inducing factors in early amphibian development. Development 99:3–14.

Smith JC (1989): Induction and early amphibian development. Curr Opin Cell Biol 1:1061–1070.

Smith JC, Price BMJ, van Nimmen K, Huylebroeck D (1990): Identification of a potent mesoderm-inducing factor as a homologue of activin A. Nature 345:729–731.

Smith JC, Slack JMW (1983): Dorsalisation and neural induction: Properties of the organizer in *Xenopus laevis*. J Embryol Exp Morphol 78:299–317.

Sokol S, Wong G, Melton DA (1990): A mouse macrophage factor induces head structures and organizes a body axis in *Xenopus laevis*. Science 249:561–564.

Spemann H (1927): Neue Arbeiten über Organisatoren in der tierischen Entwicklung. Naturwissenschaften 15:946–951.

Spemann H (1931): Uber den Anteil von Implantat und Wirkskeim an der Orientierung und Beschaffenheit der induzierten Embryoanlage. Roux Arch Entw Mech Org 123:390–517.

Spemann H (1938): "Embryonic Development and Induction." New Haven: Yale University Press.

Spemann H, Mangold H (1924): Über Induction von Embryoanlagen durch Implantation artfremder Organisatoren. Roux Arch Dev Biol 100:599–638. [The English translation can be found in Willier BH, Oppenheimer J (1974): "Foundations of Experimental Embryology," 2nd Ed. New York: Hafner Press, pp 144–184.

Sprenger F, Stevens ML, Nusslein-Volhard C (1989): The *Drosophila* gene *torso* encodes a putative receptor tyrosine kinase. Nature 338:478–483.

Sugden D, Vanecek J, Klein DC, Thomas TP, Anderson WB (1985): Activation of protein kinase C potentiates isoprenaline-induced cyclic AMP accumulation in rat pinealocytes. Nature 314:359–361.

Tannahill D, Melton DA (1989): Localized synthesis of the Vg1 protein during early *Xenopus* development. Development 106:775–785.

Thaller C, Eichele G (1987): Identification and distribution of retinoids in the developing limb bud. Nature 327:625–628.

Thaller C, Eichele G (1990): Isolation of 3,4-didehydroretinoic acid, a novel morphogenetic signal in the wing bud. Nature 345:815–819.

Thomsen G, Woolf T, Whitman M, Sokol S, Vaughan J, Vale W, Melton DA (1990): Activins are expressed in early *Xenopus* embryogenesis and can induce axial mesoderm and anterior structures. Cell (in press).

Tickle C, Alberts B, Wolpert L, Lee J (1982): Local application of retinoic acid to the limb bud mimics the action of the polarizing region. Nature 296:564–566.

Tiedemann H (1959): Ein Verfahren zur gleichzeitigen Gewinnung deuteren cephaler und mesodermaler Induktionstoffe aus Huhnereembryonen. Z. Naturf. 14:610–611.

Tiedemann H, Becker V, Tiedemann Hi (1963): Chromatographic separation of a hindbrain inducing substance into mesodermal and neural inducing subfractions. BBA 74:557–560.

Toivonen S (1979): Transmission problem in primary induction. Differentiation 15:177–181.

Toivonen S, Saxen L (1955): The simultaneous inducing action of liver and bone marrow of the guinea pig in implantation and explantation experiments with embryos of *Triturus*. Exp Cell Res 3(Suppl):346–357.

Toivonen S, Saxen L (1968): Morphogenetic interaction of presumptive neural and mesodermal cells mixed in different ratios. Science 159:539–540.

Toivonen S, Tarin D, Saxen L, Tarin PJ, Wartiovaara J (1975): Transfilter studies on neural induction in the next. Differentiation 4:1–7.

Toivonen S, Tarin D, Saxen L (1976): The transmission of morphogenetic signals from amphibian mesoderm to ectoderm in primary induction. Differentiation 5:49–55.

Van den Eijnden-Van Raaij AJM, van Zoelen EJJ, Van Nimmen K, Koster CH, Snoek GT, Durston AJ, Huylebroeck D (1990): Activin-like factor from a *Xenopus* cell line responsible for mesoderm induction. Nature 345:732–734.

Wagner M, Thaller C, Jessell T, Eichele G (1990): Polarising activity and retinoic synthesis in the floor plate of the neural tube. Nature 345:819–822.

Wahn HL, Lightbody LE, Tihen TT, Taylor JD (1975): Induction of neural differentiation in cultures of amphibian undetermined presumptive epidermis by cyclic AMP derivatives. Science 188:366–369.

Weeks DL, Melton DA (1987): A maternal mRNA localized to the vegetal hemisphere in *Xenopus* eggs codes for a growth factor related to TGF. Cell 51:861–867.

Wharton K, Johanson KM, Xu T, Artavanis-Tsakonas S (1985): Nucleotide sequence from the neurogenic locus. *Notch* implies a gene product that shares homology with proteins containing EGF-like repeats. Cell 43:567–581.

Whitman D, Melton DA (1990): Growth factors in early embryogenesis. Annu Rev Cell Biol 5:93–117.

Young S, Rothbard J, Parker PJ (1988): A monoclonal antibody recognising the site of limited proteolysis of protein kinase C. Eur J Biochem 173:247.

Cell-Cell Interactions in Early Development, pages 129–143

8. Hox-2: Gene Regulation and Segmental Patterning in the Vertebrate Head

Mai Har Sham, Stefan Nonchev, Jenny Whiting,
Nancy Papalopulu, Heather Marshall, Paul Hunt,
Ian Muchamore, Martyn Cook, and Robb Krumlauf
Division of Eukaryotic Molecular Genetics, National Institute for Medical Research, Mill Hill, London, England NW7 1AA

INTRODUCTION

The problems of pattern formation in vertebrate embryos from a molecular standpoint have received a great deal of attention recently in light of a number of gene families that are highly conserved during evolution. One particular family, homeobox-containing genes, appears to play an important role in regulating a number of processes in vertebrate development. Homeobox-containing genes are very diverse, but a subfamily of these genes, referred to as Hox, has a number of properties that suggest that they have a function in the specification of positional cues or identities along the embryonic anteroposterior axis. In this review we focus on the properties of one family, the Hox-2 genes, and detail aspects of their structure, organization, expression, and regulation in mouse development as a first step in understanding pattern specification in vertebrate development.

THE HOX-2 FAMILY

The Hox genes are distinguished from other homeobox-containing genes in that they are organized into tight clusters. We have cloned nine genes spanning 120 kb on mouse chromosome 11 (Graham et al., 1989; Rubock et al., 1990). These genes are all in the same 5′-3′ orientation with respect to the direction of transcription, and a yeast artificial chromosome (YAC) clone that contains eight members of the complex has been isolated (Rubock et al., 1990). Four related Hox clusters (Hox-1, -2, -3, and -4) have been isolated in the mouse genome by a number of groups (see review by Kessel and Gruss, 1990). Direct sequence comparisons of the predicted protein structures for the genes and their organization within each cluster reveal that these Hox complexes are related by duplication and divergence from a common ancestral cluster. Individual members display multiple regions of homology that extend outside of the homeodomain (Graham et al., 1988; Featherstone et al., 1988), and genes are more related to their counterpart in another Hox cluster than to flanking genes within the same complex. Figure 1 shows a current view of the organization and alignment of the different mouse clusters. The vertical columns represent genes related at the sequence level. It is apparent that not all clusters have the same number of genes. It must be stressed that this alignment is only preliminary, and in many cases is only based on the presence and sequence of the homeobox itself. Some clusters may have highly divergent members, which have not been identified. It is clear, however, that the overall organization of the genes in each complex is identical. The high degree of similarity at the sequence level suggests the possibility that the genes could have some degree of functional redundancy.

Hox complexes have been identified in many different vertebrates, including man, chicken, *Xenopus*, and zebrafish (for review, see Scott et al., 1989; Kessel and Gruss, 1990; Acampora et al., 1989). In general, where information is available there are highly conserved direct homologues of the mouse genes, and in some cases genes have been discovered in other species first, then identified in the mouse complexes. All vertebrates appear to have the same four clusters, which suggests that the origin of these clusters by duplication arose in an ancestor prior to the radiation of vertebrates in evolution. The

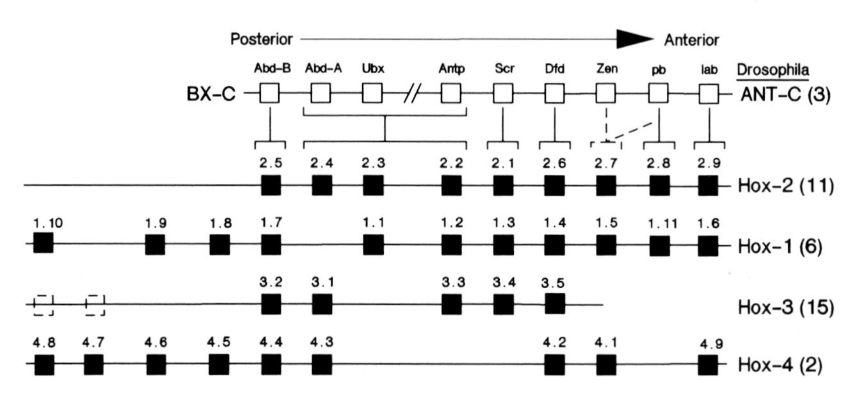

Fig. 1. Schematic alignment of the four major Hox homeobox clusters in the mouse genome. Solid boxes represent cloned and sequenced genes, and dashed boxes are genes predicted to be present based on homology to the human complexes (Acampora et al., 1989). Numbers in brackets designate the chromosomal location in the mouse genome. At the top of the diagram is an alignment with the *Drosophila* ANT-C and BX-C clusters. Vertical columns designate genes sharing extended sequence homology. Patterned after Graham et al., 1989; see also Kessel and Gruss, 1990.

maintenance of a tightly clustered organization suggests that the arrangement of the genes in the complexes is an important feature that needed to be conserved.

HOX–HOM-C HOMOLOGY, RELATION TO *DROSOPHILA* HOMEOTIC GENES

An important extension to the remarkable degree of homology exhibited among the Hox genes arose by comparisons of the mouse clusters with the ANT-C and BX-C *Drosophila* homeotic clusters. It was generally felt that while vertebrates contained genes that had homeoboxes, the roles of these genes would be drastically different in arthropods. Segmentation is thought to have evolved independently several times during evolution, and the homeobox was likely to be a conserved motif, but not an indicator of conserved processes in segmentation or patterning. In support of this, homeobox genes were found in nonsegmented organisms (Holland and Hogan, 1986) and could not be strictly linked to segmentation.

The mouse Hox genes do have multiple regions of extended homology with specific *Drosophila* homeotic genes in the HOM-C complex. A particular example is the mouse Hox-2.6 gene, which has 50% homology over its entire 269 AA length, with the *Drosophila Dfd* gene (Graham et al., 1988). Similar degrees of identity were found when comparing other members of the mouse and *Drosophila* clusters. The order of related genes was also conserved, suggesting that these homeobox clusters were derived from a common ancestor prior to the divergence between arthropods and vertebrates (Graham et al., 1989; Duboule and Dolle, 1989). Figures 1 and 2 illustrate an alignment of the arthropod and mouse clusters. This homology in structure and organization alone strongly argues for a conserved role for these gene networks in embryonic development.

COLINEARITY AND AXIAL EXPRESSION

An important feature of the *Drosophila* HOM-C clusters, colinearity, was first noted by Lewis (1978). The order of the genes along the chromosome correlates with the region of the *Drosophila* embryo they specify on the basis of genetic mutants and gene expression. Genes at one end of the cluster are involved in the specification of the most posterior segments (Abd-B) and each successive gene along the cluster regulates a more anterior region. In this way the gene order acts as a molecular representation of positions on the embryonic axis and distinguishes head from tail.

In situ hybridization studies of the mouse Hox-2 genes (Graham et al., 1989; Wilkinson et al., 1989b) revealed that there is a similar correlation or colinearity in the expression of these murine genes in the developing nervous sys-

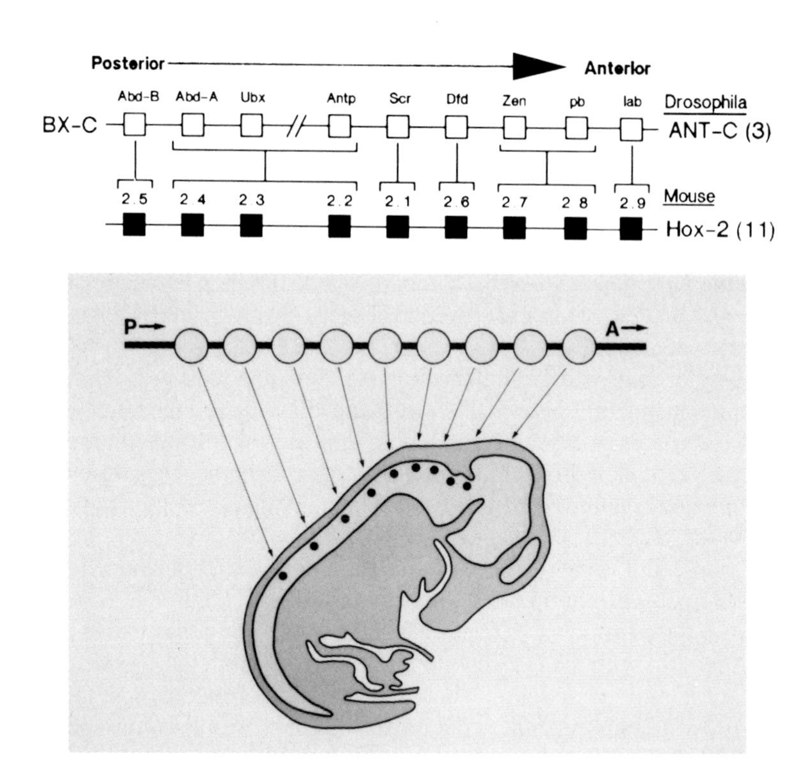

Fig. 2. Colinearity between gene position and expression along the embryonic axis. **Top**, the Hox-2 cluster and its alignment with the *Drosophila* HOM-C complex. Large arrow at the top indicates the trend in expression and specification of segmental identity from posterior to anterior in the embryo exhibited by the *Drosophila* homeotic genes. The mouse Hox-2 genes also display an ordered array of expression with the anterior boundaries of expression correlating with gene position (**bottom**).

tem. In 12.5 days post coitum (dpc) embryos, Hox-2.5 is expressed in the neural tube from the most posterior regions to a sharp anterior boundary that maps adjacent to the third prevertebra. Other Hox-2 genes are also expressed in the posterior neural tube but their anterior boundaries are distinctly different. Figure 2 shows in schematic form that the relative anterior limit for expression within the neural tube is successively more anterior for each gene in the cluster, moving from the most 5′ gene (Hox-2.5) to the most 3′ gene (Hox-2.9). Therefore, while the domains of Hox-2 expression are partially overlapping, they display a colinear relationship between gene order and expression boundaries along the anterior–posterior axis.

Other Hox clusters have similar properties. Duboule and Dolle (1989) and Gaunt et al. (1988) have shown that members of the Hox-1 and Hox-4 clusters exhibit posterior-to-anterior trends in domains of expression that correlate with position in a cluster. These correlations are not restricted to the nervous system, but have been observed in para-axial mesoderm, ectoderm, the developing gut, and organs. Recently, it has been found that there are graded domains of expression in the developing limb bud, which temporally and axially correlate with the order of genes in the Hox-4 complex (Dolle et al., 1989). In initial studies aimed at examining if the domains of gene expression for related genes in the different Hox clusters are similar, Gaunt et al. (1989) compared one subfamily. There are similar anterior boundaries of expression in the neural tube and somitic mesoderm, supporting the possibility of functional redundancy within a subgroup. However, there are clear tissue-specific and temporal differences in the expression of these genes. Therefore not all of the patterns of expression overlap, and specific functions for each gene could be related to particular tissues or times in development.

The common features of structure, organization, and expression shared between mouse and *Drosophila* homeobox clusters suggest that they may have a conserved function in axial patterning (Akam, 1989; Graham et al., 1989; Duboule and Dolle, 1989). Together these studies argue that the Hox network is part of the molecular pathway that vertebrate embryos use in specifying axial positions and regional identity in a variety of embryonic fields.

SEGMENT-RESTRICTED EXPRESSION OF HOX-2 IN THE MOUSE HINDBRAIN

Many of the boundaries of Hox-2 expression in the neural tube map within the hindbrain at 12.5 dpc. At this embryonic stage no clear morphological distinction that correlates with the anterior limits of gene expression is seen. However, in early stages of vertebrate development, periodic swellings arise in the neural tube; within the hindbrain these units are termed *rhombomeres*. The number of rhombomeres in the hindbrain of different vertebrates is identical and it has been argued that these rhombomeric swellings represent organizational or segmental units in hindbrain development. Recent experiments by Lumsden and colleagues (for review, see Lumsden, 1990) have detailed the cellular properties of rhombomeres in the chicken. Neurogenesis and axon growth arises in even-numbered rhombomeres before the odd numbers (Lumsden and Keynes, 1989). Cranial sensory ganglia and branchial motor nerve exit points are not associated with every rhombomere but display a two-segment periodicity, aligning with the even rhombomeres. Pairs of rhombomeres are also associated with specific branchial arches. These properties suggest that molecular events operating at a two-segment periodicity are important in patterning the hindbrain. Single cell dye injection experiments by Fraser et al. (1990) directly demonstrated that the rhombomeres are cellular compartments. In the mouse, a gene encoding a putative transcription factor (zinc-finger containing gene), Krox-20, is expressed in rhombomeres r3 and r5 (Wilkinson et al., 1989a). This supports the concept that the rhombomeres are organized in a two-segment periodicity, and are important segmental units in patterning the hindbrain.

We therefore investigated the possibility that the anterior limits of Hox expression at 12.5 dpc correspond with the developing rhombomeres in earlier stages of embryogenesis. Some of the Hox-2 genes do have limits of expression that coincide with rhombomeric boundaries (Wilkinson et al., 1989b; Murphy et al., 1989). Figure 3 summarizes the findings and shows an example of the segment-restricted expression of the mouse Hox-2.9 gene. The limits of expression in the hindbrain are restricted to rhombomere 4. This is an unusual pattern for members of the Hox-2 cluster because other genes have a sharp anterior limit that maps to a rhombomere boundary but is expressed in posterior domains extending to the very tail of the embryo. Hox-2.9 has both a sharp anterior and posterior limit. An important feature to the patterns of gene expression is that the boundaries of adjacent genes in the cluster vary with a two-rhombomere periodicity. The successively more anterior limits of expression or colinearity observed at 12.5 dpc are also seen at 8.5–9.0 dpc. In initial experiments in chicken embryos, we have found that the segmentally restricted expression of the Hox-2.9 gene is similar to the mouse (Maden et al., 1991). This suggests that the rhombomere patterns displayed by the Hox-2 genes are conserved in other vertebrates and are important in hindbrain development.

The timing of the segment-restricted patterns of Hox-2 expression appear slightly later than for the Krox-20 gene. This leads to the suggestion that the zinc-finger gene Krox-20 may be more analogous to a *Drosophila* segmentation gene and involved in establishing the rhombomeric units. The Hox-2 homeobox genes could be involved in the later processes of determining the phenotype or identity of the rhombomeres, which is analogous to the roles of the *Drosophila* HOM-C homeotic genes in specifying segment identity. Therefore the vertebrate and *Drosophila* Hox and HOM-C clusters of homeobox genes could share conserved roles in the processes of segmentation. However, we believe that a role for the Hox-2 genes in segmental identity in the hind-

Fig. 3. Segmental expression of Hox-2 genes in the hindbrain and branchial arches. **Right**, the ordered domains of expression that correspond to rhombomere boundaries in the hindbrain. **Bottom**, an example of the Hox-2.9 pattern of expression. Note that within the neural tube the expression is restricted to one rhombomere (r4). Crescent-shaped domains of expression lateral to the neural tube represent cranial neural crest cells. Left is anterior and right posterior. **Left**, the restricted domains of expression in the surface ectoderm and visceral arches. Note the correlation in gene order and anterior boundaries of expression.

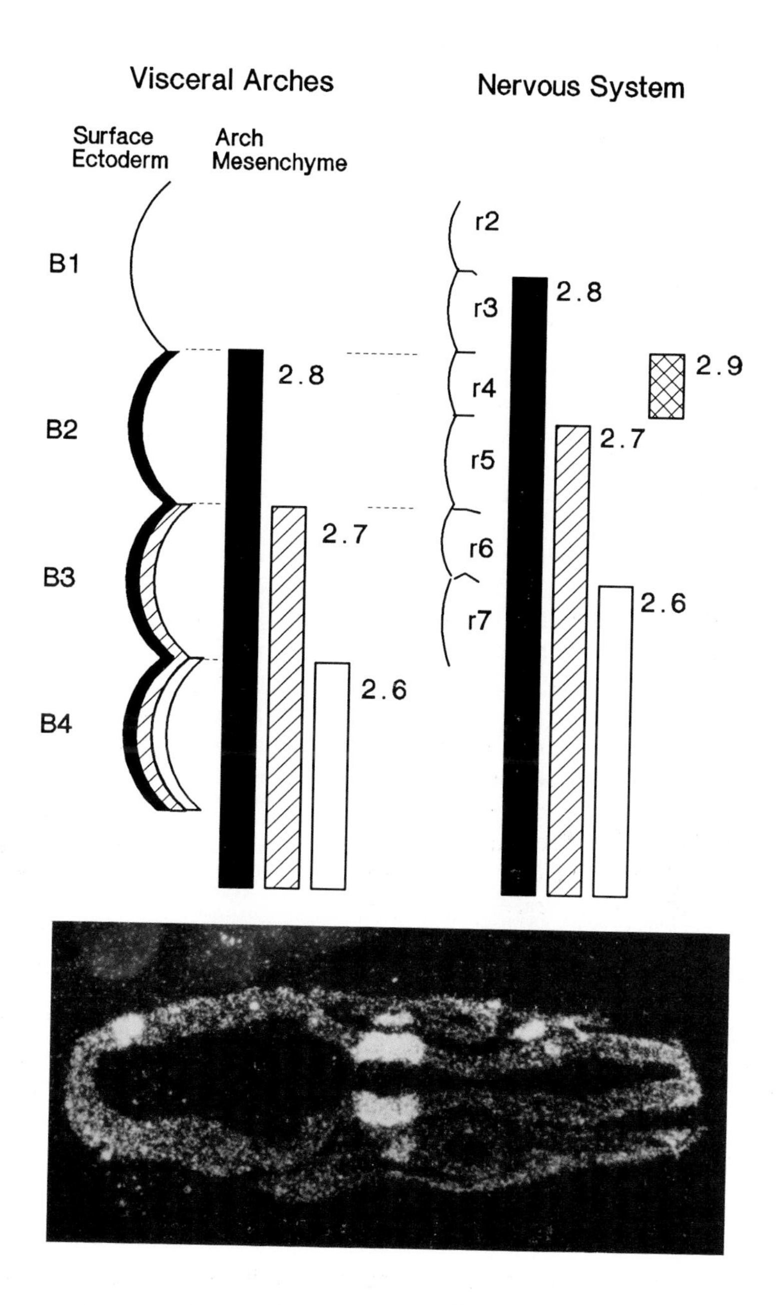
Visceral Arches
Nervous System
Surface Ectoderm
Arch Mesenchyme
B1
B2
B3
B4
2.8
2.7
2.6
r2
r3
r4
r5
r6
r7
2.8
2.9
2.7
2.6

brain is not a conserved feature of an ancient role in segmentation. We favor the idea that a role for the Hox and HOM-C complexes in positional specification of axial anterior–posterior information was the most likely basis for conservation of the ancient clusters (Wilkinson et al., 1989b; Wilkinson and Krumlauf, 1990). Roles for these genes in segmentation represent a recruitment or reutilization of the positional signaling system and arose independently in arthropods and vertebrates.

RELATIONSHIP OF THE HINDBRAIN AND HOX-2 TO HEAD DEVELOPMENT

In the head, most of the mesenchymal structures are derived from the neural crest. Unlike the trunk, para-axial mesoderm does not contribute the major components to the bone and connective tissue of the head. Somites are important segmental units in organization of the trunk, and it appears that the rhombomeres are important in the head. Several lines of evidence suggest that cranial neural crest is an imprinted or prepatterned tissue that carries the information necessary to specify structures it will form or organize. Because much of the cranial neural crest is derived from the hindbrain, we were interested in the idea that the potential positional specification system represented by the Hox network in the hindbrain might also serve to provide cues for the migratory neural crest. In Figure 3, lateral domains of expression of the Hox-2.9 gene can be observed adjacent to the neural tube. These represent neural crest cells, which will form part of the sensory ganglia. When all of the Hox-2 genes are examined, branchial arch-restricted domains of expression, which define specific subpopulations of neural crest cells, are found (Hunt et al., 1991a,b) (Fig. 3, left). The graded overlapping domains of expression found in the rhombomeres of the hindbrain are also found in the sensory ganglia, surface ectoderm, and visceral arch mesenchyme.

Our results show that the Hox genes are expressed in a manner consistent with patterning in the cranial neural crest. This suggests that the rhombomeres may have a greater role in head development, serving to pattern the premigratory neural crest and specify craniofacial structures. It appears that one molecular component of this signaling system could be the Hox genes. Our prediction is that alterations in the domain of Hox expression would provide a different address or signal that would alter the pattern of neural crest differentiation. This idea can be directly tested in transgenic mice.

RETINOIC ACID AND HOX COLINEARITY

The colinear expression of Hox-2 genes could arise from a graded signal or morphogen in the embryo, which induces a differential Hox response. Many homeobox genes are stimulated by retinoic acid (RA) in cultured cells, and

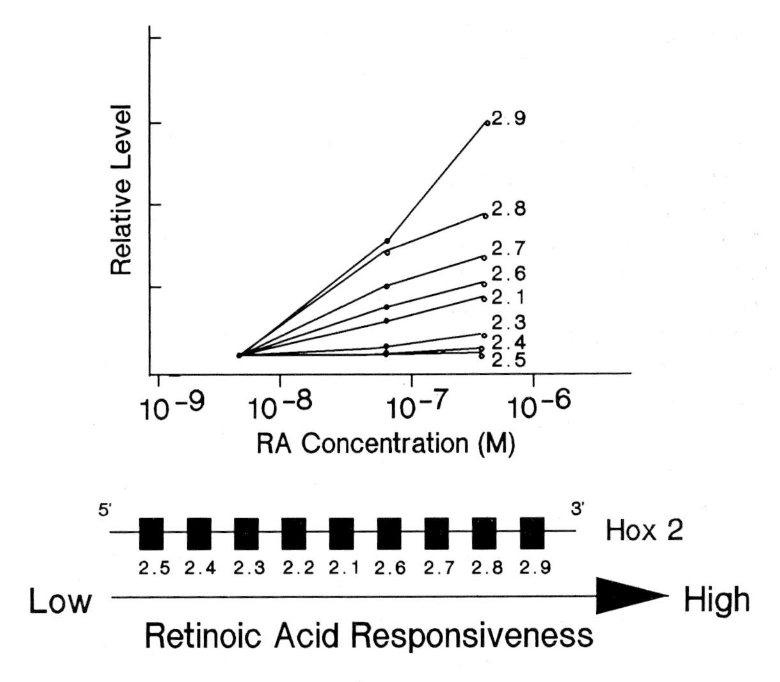

Fig. 4. Colinear responsiveness of Hox-2 genes to retinoic acid. The level of Hox expression in F9 cells treated with three different concentrations of RA 24 hr after exposure to RA (**top**). The amounts were compared to both the preinduced levels and the levels at the lowest concentration of applied RA. The most responsive gene maps to the 3′ end of the complex. **Bottom**, a summary showing the colinearity between response and gene order in the cluster.

experimental evidence in chicken embryos suggests that RA is important in patterning the limb (Brockes, 1989). We previously demonstrated that the Hox-2.1 and Hox-2.6 genes can be induced in the RA-dependent differentiation of mouse F9 teratocarcinoma cells (Graham et al., 1988). We therefore compared the degree of responsiveness of all the Hox-2 genes to RA at three different concentrations in F9 cells (Fig. 4). There is a wide variation in the relative response for different Hox-2 genes. The most responsive genes are located at the 3′ end of the cluster and the least responsive at the 5′ end of the complex. These results are remarkably analogous to the colinear patterns we detected in the mouse embryo. Recently, Simeone et al. (1990) have shown that the human Hox-2 cluster also has a colinear differential sensitivity to RA in human NTera-2 cells. While these results were obtained in cultured cells, a similar set of results is seen when *Xenopus* embryos are treated with RA (Papalopulu et al., 1990). It therefore appears that Hox-2 genes can differentially respond to RA, and this could be one of the means that is used during embryogenesis to set up a gradient of homeobox expression and ordered, partially overlapping domains of expression.

GENE REGULATION AND CLUSTER ORGANIZATION AND CONSERVATION

The Hox proteins have been highly conserved in evolution, yet it is not necessary to maintain gene clustering to coordinately regulate a gene family. The ordered spatial and temporal patterns of Hox expression may be very important for their function, and the clustered organization a critical part of regulating the restricted domains of expression. The conserved organization of the clusters may therefore reflect selection based on gene regulation rather than simply protein structure. Regulatory elements distributed over a large region of the complex could be required to generate the appropriate patterns of expression, and disruption of the complex would prevent normal control elements from functioning. To investigate the molecular basis of Hox-2 expression, we have used cultured cells and transgenic mice to map promoters and regulatory elements required for spatial and temporal control.

Initially we examined the Hox-2.1 gene. The intergenic distance between Hox-2.1 and the immediate 5′ flanking gene is about 2 kb. When a clone containing 5′ flanking DNA and the entire Hox-2.1 gene marked with an oligonucleotide is transfected into F9 cells, a Hox-2.1 mRNA of the proper size is generated. Therefore, to map the regulatory regions, we have generated a series of deletion constructs containing 5′ flanking sequences of the Hox-2.1 gene fused to a CAT reporter (Fig. 5). Transfection of these constructs into F9 teratocarcinoma cells has allowed us to identify a number of regions that work in both a positive and negative manner to modify CAT activity. A number of fragments stimulated CAT activity in transient assays (Fig. 5), but only two of these regions stimulated CAT activity in stably transfected cell lines. It is interesting to note that in the transient assays, combinations of positive elements did not act in an additive manner. In some cases combinations actually reduced or eliminated expression, suggesting that negative elements exist in the intergenic region.

In contrast to the results obtained in cultured cells, constructs that contained the entire intergenic region between Hox-2.2 and Hox-2.1 did not show any expression when tested in transgenic mice. We have also tested a range of further constructs that contained 5′ flanking DNA extending through and upstream of the Hox-2.2 gene. None of these DNA samples were functional in transgenic mice. It therefore appears that the elements we identified in cultured cells represent part of the basic promoter requirements for the Hox-2.1 gene, but that other elements are needed to generate the normal pattern of Hox-2.1 expression in embryogenesis.

Constructs extending in the 3′ direction from Hox-2.1 were prepared to test regulatory elements in this region. We have identified a small region that drives expression of the transgene in part of the neural tube. This represents a limited subset of the Hox-2.1 pattern. This 3′ region which can function on Hox-2.1

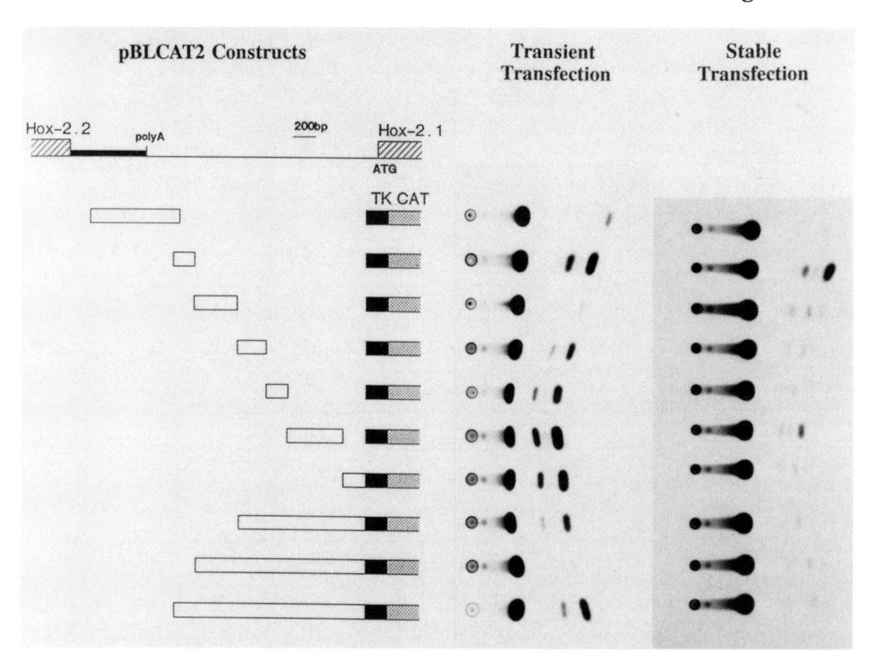

Fig. 5. Regulatory elements for the Hox-2.1 gene. **Left**, a diagram showing the relative distance between Hox-2.1 and the 5′ flanking Hox-2.2 gene. Open rectangles represent the DNA from the intergenic region attached to a CAT reporter gene and transfected into F9 cells. **Right**, transient expression (48 hr after transfection) and stable expression for the constructs.

is also 5′ of the Hox-2.6 gene, and it is possible that a single element such as this could regulate multiple genes in the complex.

Transgenic experiments examining the Hox-2.6 gene marked with a lacZ reporter have produced different results. In collaboration with Drs. P. Rigby and R. Allemann, we have found that only a small region of the Hox-2.6 gene and flanking DNA is required to reconstruct the normal pattern of Hox-2.6 expression in transgenic mice. An important regulatory region maps to the 3′ flanking DNA. With Hox-2.6 it therefore appears that the overall organization of the Hox-2 cluster is not needed for spatial control.

The next gene in the Hox-2 cluster flanking Hox-2.6 on the 3′ side is Hox-2.7. The intergenic distance is approximately 30 kb. However, in recent experiments, evidence has been found that two transcripts for the Hox-2.7 gene are produced from two different promoters. One promoter (P1) maps immediately upstream of the ATG in the protein. The second promoter (P2), identified by polymerase chain reaction (PCR) analysis, is located 25 kb upstream of P1. The positions of the two Hox genes and the different Hox-2.7 promoters is illustrated in Figure 6. The upstream P2 promoter maps near the 3′ end of the Hox-2.6 gene in a region where we have shown that elements for the appropriate

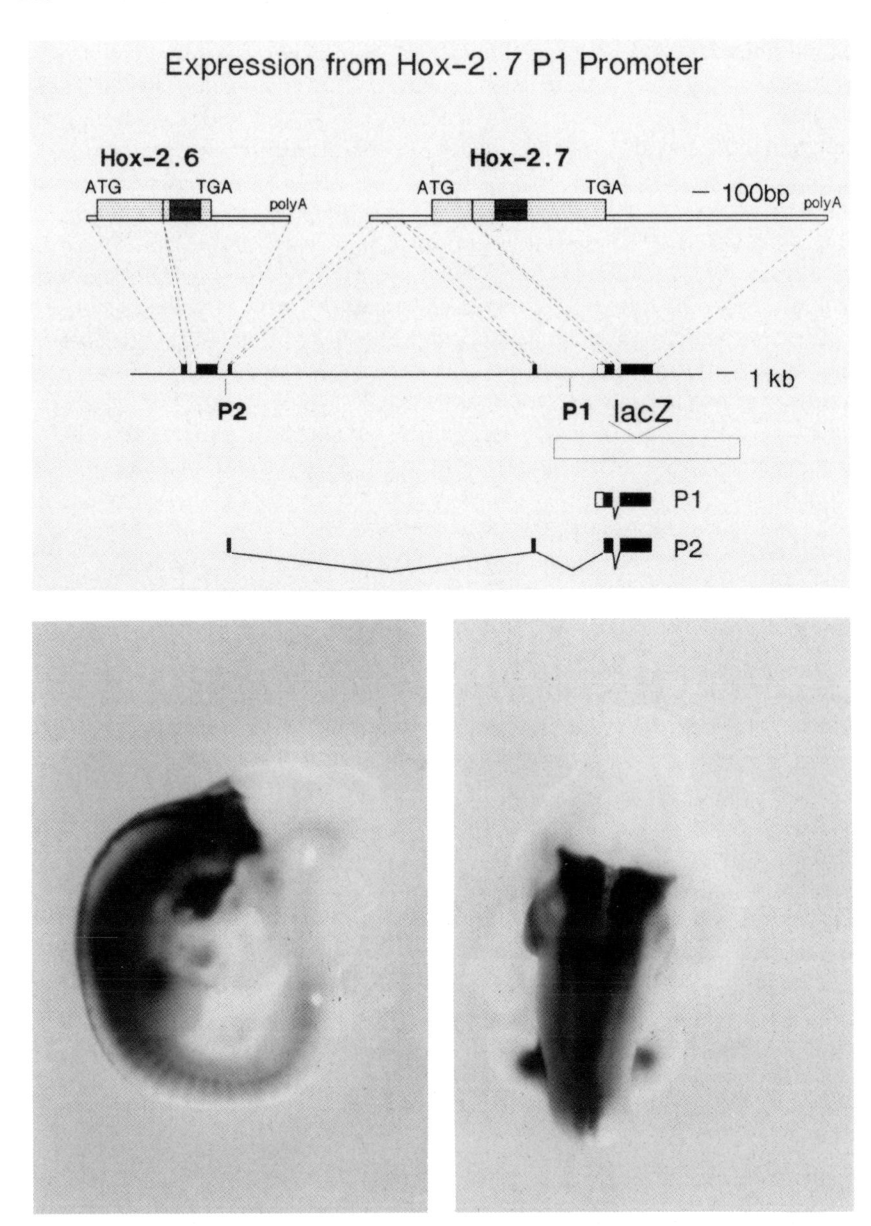

Fig. 6. Multiple promoters and the potential for shared regulatory regions. Two promoters for the Hox-2.7 gene are designated P1 and P2. The upstream P2 site maps near the 3′ end of the Hox-2.6 gene within a region containing an enhancer that is required for Hox-2.6 expression (**top**). **Bottom**, a lacZ staining pattern of transgenic mice generated with constructs containing only the P1 promoter.

expression of the Hox-2.6 gene are located. The 3′ Hox-2.6 regulatory region is therefore positioned next to one of the two Hox-2.7 start sites and may modulate the expression of both Hox-2 genes. To test if the P1 region can function independently from the P2 region, lacZ constructs were used to make transgenic mice. A staining pattern for a 10.5 dpc embryo shows that part of the normal expression of the Hox-2.7 gene is generated by the P1 constructs (Fig. 6, bottom). Note the sharp anterior boundary of lacZ expression in the hindbrain. We are currently constructing larger versions of this region to examine the role of the P2 regions on expression.

The transgenic experiments support the idea that regulatory elements for the Hox-2 complex are spread over a broad region. It appears that several genes share some of the same control regions and that multiple promoters are used to express the genes. In some cases, such as Hox-2.6 or Hox-1.1 (Puschel et al., 1990), isolated members appear to faithfully reconstruct the normal patterns of expression outside of the context of the cluster. However, we cannot be certain that all transcripts are reconstructed, and further work is needed to fully substantiate that all aspects of the pattern are normal. Some of the regions utilized by the Hox-2.6 gene could be necessary for another gene (Hox-2.7). Based on the limited numbers of genes examined, the clustered Hox organization is important to maintain regulatory patterns and is very likely a key ingredient in the conservation of this network in vertebrates.

CONCLUSIONS

The patterns of expression for Hox genes strongly supports the idea that they play a role in positional specification in vertebrate embryogenesis. The genes are expressed in many developing systems, such as the axial skeleton, nervous system, head, limb, and gut. They appear to have a central role in embryonic fields where the specification of an axis is important. The emerging concept is that the genes provide a combinatorial code to establish an address or particular identity. Homology between different Hox clusters points to potential problems in redundancy of function. In the absence of naturally occurring genetic mutants, reverse genetics will be essential in delineating how this network of genes functions. The indentification of regulatory regions involved in the spatial regulation of the Hox genes provides an important tool that can be used to alter the normal domains of Hox expression and directly test their role in processes such as segmentation.

ACKNOWLEDGMENTS

Dr. S. Nonchev was supported by a Wellcome grant. We thank our collaborators Drs. David Wilkinson and Andrew Lumsden for their interactions and help in this work.

REFERENCES

Acampora D, D'Esposito M, Faiella A, Pannese M, Migliaccio E, Morelli F, Stornaiuolo A, Nigro V, Simeone A, Boncinelli E (1989): The human HOX gene family. NAR 17:10385–10402.

Akam M (1989): Hox and HOM: Homologous gene clusters in insects and vertebrates. Cell 57:347–349.

Brockes JP (1989): Retinoids, homeobox genes and limb morphogenesis. Neuron 2:1285–1294.

Dolle P, Izpisua-Belmonte JC, Falkenstein H, Renucci A, Duboule D (1989): Coordinate expression of the murine Hox-5 complex homeobox containing genes during limb pattern formation. Nature 342:767–772.

Duboule D, Dolle P (1989): The murine Hox gene network: Its structural and functional organisation resembles that of *Drosophila* homeotic genes. EMBO J 8:1507–1508.

Featherstone M, Baron A, Gaunt S, Mattei M, Duboule D (1988): Hox-5.1 defines a homeobox containing gene locus on mouse chromosome 2. Proc Natl Acad Sci USA 85:4760–4764.

Fraser S, Keynes R, Lumsden A (1990): Segmentation in the chick embryo hindbrain is defined by cell lineage restrictions. Nature 344:431–435.

Gaunt SJ, Sharpe PT, Duboule D (1988): Spatially restricted domains of homeogene transcripts in mouse embryos: Relation to a segmented body plan. Development 104(suppl): 169–179.

Gaunt SJ, Krumlauf R, Duboule D (1989): Mouse homeobox genes within a subfamily, Hox 1.4, 2.6, and 5.1, display similar anteroposterior domains of expression in the embryo, but show stage- and tissue-dependent differences in their regulation. Development 107:131–141.

Graham A, Papalopulu N, Lorimer J, McVey JH, Tuddenham EGD, Krumlauf R (1988): Characterisation of a murine homeobox gene, Hox 2.6, related to the *Drosophila Deformed* gene. Genes Dev 2:1424–1438.

Graham A, Papalopulu N, Krumlauf R (1989): The murine and *Drosophila* gene complexes have common features of organisation and expression. Cell 57:367–378.

Holland PWH, Hogan BLM (1986): Phylogenetic distribution of *Antennapedia*-like homeoboxes. Nature 321:251–253.

Hunt P, Whiting J, Muchamore I, Marshall H, Krumlauf R(1991a):Homeobox genes and models for patterning the hindbrain and branchial arches. Development (Suppl) 186–195.

Hunt P, Wilkinson D, Krumlauf R (1991b): Patterning the vertebrate head: Murine Hox-2 genes mark distinct subpopulations of premigratory and migrating cranial neural crest. Development 112:43–50.

Kessel M, Gruss P (1990): Murine developmental control genes. Science 249:374–379.

Lewis E (1978): A gene complex controlling segmentation in *Drosophila*. Nature 276:565–570.

Lumsden A (1990): The cellular basis of segmentation in the developing hindbrain. TINS 13:329–335.

Lumsden A, Keynes R (1989): Segmental patterns of neuronal development in the chick hindbrain. Nature 337:424–428.

Maden M, Hunt P, Eriksson U, Kuroiwa A, Krumlauf R, Summerbell D (1991): Retinoic acid-binding protein, rhombomeres and the neural crest. Development 111:35–43.

Murphy P, Davidson D, Hill R (1989): Segment-specific expression of a homeobox containing gene in the mouse hindbrain. Nature 341:156–159.

Papalopulu N, Hunt P, Wilkinson D, Graham A, Krumlauf R (1990): Hox-2 homeobox genes and retinoic acid: Potential roles in patterning the vertebrate nervous system. In "Advances in Neural Regeneration Research," Vol 60. New York: Wiley-Liss, pp 291–308.

Puschel AW, Ballin R, Gruss P (1990): Position-specific activity of the Hox 1.1 promoter in transgenic mice. Development 108:435–442.

Rubock M, Larin Z, Cook M, Papalopulu N, Krumlauf R, Lehrach H (1990): A yeast artificial chromosome containing the mouse homeobox cluster Hox-2. Proc Natl Acad Sci USA 87:4751–4755.

Scott M, Tamkun J, Hartzell G (1989): The structure and function of the homeodomain. Biochim Biophys Acta 989:25–48.

Simeone A, Acampora D, Arcioni L, Andrews P, Boncinelle E, Mavilio F (1990): Sequential activation of HOX2 homeobox genes by retinoic acid in human embryonal carcinoma cells. Nature 346:763–766.

Wilkinson D, Bhatt S, Chavrier P, Bravo R, Charnay P (1989a): Segment-specific expression of a zinc finger gene in the developing nervous system of the mouse. Nature 337:461–464.

Wilkinson DG, Bhatt S, Cook M, Boncinelli E, Krumlauf R (1989b): Segmental expression of Hox-2 homeobox-containing genes in the developing mouse hindbrain. Nature 341:405–409.

Wilkinson DG, Krumlauf R (1990): Molecular approaches to the segmentation of the hindbrain. TINS 13:335–339.

Cell-Cell Interactions in Early Development, pages 145–162

9. Development of the Terminal Anlagen of the *Drosophila* Embryo Depends Upon Interactions Between the Germline and the Somatic Follicle Cells

Leslie M. Stevens and Christiane Nüsslein-Volhard
Max-Planck-Institut für Entwicklungsbiologie, Spemannstrasse 35/III, D-7400 Tübingen, Germany

INTRODUCTION

The initiation of embryonic development in *Drosophila melanogaster* is dependent upon maternal gene products that are present in the freshly laid egg. These products are responsible for establishing polarity along the anterior–posterior and dorsal–ventral axes of the embryo through the restricted spatial activation of zygotic target genes. Subsequently, the zygotic genome, through a series of hierarchical interactions involving the gap, segmentation, and homeotic genes, subdivides the embryo and gives the individual segments their identity. Genetic analysis has indicated that for the establishment of the anterior–posterior axis, three classes of maternal genes are required (Nüsslein-Volhard et al., 1987). Each system acts independently of one another to bring about the development of anterior structures (head and thorax), posterior structures (abdomen), and the unsegmented terminal regions (acron and telson).

Among these three groups of genes, two different mechanisms are employed to establish polarity within the region for which they are responsible. For both anterior and posterior development, this is achieved by the localization of a maternal gene product within the egg cytoplasm. In contrast, terminal development appears to be mediated by the localized activation of a uniformly distributed gene product. The activating signal is transmitted from the somatic follicle cells to the germline-derived oocyte. Thus, the structures that form at the poles of the embryo appear to arise from an inductive event between the germline and the soma.

THE TERMINAL CLASS GENES

Six genes have currently been identified that are required maternally for the development of the acron and telson at the ends of the embryo: *torso (tor)*

(Schüpbach and Wieschaus, 1986a); *torsolike (tsl)* (Stevens et al., 1990); *trunk (trk)* (Schüpbach and Wieschaus, 1986a); *fs(1)Nasrat* [*fs(1)N*] (Degelmann et al., 1986); *fs(1)pole hole* [*fs(1)ph*] (Perrimon et al., 1986); and *l(1)pole hole* [*l(1)ph*] (Perrimon et al., 1985; Ambrosio et al., 1989a). Embryos derived from mothers that are homozygous mutants for any one of these genes exhibit a virtually identical phenotype (Fig. 1b,c). At the anterior, the labrum is missing and the head skeleton is collapsed. Posteriorly, everything posterior to abdominal segment 7 is deleted, including abdominal segment 8 and the telson, which includes the anal plates, tuft, and the *filzkörper*, posterior tracheal specializations. Internal organs that derive from the terminal anlagen, such as the hind and posterior midgut, are also absent.

tor is distinct among the terminal class genes in that it can also be mutated to gain-of-function alleles in which the embryos exhibit a phenotype that is complementary to the lack-of-function phenotype (Klingler et al., 1988; Strecker et al., 1989). In embryos derived from mothers carrying a weak gain-of-function *tor* allele, the head and telson are normal, but the segmented region is disrupted and segments can be lost to variable degrees (Fig. 1d). In the intermediate case, the entire segmented region that will form the head, thorax, and abdomen is affected; segments no longer form and the telson is enlarged. In the extreme gain-of-function phenotype, no cuticle is formed. However, internal organs do develop, and these are primarily derivatives of the terminal anlagen, such as the posterior midgut.

The existence of *tor* gain-of-function alleles that exhibit a phenotype different from that produced by mutations in the other terminal class genes makes it possible, through the analysis of double mutant combinations, to determine the epistatic relationships between *tor* and the other members of this gene group. This type of analysis assumes that the manifestation of the gain-of-function phenotype will be independent of genes that act upstream in the same developmental pathway, but will require the function of genes that act downstream. Accordingly, when mutations in terminal class genes that function

---

Fig. 1. The *torso* lack and gain-of-function phenotypes. Dark-field photographs of cuticular preparations. Anterior is up. (**a**) Wild-type *Drosophila* larva. Ventral is to the left. Thoracic (T1–3) and abdominal (A1–8) segments are demarcated by the ventral denticle belts. Although most of the head skeleton (hs) originates from segmented anlagen, the labrum (lb) derives from the unsegmented acron (Jürgens et al., 1986). At the posterior, the unsegmented telson includes the *filzkörper* (fk), anal plates, and tuft (t) (Jürgens, 1987). (**b,c**) *tsl* and *tor* phenotypes, respectively. Ventral views. At the anterior, the labrum is lacking and the head skeleton is reduced in size. Posteriorly, A8 and the telson are deleted. (**d–f**) Ventral is to the left. (**d**) *tor* gain-of-function phenotype of intermediate strength. The head and telson are normal or even enlarged, while the segmented region is disrupted and segments deleted to variable degrees. (**e,f**) Two focal planes of a double mutant between a *tor* gain-of-function allele, tor^{Y9}, and *tsl*. The phenotype is equivalent to that of tor^{Y9}/tor^{+} alone, indicating that *tsl* functions upstream of *tor*. (Reprinted by permission from Nature, vol. 346, pp. 660–663. Copyright © 1990 Macmillan Magazines Ltd.)

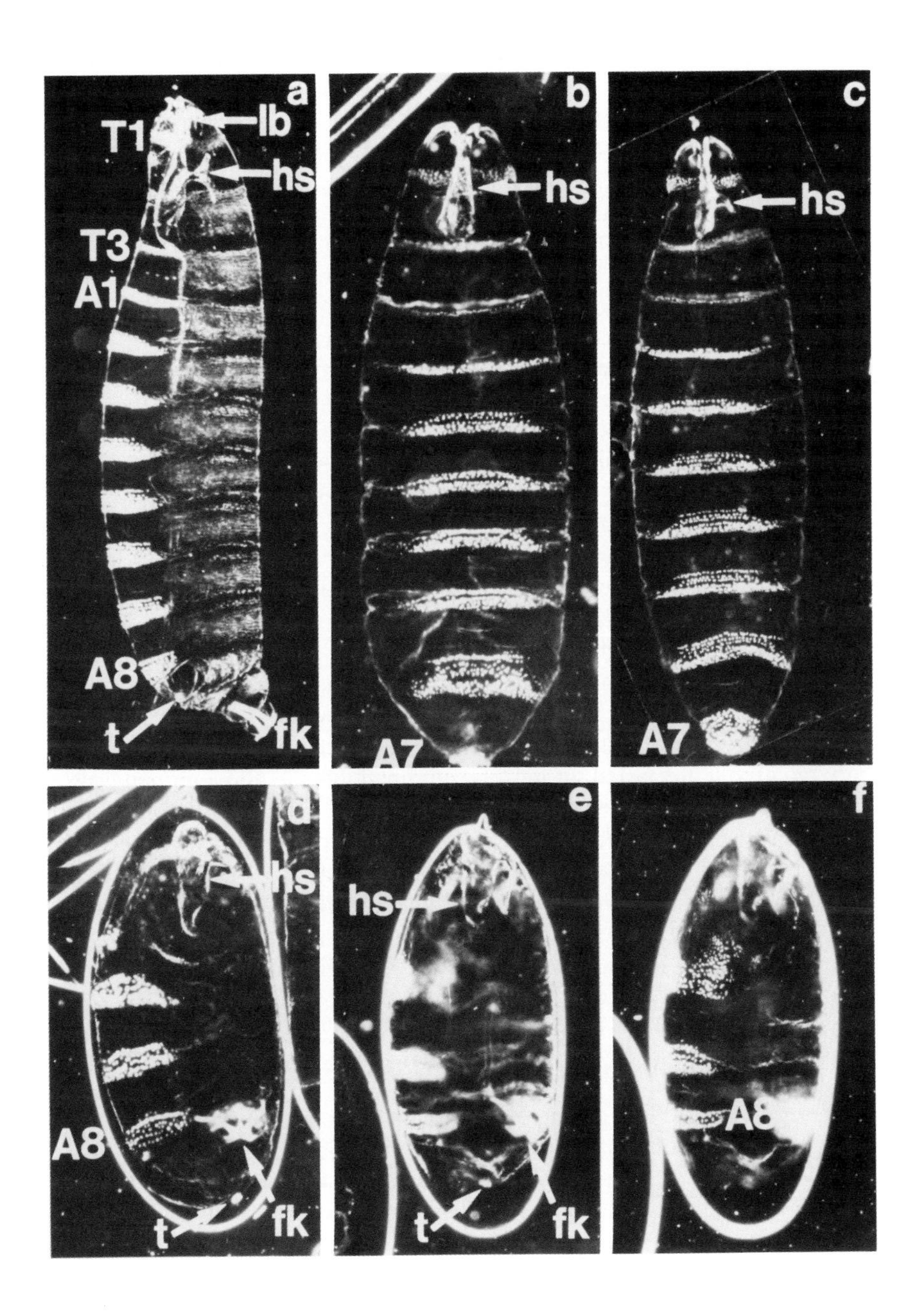
a
T1
lb
hs
T3
A1
A8
t
fk
b
hs
A7
c
hs
A7
d
hs
A8
t
fk
e
hs
t
fk
f
A8

upstream of *tor* are combined with *tor* gain-of-function alleles, the gain-of-function phenotype will prevail. In contrast, mutations in genes that act downstream of *tor* will block the expression of the gain-of-function phenotype, and the embryos will instead exhibit the phenotype of the downstream mutation. Through this type of analysis, it has been determined that four of the terminal class genes, *tsl, trk, fs(1)ph,* and *fs(1)N*, act upstream of *tor* (Fig. 1e,f) (Casanova and Struhl, 1989; Stevens et al., 1990), while *l(1)ph* functions downstream (Klingler, 1989; Ambrosio et al., 1989b).

A similar analysis has been carried out to examine the relationship between *tor* and the zygotic gap gene *tailless (tll)* (Strecker et al., 1986). Embryos that are homozygous mutants for *tll* exhibit a phenotype that is similar to the maternal terminal phenotype in that parts of the head skeleton and posterior cuticular derivatives are deleted, suggesting that *tll* might be a target for the action of the maternal genes of the terminal class. To investigate the relationship between *tll* and *tor*, embryos were analyzed that derived from mothers carrying *tor* gain-of-function alleles, and were themselves homozygous mutants for *tll* (Klingler et al., 1988; Strecker et al., 1989). In these embryos, the *tor* gain-of-function phenotype is suppressed and the embryos exhibit a phenotype very similar to that of *tll* alone. The telson and parts of the head skeleton are deleted due to the lack of *tll* activity, but the thoracic and abdominal segments that are disrupted by *tor* gain-of-function mutations are completely restored. This implies that the *tll* gene product mediates the suppression of central structures seen in *tor* gain-of-function embryos.

These observations suggest that in wild-type embryos, *tll* activity is not only required at the poles but must be confined there to allow normal central development. This implies that a primary role of *tor* activity is to bring about the spatially restricted activation of *tll* at the poles of the embryo. Indeed, the initial transcription domain of *tll* is confined to the most anterior and most posterior 20% of wild-type embryos (Pignoni et al., 1990). Further, as predicted by the genetic data, in embryos derived from *tor* gain-of-function mothers, *tll* expression spreads toward the middle region of the embryo (Frank Sprenger, personal communication).

THE MOLECULAR ANALYSIS OF *torso*

There are at least two mechanisms by which the spatially restricted activation of *tll* expression at the ends of the embryo could be achieved. A simple scheme would be if *tor* protein itself were confined to the poles of the embryo. Alternatively, *tor* could be present throughout the embryo, but active only in the terminal regions. Information that enables us to discriminate between these two models has been obtained from a molecular analysis of the *tor* gene (Sprenger et al., 1989). *tor* encodes a 923 amino acid protein with a predicted molecular weight of 105 kD. The DNA sequence reveals that the *tor* protein exhibits significant homology to the tyrosine kinase family of proteins. In addition to amino acid identity within the kinase domain, *tor* exhibits structural similarities to a subclass of tyrosine kinases, the receptor tyrosine kinases. These features include a relatively large extracellular domain, a single transmembrane region, and a kinase-containing cytoplasmic domain (Ullrich and Schlessinger, 1990).

The distribution of *tor* in the embryo has been analyzed using antibodies directed against the *tor* protein (Fig. 2). *tor* is not localized to the poles of the embryo, but rather is distributed uniformly along the anterior–posterior axis (Fig. 2b), and appears to be associated with the membrane of the embryo (Fig. 2d), although the nature of that association cannot be determined at this level of analysis. Similar conclusions have been reached by Casanova and Struhl (1989). These observations clearly demonstrate that *tor* protein is not localized to the polar regions of the embryo, and thus a restricted pattern of *tor* protein distribution cannot be responsible for the confinement of *tll* expression to the terminal regions of the embryo.

Instead, the protein distribution and sequence data suggest a model in which *tor* protein is uniformly distributed along the anterior–posterior axis of the embryo and is locally activated only at the poles. By analogy to other receptor tyrosine kinases, it has been proposed that *tor* is activated by the binding of a ligand molecule (Sprenger et al., 1989; Stevens et al., 1990). Restriction of that ligand to the polar regions of the embryo could then explain the localized activation of *tor* and hence of *tll*. This model also suggests an explanation for the gain-of-function *tor* alleles, in that these could encode proteins that are constitutively active, independent of ligand binding. Since *tor* is present everywhere, this would result in ectopic *tor* activity in the middle regions of the embryo.

EVENTS DOWNSTREAM OF *torso*

The model described above proposes that *tor* is locally activated at the poles of the embryo by the binding of a ligand molecule, and that the activation of *tor* leads ultimately to transcription of the *tll* gene at the ends of the embryo. Genetic data presented earlier suggest that the activation of *tll* expression is mediated via the product of the *l(1)ph* gene. *l(1)ph* has also been cloned and sequenced and determined to be the *Drosophila* homologue of the vertebrate proto-oncogene Raf-1, a serine/threonine kinase (Mark et al., 1987; Nishida et al., 1988). In mammalian cells, Raf-1 has been shown to be activated by the platelet-derived growth factor B-receptor (Morrison et al., 1989), and by analogy, this would suggest that *tor* directly phosphorylates and consequently activates *l(1)ph*. Although this is consistent with the idea that the terminal pathway involves a kinase cascade mechanism, the identification of *l(1)ph* as a kinase also suggests that it is

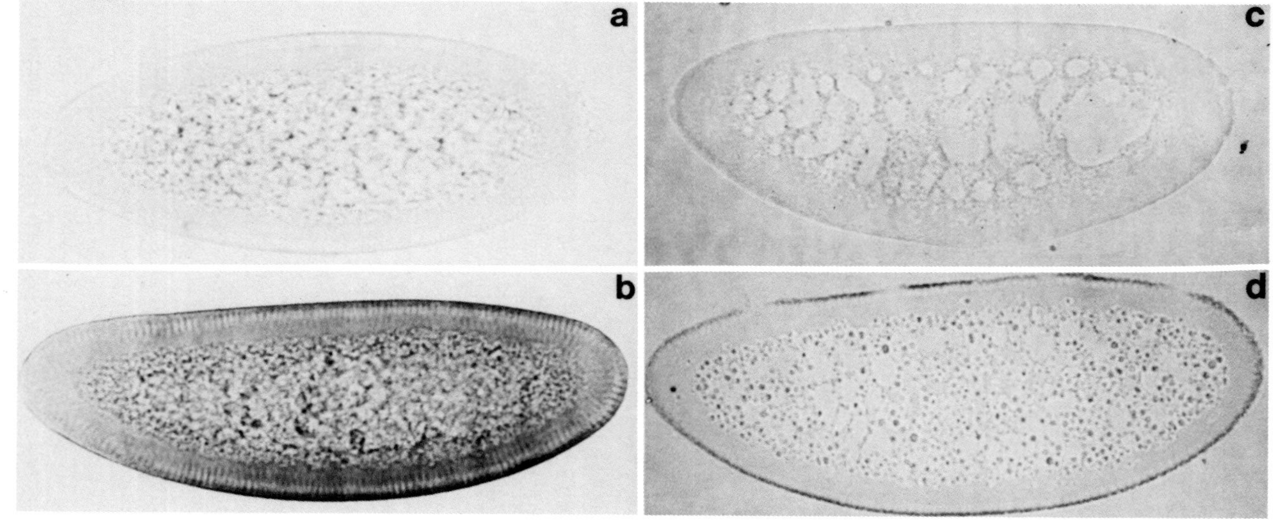

Fig. 2. *torso* protein distribution in whole mounts and sections of embryos. The distribution of *tor* was analyzed by staining embryos with antisera against the *tor* protein. (**a,b**) Whole amounts of embryos. (**c,d**) Ten-micron araldite sections. (a,c) Control embryos derived from mothers homozygous for tor^{XR1}, a protein null mutation. No staining is detectable. (b,d) Embryos derived from tor^{+} mothers. Staining is detectable all along the anterior–posterior axis, and in the section appears to be associated with the membrane of the embryo. Antisera against β-galactosidase fusion proteins (Rüther and Müller-Hill, 1983) containing either the extracellular or cytoplasmic regions of the *tor* protein were generated in rabbits. The binding of affinity-purified antiserum was visualized by staining embryos with biotinylated horseradish peroxidase-avidin complexes bound to biotinylated second antibody. tor^{+} mothers were homozygous mutants for *oskar*, a maternal mutation in which the embryos lack pole cells (Lehmann and Nüsslein-Volhard, 1986). Control and experimental embryos were stained together, and tor^{+} embryos were identified by the presence of the *oskar* phenotype.

unlikely to be directly responsible for the activation of *tll* transcription. Thus, the product of at least one other gene must function in this part of the pathway.

ZYGOTIC TARGET GENES

As suggested above, a primary function of the maternal terminal class genes is the proper spatial activation of the expression of the gene *tll*. However, although the phenotype exhibited by *tll* mutant embryos is very similar to that produced by mutations in maternal genes of the terminal class, the most anterior and posterior regions of the larva, which are deleted by the maternal mutations, are present in *tll* embryos. To account for this discrepancy, at least one other zygotic gene must be postulated to be activated by *tor*. One such gene is *hückebein (hkb)*, a zygotic gap gene that when mutated results in the loss of the posterior midgut (Weigel et al., 1990). Together, *tll* and *hkb* appear to account for all of the function of *tor* in its posterior domain. *hkb* does not fulfill this role at the anterior pole, however, indicating that at least one more gene must be activated by *tor* in this region.

Sequence analysis of the *tll* gene indicates that it exhibits homology to the steroid hormone receptor family (Pignoni et al., 1990). This suggests that *tll* may act as a transcription factor and be involved in regulating the expression of other zygotic genes. Thus, the spatially restricted activation of *tll* expression by the maternal pathway will result in the initiation of a specific program of zygotic gene expression in the terminal regions of the embryo.

EVENTS UPSTREAM OF *torso*

To investigate the possibility that the genes acting upstream of *tor* in this developmental pathway are involved in regulating the expression or modification of the *tor* protein, Western blot analysis (Fig. 3) and whole-mount staining of embryos using antibodies against *tor* have been carried out (Casanova and Struhl, 1989; unpublished observations). In embryos derived from mothers homozygous for mutations in *tsl, trk, fs(1)ph* and *fs(1)N*, the spatial distribution, quantity, and electrophoretic mobility of *tor* protein are indistinguishable from that in embryos derived from wild-type mothers. Thus, these genes do not appear to be involved in regulating the expression of *tor* protein itself. Rather, it seems likely that the role of the upstream genes is to generate the localized signal that has been postulated to activate *tor*. Indeed, several lines of evidence suggest that *tsl* may encode a ligand for *tor*, and that its expression during oogenesis may determine the spatial parameters of *tor* activation in the embryo.

Oogenesis in *Drosophila* requires the interaction of two distinct tissues (King, 1970). The egg derives from a germline precursor cell that divides in the germarium to form the 16-cell cluster. One of these cells will develop into the oocyte while the others serve as the accessory nurse cells, providing the growing oocyte with the products of maternally expressed genes. The 16-cell cluster is ensheathed by an epithelium of somatically derived follicle cells that synthesize the vitelline membrane and the chorion, or eggshell. Significantly, the predicted structure of *tor* (Sprenger et al., 1989) indicates that the receptor portion of the molecule lies outside the membrane of the embryo, in the perivitelline space underneath the vitelline membrane. This suggests that a ligand for *tor* must at some time also be present in this space.

If we assume that the spatial restriction of *tor* function is mediated through localized activation by ligand binding, there are at least two requirements that must be fulfilled by a ligand for *tor*. It must be present in the perivitelline space that surrounds the embryo, and as discussed earlier, it must be localized to the poles of the embryo. A mechanism that would fulfill both of these requirements was suggested by observations made on the tissue specificity of the terminal class genes. By constructing mosaic females, in which the germline is mutant for a particular maternal gene but the somatic tissue is wild type, or vice versa, it is possible to determine whether a maternal gene is required in the germline or the soma. Through this type of analysis, it has been shown that *trk*, *fs(1)ph*, and *fs(1)N* are germline-dependent and are thus required within the oocyte/nurse cell complex (Perrimon and Gans, 1983; Schüpbach and Wieschaus, 1986b). In contrast, *tsl* is unique among the terminal class genes in that it is required to be expressed by the somatic follicle cells (Stevens et al., 1990).

The somatic dependence of *tsl* expression suggests a mechanism by which the *tsl* product can be deposited into the perivitelline space and localized near the polar regions of the oocyte. This could occur if *tsl* is expressed by a subpopulation of follicle cells present only at the poles of the follicle. In fact, the existence of specialized follicle cells at the poles has been implied both by the structure of the chorion they secrete (Margaritas et al., 1980), and by their differential expression of an antigen recognized by a monoclonal antibody (Brower et al., 1981). More recent evidence has been obtained through the use of enhancer detectors or ''enhancer traps'' (O'Kane and Gehring, 1987). This method involves the use of germline-mediated transformation to randomly introduce the bacterial enzyme β-galactosidase as a reporter gene into the genome

Fig. 3. Western blot analysis of *torso* protein in terminal class mutants. Embryonic extracts from 0–4-h-old embryos derived from wild-type and mutant mothers were separated by SDS-PAGE and transferred to nitrocellulose. The Western blot was incubated with antiserum against *tor* protein, and staining visualized with alkaline phosphatase-coupled goat antirabbit antibodies. Apparent molecular weights of marker proteins in kd are indicated at the left. Extracts of embryos obtained from tor^{XR1} mothers (−) serve as controls for the specificity of the staining. The apparent mobility and quantity of *tor* protein in embryos derived from mothers mutant for *tsl*, *fs(1)ph*, and *trk* and are not significantly different from wild type (+).

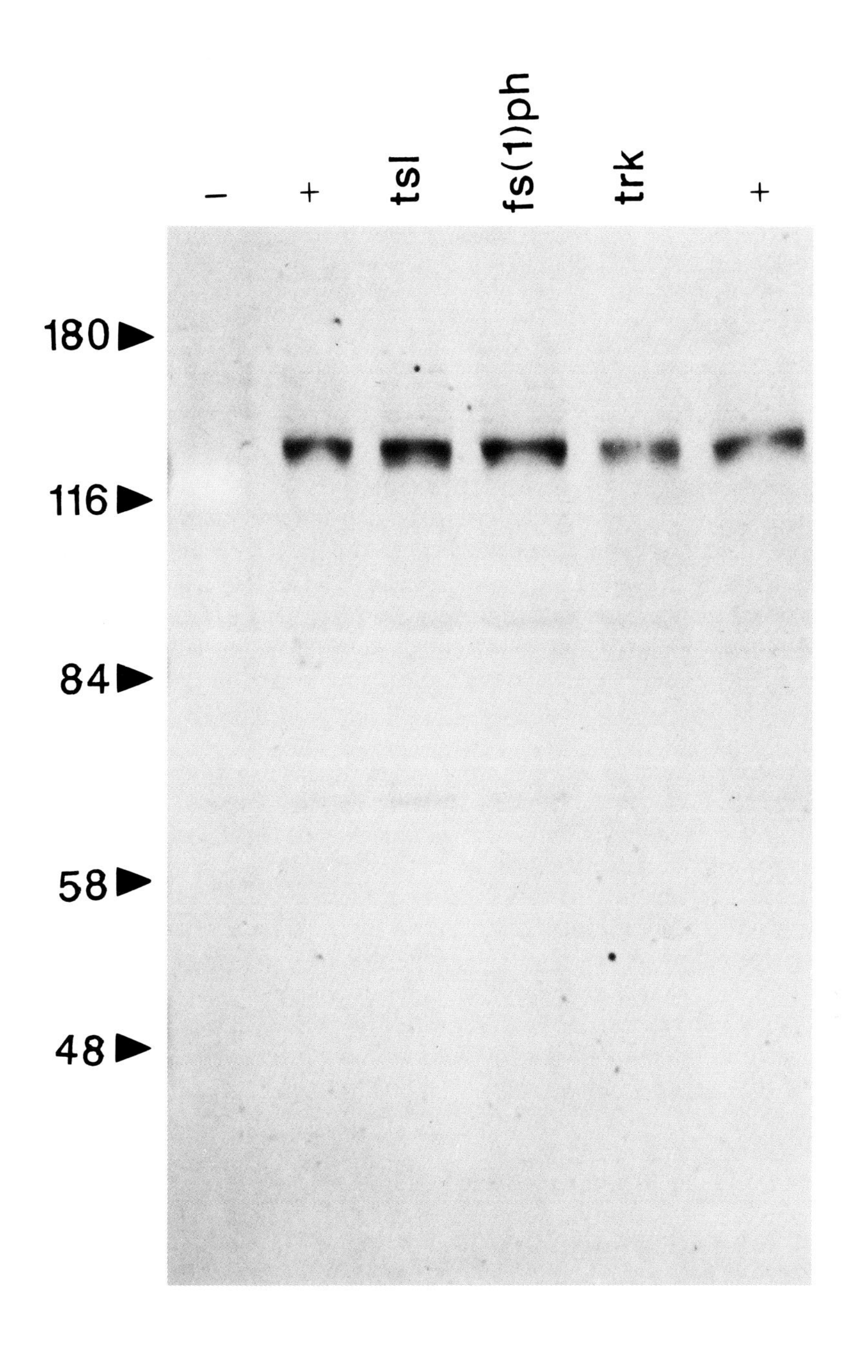
–
+
tsl
fs(1)ph
trk
+
180
116
84
58
48

of the fly, where it may come under the control of an endogenous regulatory element. Such an insertion can result in tissue-specific expression of β-galactosidase, and a number of lines have been isolated in which the reporter gene is expressed in the ovary. In several cases, the reporter gene is expressed only by a small number of cells present at the poles of the oocyte (Fasano and Kerridge, 1988; Grossniklaus et al., 1989). Using this technique as a measure of differential gene expression, these observations suggest that a specialized subpopulation of follicle cells exists at the poles.

MOSAIC ANALYSIS OF *tsl* EXPRESSION

The existence of specialized cells at the poles of the follicle raises the possibility that *tsl* is expressed by such a polar subpopulation. To test this hypothesis, a mosaic analysis was carried out to identify the follicle cells required to express *tsl* (Stevens et al., 1990). Clones of *tsl* follicle cells were created by mitotic recombination in the ovaries of females heterozygous for *tsl*, and their progeny were examined for the influence of the clones on the embryonic phenotype. If *tsl* expression is required only in a subpopulation of follicle cells present at the poles of the oocyte, clones of *tsl* follicle cells that include these cells, even if they are quite small, will produce a complete *tsl* phenotype at that end of the future embryo, while the other end will be unaffected. In contrast, if *tsl* is required throughout the follicular epithelium, small clones will not influence the embry onic phenotype, while large clones will affect both ends of the embryo equiv alently. Thus, three parameters were examined in this experiment: 1) Whether small clones are capable of influencing the embryonic phenotype, or if an effect is observed only after the induction of large clones; 2) whether clones that produce an embryonic phenotype are localized to the poles of the oocyte; and 3) whether the development of the two ends of the embryo can be separated from one another, or if they are always simultaneously affected.

The size of follicle cell clones was regulated by irradiating the females (to induce mitotic recombination) at different stages of development. During oogenesis, when the germline-derived 16-cell cluster leaves the germarium, it is accompanied by approximately 80 somatic follicle cells (King, 1970). During the first six stages of oogenesis, the follicle cells undergo four rounds of cell division to reach a final population of about 1,200. At stage 7 they stop dividing and become polyploid, and later secrete the vitelline membrane and chorion. The minimum time required after leaving the germarium until egg deposition is about four days at room temperature (22–23°C). By irradiating adult females, in which the process of oogenesis is already taking place, recombination was induced in follicle cells that were just leaving the germarium and would undergo at most only four additional rounds of cell division, resulting in clones as small as 16 cells. To restrict the analysis to such small clones, only progeny derived within the first four days of adult irradiation were exam-

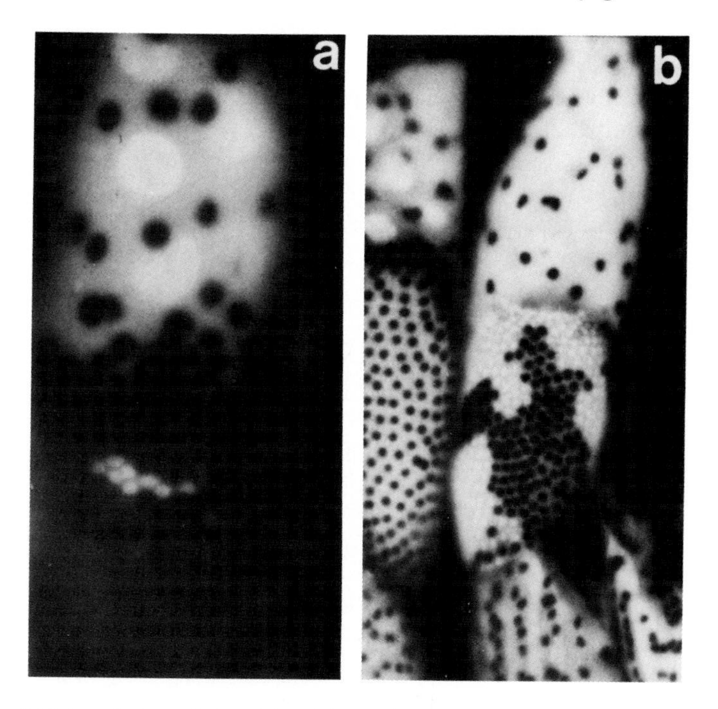

Fig. 4. Follicle cell clones induced by adult and larval irradiation. Fluorescent micrographs of follicle cell clones in ovaries taken from females heterozygous for an enhancer trap insertion in which all follicle cells express β-galactosidase (Fasano and Kerridge, 1988). Clones of non-expressing cells were generated by irradiating the females with 1,050 R to induce mitotic recombination. Ovaries were fixed in 1% glutaraldehyde in phosphate-buffered saline, then stained overnight at 37°C for β-galactosidase activity according to Simon et al. (1985). Ovaries were then counterstained with 4′,6-diamidino-2-phenylindole (DAPI, 0.5 μg/ml). β-galactosidase staining quenches the DAPI fluorescence, such that only cells contained within the clone appear fluorescent. (**a**) Follicle from adult female irradiated four days previously. The clone is quite small, consisting of eight cells. (**b**) Follicle from female that was irradiated during larval stages. A high percentage of the cells are clonally derived.

ined. Larger clones were produced by irradiating females during larval stages, before ovary formation, when the precursors to the somatic cells in the germarium are still proliferating (King, 1970). When mitotic recombination is induced in follicle cell precursors, the resulting adult females contain germaria, and consequently follicles, in which a high proportion of the cells are clonally derived (Wieschaus et al., 1981; Szabad and Hoffman, 1989).

Examples of the different sizes of follicle cell clones that are produced by adult and larval irradiation are shown in Figure 4. It should be noted that these are not *tsl* clones. Rather, these ovaries were taken from females that were heterozygous for an enhancer-trap insertion that causes all follicle cells to express β-galactosidase (Fasano and Kerridge, 1988). Clones of cells lacking the marker were produced by inducing mitotic recombination with X-irrradiation. Thus, the clonally derived cells appear white, while the cells that still carry the marker are darkly stained. The follicle shown in Figure 4a is derived from a female that was irradiated four days prior to examination. The size of the clone is quite small, consisting of only eight cells. In contrast, Figure 4b shows a follicle taken from a female that was irradiated during larval stages. A large number of the cells are clonally derived. This demonstrates the validity of using adult and larval irradiation for a comparison of the effects of small and large *tsl* clones on the embryonic phenotype.

Although the loss of β-galactosidase expression provides a convenient marker for the location of follicle cell clones in oogenic stages, it could not be used for the *tsl* mosaic analysis because the follicle cells are no longer associated with the egg after it is laid. Instead, the chorion that is secreted by the follicle cells was marked using a mutation called *fragile chorion (fch)*. Mothers that are homozygous mutants for *fch* produce embryos in which the chorion is very thin and easily disrupted. The presence of the *fch* phenotype in the chorion can then be used to determine the location within the follicular epithelium of the cells that synthesized that part of the chorion.

Finally, although *tsl* is required for the development of both anterior and posterior terminal regions, the structures formed at the two ends of the embryo are quite different. In order to compare the two poles directly, the experiment was done in a *bicoid (bcd)* background. *bcd* is required maternally for the development of anterior structures, and acts independently of the terminal class genes (Frohnhöfer and Nüsslein-Volhard, 1986). In embryos derived from mothers that are homozygous mutants for *bcd*, the head and thorax are deleted and replaced by a duplicated telson (Fig. 5a). Thus, these embryos consist of an abdomen with a telson at each end, making it possible to compare the development of two equivalent structures in the same embryo. It is worth noting that the mother of the embryo pictured in Figure 5a was also homozygous for the marker mutation, *fch*, indicating that the presence of *fch* in the maternal genotype does not interfere with the development of the telson.

The scheme for the mitotic recombination is shown below Table I. *bcd, fch*, and *tsl* lie on the right arm of the third chromosome, at map positions 48, 55, and 71, respectively. Irradiated females were homozygous for *bcd*, so all embryos exhibited the *bcd* phenotype. They were heterozygous for *fch* and *tsl*. After the induction of mitotic recombination by X-irradiation, which preferentially occurs near the centromere (García-Bellido, 1972), one progeny cell became wild type at the *fch* and *tsl* loci and was indistinguishable from the majority of cells that were heterozygous for these two recessive mutations. Its sister cell became homozygous for all three mutations, and its progeny formed a clone of *fch tsl* cells. Control embryos were derived from nonirradiated females of the same genotype. Embryos were scored for the presence of *filzkörper*, a telson structure that is dependent upon *tsl* function.

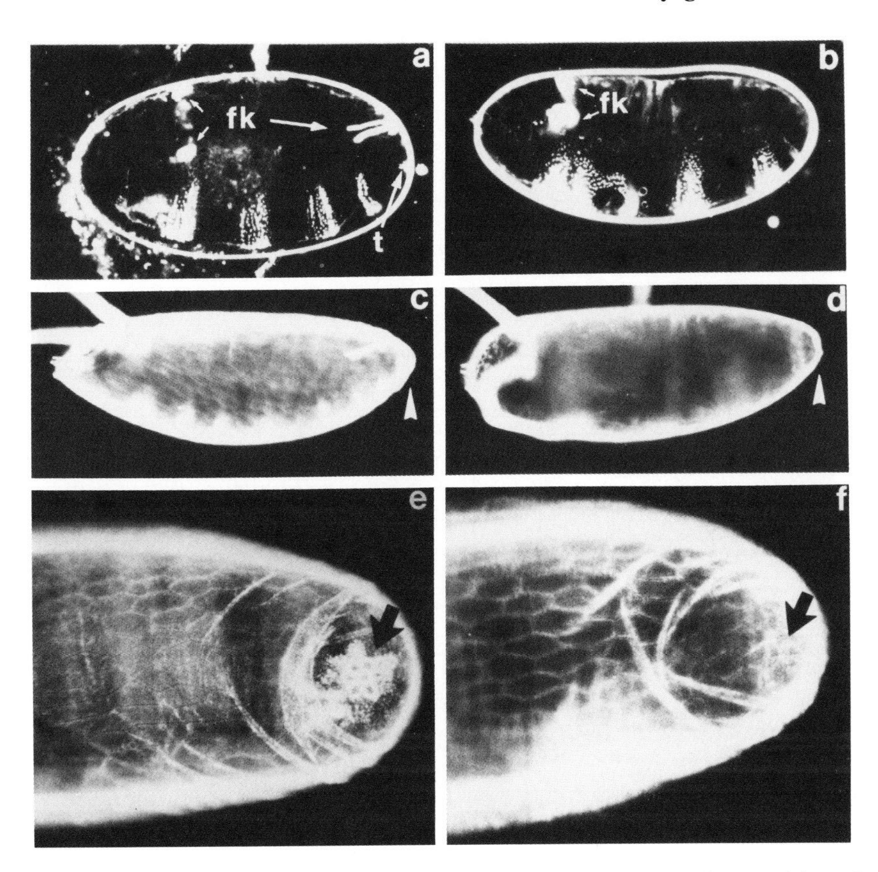

Fig. 5. Cuticular and chorion phenotypes of embryos derived from follicles containing *tsl* clones. Dark-field photographs. Anterior is to the left. Maternal genotypes: (**a**) $bcd^{E1}fch^{055}/bcd^{E1}fch^{055}$, (**b–f**) $bcd^{E1}fch^{267}tsl^{691}/bcd^{E1}fch^{+}tsl^{+}$. Embryos in b, d, and f were produced by mothers irradiated as larvae (b,d), or as adults (f). Embryos in c and e are controls obtained from nonirradiated females. (**a**) *bcd fch* phenotype. The head and thorax are deleted and replaced by a duplicated telson. *Filzkörper* (fk) are present at both ends of the embryo. (**b**) Cuticular phenotype of experimental embryo. Telson structures (fk) are present at the anterior, but not at the posterior pole. (**c,e**) Chorion of control embryos. Telson and aeropyle (c, arrowhead) are present. Forming the aeropyle are the bright imprints of seven central cells (e, arrow). (**d**) Chorion of experimental embryo in which the posterior telson is deleted. Aeropyle is absent (arrowhead). (**f**) Chorion of experimental embryo in which the posterior telson is deleted. The imprints of the central cells are absent (arrow). (Reprinted by permission from Nature, vol. 346, pp. 660–663. Copyright © 1990 Macmillan Magazines Ltd.)

This analysis indicated that it is possible to independently influence the development of the two ends of the embryo after the induction of *tsl* clones in the follicular epithelium. In the embryo shown in Figure 5b, which was derived from an irradiated mother, the anterior telson is normal but the posterior telson is completely absent. This result is consistent with a model in which there

TABLE I.
The Induction of *fch tsl* Follicle Cell Clones

Type of irradiation[a]	No. of embryos scored	No. of clones identified	No. of clones with chorion defects
Larval	7,940	5	4
Adult	3,020	2	2
None	7,180	0	—

[a]Females were irradiated with 1,050 R.

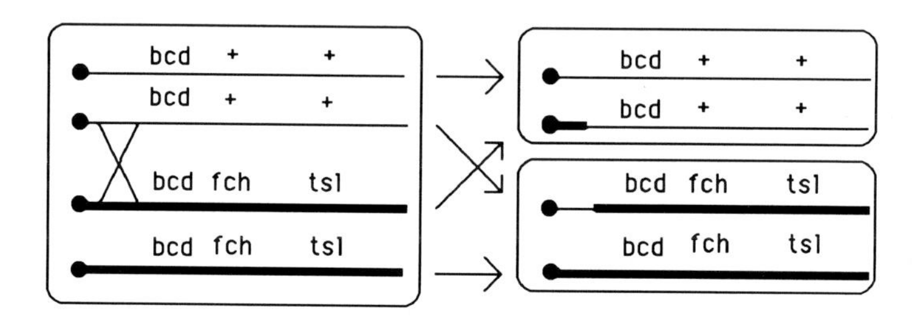

is a separate source of *tsl* for each end of the embryo, and argues against the idea that the entire follicular epithelium is required to synthesize *tsl*.

Among almost 11,000 experimental embryos scored (Table I), seven were detected in which the posterior end exhibited a complete *tsl* phenotype, while the anterior end was wild type with respect to *tsl*. Importantly, two of these seven embryos were derived from mothers that were irradiated as adults, indicating that quite small clones, perhaps consisting of 16 or fewer cells, are capable of producing a complete *tsl* phenotype at one end of the embryo. This implies that only a small number of follicle cells are required to express *tsl*.

Finally, to determine the location of the clones within the follicular epithelium, the chorions of these seven embryos were examined for the presence of the *fch* phenotype. Six were associated with chorion defects, indicating the presence of a *fch* clone in the follicle. In three of the embryos, the clones were quite large and the chorion had become disrupted. In the remaining three cases, however, including the two produced by adult irradiation, the clones were small and the chorion remained associated with the embryo. In all three, the chorion phenotype indicated that the clones had been located at the posterior end of the follicle. The posterior pole of the wild-type *Drosophila* chorion is characterized by the aeropyle, a protruberant structure formed by the imprints of the 6–15 small central cells surrounded by the imprints of 10–15 peripheral cells (Fig. 5c,e) (Margaritas et al., 1980). The central and peripheral cell imprints are distinct both from one another and from the imprints of the cells that form the main body of the chorion. In the three embryos with small clones, the central cell imprints were absent (Fig. 5d,f), and the peripheral cell imprints were deleted to variable degrees, while the main body of the chorion appeared unaffected. Based on the number of central and peripheral cell imprints remaining, these observations suggest a clone size varying from a minimum of 6 to a maximum of 30 cells. This result confirms the interpretation of the adult irradiation data, implying that *tsl* expression is required in only a small number of follicle cells (<2.5% of the follicular epithelium). Further, it indicates that this small subpopulation of follicle cells must be located at the poles of the follicle.

The results of this mosaic study support a model in which *tsl* is expressed by a small subpopulation of follicle cells present at the poles of the oocyte, and thus *tsl* may encode a localized signal that activates the *tor* receptor. This type of analysis examines the requirement for *tsl*, and does not necessarily reflect its pattern of expression. However, based on our present knowledge, the simplest interpretation of these results suggests that the synthesis of *tsl* is spatially regulated. Further, although these results are consistent with the idea that *tsl* itself encodes a ligand for *tor*, they are also compatible with a model in which *tsl* acts at another level, perhaps to regulate expression of the postulated ligand. Further investigation will be required to distinguish these possibilities.

The functions of *trk*, *fs(1)ph*, and *fs(1)N* in this pathway have not been determined. One possibility that we have considered is that they may be involved in promoting the formation of the polar subpopulations in the follicle. If these cells failed to differentiate, it is possible that the progeny of mutant females would exhibit a *tsl* phenotype. However, using specific enhancer trap insertions to mark the polar subpopulations, we have found that these cells are present in the ovaries of females homozygous for *trk*, *fs(1)ph*, and *fs(1)N* (unpublished observations). This suggests that these genes are likely to function further downstream, and may be involved in transmitting the signal from the follicle cells to the egg.

CONCLUSIONS

Although the individual roles played by all the genes in this pathway have not yet been elucidated, current information permits us to propose the following model. During oogenesis a localized signal, encoding spatial information required for the development of the terminal anlagen of the *Drosophila* embryo, is transmitted from the polar follicle cells to the germline-derived oocyte. It is presumably stored, perhaps in the vitelline membrane, until embryogenesis commences and it is released and becomes competent to activate the *tor* receptor. The signal is then transduced across the embryonic membrane by *tor*, resulting in a localized cytoplasmic kinase cascade involving the product of the gene *l(1)ph*. The function of *l(1)ph* is to activate, either directly or indirectly, downstream gene products that are responsible for the expression of zygotic target genes such as *tll* and *hkb*. These zygotic genes, which may themselves act as transcriptional regulators, initiate a specific program of zygotic gene expression that ultimately leads to the formation of head and tail structures of the larva. An intriguing aspect of this pathway of embryonic development is that spatial information, contained within the organization of the follicle, is communicated from the follicle cells to the oocyte. In this respect, terminal differentiation differs from the other two systems of anterior–posterior development and more closely resembles the establishment of the dorsal–ventral axis (see Schüpbach et al., Chapter 10, this volume). Thus, the formation of pattern at the ends of the embryo is dependent upon the establishment of polarity and the differentiation of subpopulations within the somatic follicular epithelium.

ACKNOWLEDGMENTS

We are grateful to Stephen Kerridge for providing transformant lines with β-galactosidase expression in follicle cells. We thank Peter Krolla for expert technical assistance, Wolfgang Driever for generous advice on antibody staining, Helen Doyle and Frank Sprenger for comments on the manuscript, and Angela Dressel and Roswitha Grömke-Lutz for photographic assistance. This work was supported by the Deutsche Forschungsgemeinschaft.

REFERENCES

Ambrosio L, Mahowald AP, Perrimon N (1989a): *l(1)pole hole* is required maternally for pattern formation in the terminal regions of the embryo. Development 106:145–158.

Ambrosio L, Mahowald AP, Perrimon N (1989b): Requirement of the *Drosophila raf* homologue for *torso* function. Nature 342:288–291.

Brower DL, Smith RJ, Wilcox M (1981): Differentiation within the gonads of *Drosophila* revealed by immunofluorescence. J Embryol Exp Morphol 63:233–242.

Casanova J, Struhl G (1989): Localized surface activity of *torso*, a receptor tyrosine kinase, specifies terminal body pattern in *Drosophila*. Genes Dev 3:2025–2038.

Degelmann A, Hardy PA, Perrimon N, Mahowald AP (1986): Developmental analysis of the torso-like phenotype in *Drosophila* produced by a maternal-effect locus. Dev Biol 115:479–489.

Fasano L, Kerridge S (1988): Monitoring positional information during oogenesis in adult *Drosophila*. Development 104:245–253.

Frohnhöfer HG, Nüsslein-Volhard C (1986): Organization of anterior pattern in the *Drosophila* embryo by the maternal gene *bicoid*. Nature 324:120–125.

García-Bellido A (1972): Some parameters of mitotic recombination in *Drosophila melanogaster*. Mol Gen Genetics 115:54–72.

Grossniklaus U, Bellen HJ, Wilson C, Gehring WJ (1989): P-element-mediated enhancer detection applied to the study of oogenesis in *Drosophila*. Development 107:189–200.

Jürgens G (1987): Segmental organisation of the tail region in the embryo of *Drosophila melanogaster*. Roux Arch Dev Biol 196:141–157.

Jürgens G, Lehmann R, Schardin M, Nüsslein-Volhard C (1986): Segmental organisation of the head in the embryo of *Drosophila melanogaster*. Roux Arch Dev Biol 195:359–377.

King RC (1970): "Ovarian Development in *Drosophila melanogaster*." New York: Academic Press.

Klingler M (1989): Die Funktion des Gens *torso* bei der Determination terminaler Anlagen im *Drosophila* Embryo. Thesis, Eberhard-Karls-Universität, Tübingen.

Klingler M, Erdélyi M, Szabad J, Nüsslein-Volhard C (1988): Function of *torso* in determining the terminal anlagen of the *Drosophila* embryo. Nature 335:275–277.

Lehmann R, Nüsslein-Volhard C (1986): Abdominal segmentation, pole cell formation, and embryonic polarity require the localized activity of *oskar*, a maternal gene in *Drosophila*. Cell 447:141–152.

Margaritas LH, Kafatos FC, Petri WH (1980): The eggshell of *Drosophila melanogaster*. I. Fine structure of the layers and regions of the wild-type eggshell. J Cell Sci 43:1–35.

Mark GE, MacIntyre RJ, Digan ME, Ambrosio L, Perrimon N (1987): *Drosophila melanogaster* homologs of the raf oncogene. Mol Cell Biol 7:2134–2140.

Morrison DK, Kaplan DR, Escobedo JA, Rapp UR, Roberts TM, Williams LT (1989): Direct activation of the serine/threonine kinase activity of raf-1 through tyrosine phosphorylation by the PDGF B-receptor. Cell 58:649–657.

Nishida Y, Hata M, Ayaki T, Ryo H, Yamagata M, Shimizu K, Nishizuka Y (1988): Proliferation of both somatic and germ cells is affected in the *Drosophila* mutants of raf proto-oncogene. EMBO J 7:775–781.

Nüsslein-Volhard C, Frohnhöfer HG, Lehmann R (1987): Determination of anteroposterior polarity in *Drosophila*. Science 238:1675–1681.

O'Kane CJ, Gehring WJ (1987): Detection *in situ* of genomic regulatory elements in *Drosophila*. Proc Natl Acad Sci USA 84:9123–9127.

Perrimon N, Engstrom L, Mahowald AP (1985): A pupal lethal mutation with a paternally influenced maternal effect on embryonic development in *Drosophila melanogaster*. Dev Biol 110:480–491.

Perrimon N, Gans M (1983): Clonal analysis of the tissue specificity of recessive female-sterile mutations of *Drosophila melanogaster* using a dominant female-sterile mutation *Fs(1)K1237*. Dev Biol 100:365–373.

Perrimon N, Mohler D, Engstrom L, Mahowald AP (1986): X-linked female-sterile loci in *Drosophila melanogaster*. Genetics 113:695–712.

Pignoni F, Baldarelli RM, Steingrimsson E, Diaz RJ, Patapoutian A, Merriam JR, Lengyel JA (1990): The *Drosophila* gene *tailless* is expressed at the embryonic termini and is a member of the steroid receptor superfamily. Cell 62:151–163.

Rüther U, Müller-Hill B (1983): Easy identification of cDNA clones. EMBO J 2:1791–1794.

Schüpbach T, Wieschaus E (1986a): Maternal-effect mutations altering the anterior–posterior pattern of the *Drosophila* embryo. Roux Arch Dev Biol 195:302–317.

Schüpbach T, Wieschaus E (1986b): Germline autonomy of maternal-effect mutations altering the embryonic body pattern of *Drosophila*. Dev Biol 113:443–448.

Simon JA, Sutton CA, Lobell RB, Glaser RL, Lis JT (1985): Determinants of heat shock-induced chromosome puffing. Cell 40:805–817.

Sprenger F, Stevens LM, Nüsslein-Volhard C (1989): The *Drosophila* gene *torso* encodes a putative receptor tyrosine kinase. Nature 338:478–483.

Stevens LM, Frohnhöfer HG, Klingler M, Nüsslein-Volhard C (1990): Localized requirement for *torso-like* expression in follicle cells for development of terminal anlagen of the *Drosophila* embryo. Nature 346:660–663.

Strecker TR, Halsell SR, Fisher WW, Lipshitz HD (1989): Reciprocal effects of hyper- and hypoactivity mutations in the *Drosophila* pattern gene *torso*. Science 243:1062–1066.

Strecker TR, Kongsuwan K, Lengyel JA, Merriam JR (1986): The zygotic mutant tailless affects the anterior and posterior ectodermal regions of the *Drosophila* embryo. Dev Biol 113:64–76.

Szabad J, Hoffman G (1989): Analysis of follicle-cell functions in *Drosophila*: The *Fs(3)Apc* mutation and the development of chorionic appendages. Dev Biol 131:1–10.

Ullrich A, Schlessinger J (1990): Signal transduction by receptors with tyrosine kinase activity. Cell 61:203–212.

Weigel D, Jürgens G, Klingler M, Jäckle H (1990): Two gap genes mediate maternal terminal pattern information in *Drosophila*. Science 248:495–498.

Wieschaus E, Audit C, Masson M (1981): A clonal analysis of the roles of somatic cells and germline during oogenesis in *Drosophila*. Dev Biol 88:92–103.

Cell-Cell Interactions in Early Development, pages 163–174

10. Dorsoventral Signaling Processes in *Drosophila* Oogenesis

Trudi Schüpbach, Robert J. Clifford, Lynn J. Manseau, and James V. Price
Department of Molecular Biology, Princeton University, Princeton, New Jersey 08544

INTRODUCTION

The mature egg of *Drosophila melanogaster* has a well-defined anterior–posterior and dorsoventral pattern. When the egg is laid, asymmetries are visible both in the shape of the egg itself and in the patterning of its eggshell, an extracellular multilayered structure that is secreted by the follicle cells during oogenesis. The embryo which develops inside the egg forms its anterior–posterior and dorsoventral axis in exact alignment with the axes of the eggshell. Embryonic patterning in *Drosophila* depends on the localized activities of certain maternal gene products such as *bicoid, torso, nanos*, and *dorsal*, all of which are produced in the germline during oogenesis (Nüsslein-Volhard et al., 1987; Driever and Nüsslein-Volhard, 1988a,b; Klingler et al., 1988; Roth et al., 1989; Steward et al., 1988; Strecker et al., 1989). The correspondence between embryonic development and eggshell pattern reflects therefore an alignment between the localization or local activation of the maternal embryonic determinants and the secretion pattern of the overlying follicle cell epithelium. Given this invariant alignment, it is very likely that a mechanism operates during oogenesis that coordinates the two patterns. It is theoretically possible that a single event simultaneously governs patterning in the follicle cell epithelium and localization processes in the germline (e.g., localized factors produced by the maternal organism might polarize both the follicle cell epithelium and the egg). An alternate, and more likely, possibility is that primary patterning processes occur first either in the germline or follicle cell epithelium, and the polarity of that cell type subsequently controls or orients the patterning processes in the other cell type.

Much of the primary patterning steps in anterior–posterior pattern occur early in oogenesis, at a stage when the oocyte first assumes its characteristic posterior position relative to the nurse cells. The initial anterior–posterior arrangement of nurse cells and oocyte might provide the basis for the polarized localization of anterior–posterior maternal determinants within the egg.

It might also control the anterior–posterior differences in the follicle cell epithelium. Even less is known about the time and the primary mechanism that first establishes dorsoventral asymmetries in the developing egg chamber. The first morphological difference between dorsal and ventral side of the egg chamber becomes apparent when the oocyte nucleus assumes a dorsal position in midoogenesis (stage 8/9; King, 1970). Subsequently, the follicle cells on the dorsal side, which are located directly over the nucleus, assume a more highly columnar shape and become more densely packed, as compared to the follicle cells on the ventral side. These differences in follicle cell shape and arrangement become further elaborated at later stages of oogenesis, and are reflected in the pattern of the mature eggshell.

MUTATIONS THAT AFFECT THE DORSOVENTRAL PATTERN DURING OOGENESIS

Screens for female-sterile mutations have lead to the isolation of mutations at several loci that simultaneously alter the dorsoventral pattern of both the eggshell and the embryo. Mutations at the loci *gurken, torpedo* (Schüpbach, 1987), and *cornichon* (Ashburner et al., 1990) cause a ventralization of eggshell and embryo. This phenotype is characterized by an expansion of ventral cell fates and a loss of dorsal cell fates (Fig. 1), both in the follicle cell epithelium as well as in the embryos forming inside these eggs. Mutations at the loci *fs(1)K10* (Wieschaus et al., 1978) *spire*, and *cappuccino* (Manseau and Schüpbach, 1989) dorsalize the eggshell and embryo (Fig. 1).

Mosaic animals were used to determine the ovarian cell type in which wild-type activity of these six genes is required for normal development. Transplantation of germ cells or the induction of mitotic recombination in the germline yielded mosaic egg chambers in which the oocyte and its sister nurse cells were of either mutant or wild-type genotype, surrounded by follicle cells of the reciprocal genotype. The analysis of such germline mosaics demonstrated that the genes *fs(1)K10, gurken, spire, cappuccino*, and *cornichon* are required in the germline (Wieschaus et al., 1978; Schüpbach, 1987; Manseau and Schüpbach, 1989; see also Table I). In those cases, the phenotype of both the eggshell and the embryo was dictated by the genotype of the germline. The follicle cells in these mosaics secreted a ventralized eggshell if the underlying germline was mutant for *gurken* or *cornichon*, or a dorsalized eggshell, if the underlying eggshell was mutant for *fs(1)K10, cappuccino*, or *spire*, although in all cases the follicle cells themselves were of wild-type genotype. These genes identify dorsoventral patterning processes that take place in the germline, yet somehow control the secretion and migration pattern of the overlying follicle cells.

The gene *torpedo*, in contrast, tested as somatic dependent (Schüpbach, 1987). Phenotypically mutant eggs and embryos were produced by mosaics in which the germline was wild-type but the surrounding soma, including the

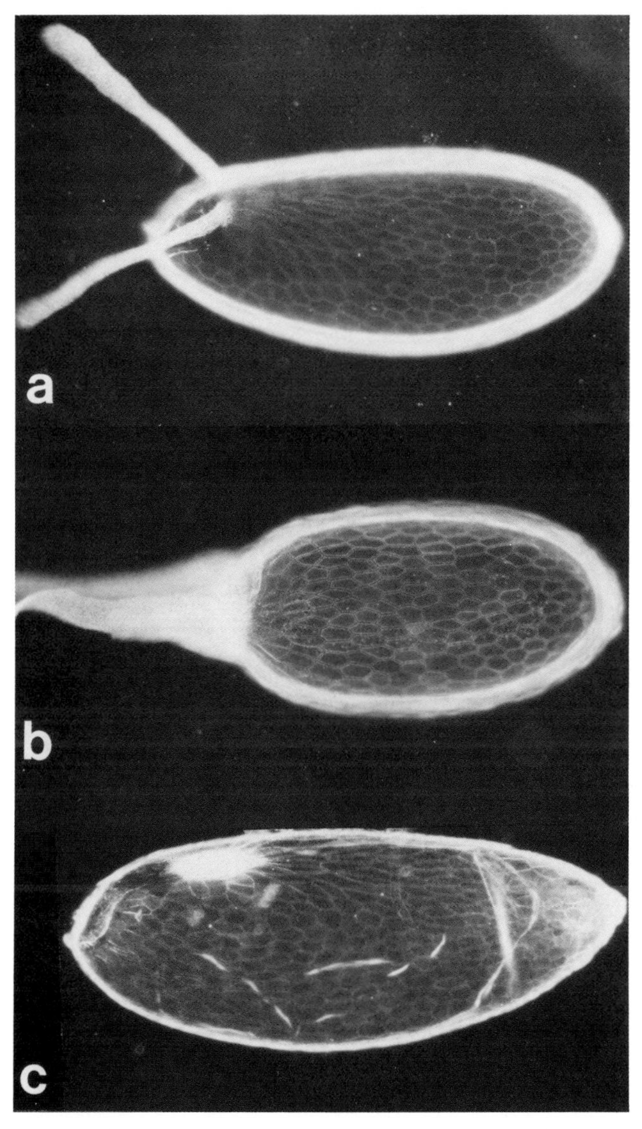

Fig. 1. Wild-type, dorsalized, and ventralized eggshell phenotypes. (**a**) Wild type. (**b**) Egg from female homozygous for *fs(1)K10*. (**c**) Egg from female homozygous for *gurken*^{HK}. During oogenesis, the eggshell is secreted by the follicle cell epithelium. The imprints left by the individual follicle cells can be seen as hexagons on the final eggshell. In the wild type, two specialized groups of follicle cells migrate out of the dorsal surface and produce the two dorsal appendages. In the extreme dorsalized egg chamber, too many follicle cells migrate out of the epithelium and produce dorsal appendage material. The resulting egg (b) has therefore an excess of dorsal appendage material around its anterior circumference and a small number of follicle cell imprints on its main body. In the ventralized egg chamber, the two migrating populations are fused on the dorsal midline and drastically reduced in cell number. The resulting egg (c) shows only a patch of dorsal appendage material, whereas an excess of follicle cells have contributed to the main body of the eggshell.

TABLE I.
Germline Mosaics Produced by Radiation-Induced Mitotic Recombination

Gene tested	Genotype of irradiated females[a]	No. of females tested	No. of females with homozygous germline clones	No. of eggs derived from homozygous clones	Phenotype of eggs and embryos derived from clones
windbeutel	$wind^{RP}/$ $Fs(2)D$	573	8	391	Normal eggs, normal embryos
cactus	$cact^{PD}/$ $Fs(2)1^{b}$	443	4	50	Normal eggs, ventralized embryos
cornichon	$cni^{AA112}/$ $Fs(2)X10$	1,012	6	15	Ventralized eggs, ventralized embryos

[a]Females were irradiated with approximately 1,300 R during larval stages and tested as described in Schüpbach and Wieschaus (1986) and Schüpbach (1987).
[b]Szabad et al. (1987).

follicle cells, were mutant. In the complementary class of mosaics, where mutant germ cells were surrounded by wild-type follicle cells, normal eggs and embryos were produced. These results demonstrated that genetic changes in the soma, including the follicle cells, can not only affect the behavior of the follicle cells themselves, but can also alter the pattern of the embryo that later develops inside the resulting eggs. Some form of dorsoventral pattern information must therefore be transmitted from the soma to the germline and mutations at *torpedo* interfere at some level with this process.

Molecular analysis of the *torpedo* locus demonstrated that it encodes the *Drosophila* homologue of the vertebrate epidermal growth factor receptor (DER; Schejter and Shilo, 1989; Price et al., 1989). This gene product belongs therefore to the larger family of receptor tyrosine kinases, transmembrane proteins that have large extracellular domains with cystein-rich motifs, and intracellular domains with tyrosine kinase activity. This structure suggests that *torpedo* plays a very direct role in the cell communication processes between germline and follicle cells during *Drosophila* oogenesis. In situ hybridization to tissue sections had shown that the gene is expressed at a relatively high level in follicle cells during oogenesis (Kammermeyer and Wadsworth, 1987), a result that corresponded very well with the somatic requirement for *torpedo* activity observed in mosaic egg chambers.

The female-sterile alleles of *torpedo* are not complete loss-of-function mutations. Genetic analysis of the locus revealed that this gene is required at several stages of the life cycle, in different tissues, presumably mediating several different cell communication processes (Clifford and Schüpbach, 1989; Schejter and Shilo, 1989; Baker and Rubin, 1989; Zak et al., 1990). Among the mutant alleles, several classes can be distinguished that preferentially (although not exclusively) affect one tissue type. Further molecular analysis should reveal what specific alterations in the gene cause this observed specificity.

Given that the *torpedo/DER* gene encodes a transmembrane receptor with an intracellular tyrosine kinase domain, the simplest model for its action in oogenesis would be to assume that the protein product is present in the membrane of the follicle cells during oogenesis and binds a ligand molecule that is produced by the germline. Mutations at *gurken, cornichon*, and *torpedo* produce a very similar, ventralized phenotype, which remains unaltered in double mutants between these genes. This suggests that the three genes participate in the same dorsoventral patterning process, since loss of one individual component produces the same effect as losing all three. Given that *gurken* and *cornichon* act in the germline, one possibility would be that those two genes might be directly involved in the production of a dorsalizing signal that is secreted by the germline and binds to, and thus activates, the *torpedo* receptor in the follicle cells. Reductions in the signal, or the receptor, will result in the same ventralized phenotype.

Mutations at *fs(1)K10* produce a very constant, dorsalized phenotype in eggshell and embryo. In double mutants between *fs(1)K10* and either *gurken, torpedo*, or *cornichon*, a ventralized phenotype is produced. Therefore, for a mutation at *fs(1)K10* to produce a dorsalized eggshell and a dorsalized embryo, the wild-type activity of the genes *gurken, torpedo*, and *cornichon* is required. These results can be most easily explained in a model where *fs(1)K10* acts as an upstream negative regulator of the dorsal signal-producing genes. In the absence of wild-type *fs(1)K10* product, the germline signal becomes deregulated and would overactivate *torpedo*, and thus dorsalize eggshell and embryo.

Mutations at *spire* and *cappuccino* (Manseau and Schüpbach, 1989) can also give rise to strongly dorsalized eggshells and embryos, although their eggshell phenotype is always variable. Usually some of the eggs produced by females mutant for *spire* and *cappuccino* are of normal morphology, and a fraction of the eggs may even be ventralized. In addition to this variable effect on dorsoventral patterning, *spire* and *cappuccino* also affect anterior–posterior patterning in the embryo. Eggs from mutant mothers always lack the posterior pole plasm, which contains the determinants for the germline and a particular class of determinants for abdominal segmentation. Such embryos therefore lack pole cells and usually develop severe defects in abdominal segmentation in addition to the more variable alterations in dorsoventral patterning. Genetic studies have suggested that the abdominal segmentation phenotype of *spire* and *cappuccino* reflects a mislocalization of posterior determinants rather than a failure to synthesize them (Manseau and Schüpbach, 1989). It is therefore well possible that the dorsoventral effects of *cappuccino* and *spire* also reflect a role of these genes in localization processes that restrict dorsoventral pattern information to one side of the developing egg chamber. These two loci are required in the germline, and, in genetic tests, appear to function upstream of *gurken* and *torpedo* with respect to their effect on dorsoventral patterning.

CURRENT MODEL

These observations have led to a current working model (Fig. 2), in which the signal that activates the *torpedo* receptor in the follicle cells, and acts to promote dorsal cell fates in eggshell and embryo, is produced with the direct help of *gurken* and *cornichon*. The activity of these genes is (directly or indirectly) negatively controlled by the action of *fs(1)K10*, such that only a limited amount of active signal is produced in the egg chamber. Molecular analysis of *fs(1)K10* has shown that this gene encodes a protein of unknown function that localizes almost exclusively to the oocyte nucleus (Proust et al., 1988). It might therefore function as a direct negative transcriptional regulator of any of the genes involved in producing the dorsal signal. Given that the *fs(1)K10* protein product is strictly localized to the oocyte nucleus, it is attractive to speculate that the oocyte nucleus might be the source of the dorsal signal. The dorsalization seen in mutations at *fs(1)K10* is always more extensive in the anterior part of the embryo (Wieschaus, 1979), which is close to the oocyte nucleus, and thus may reflect a local high level of dorsal signal produced in the absence of *fs(1)K10*. The negative action of *spire* and *cappuccino*, on the other hand, might serve to restrict the active signal to the dorsal side of the egg chamber. In the absence of these two genes, the signal from the oocyte nucleus, might, for instance, be more free to diffuse to the ventral side. A more random diffusion of the signal might account for the variable dorsalization phenotype seen in these mutants.

The *torpedo/DER* receptor gene product is initially present in all follicle cells (Kammermeyer and Wadsworth, 1987). With a limited amount of dorsal

---

Fig. 2. Model for signaling processes during oogenesis. As indicated in the small schematic drawing at the top of the figure, only a section of an egg chamber with follicle cells and oocyte is represented in detail. The oocyte nucleus is acentrically located on the dorsal side of the oocyte. A dorsal signal is produced in the oocyte nucleus with the help of *gurken* and *cornichon*. The product of the *fs(1)K10* locus negatively controls the production of this signal. The products of *spire* and *cappuccino* help to restrict the active signal to the dorsal side of the egg chamber. The signal is secreted by the oocyte and is bound by the *torpedo/DER* receptor tyrosine kinase. The activated kinase promotes a dorsal follicle cell fate. It also negatively regulates the activity of the three somatically active genes *windbeutel, nudel*, and *pipe*. On the ventral side, in the absence of kinase activity, these three genes produce a new signal in the follicle cells. This new signal is transmitted back to the egg and promotes ventral embryonic cell fates via the activities of the *Toll-dorsal* cascade of genes.

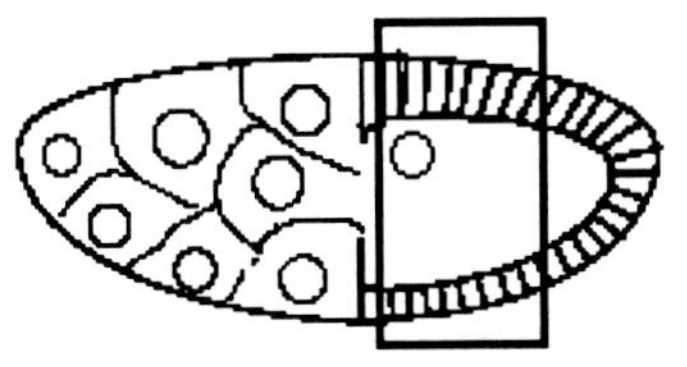

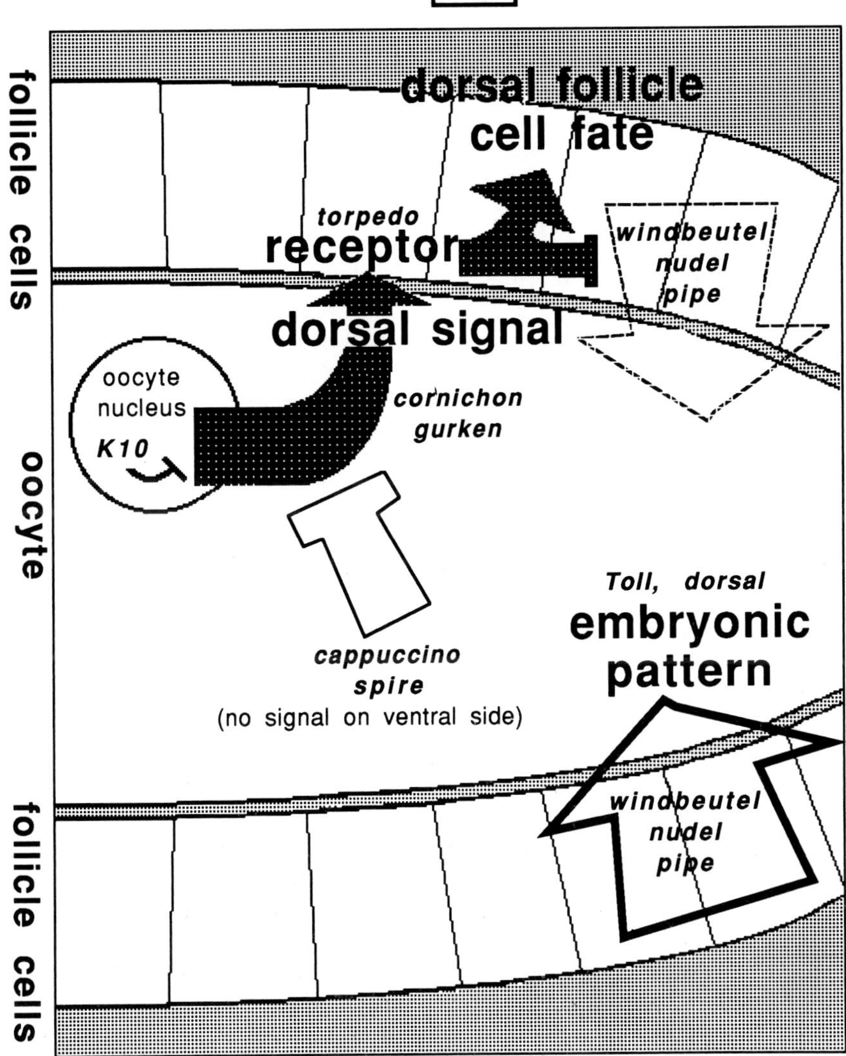
follicle cells
oocyte
follicle cells
dorsal follicle
cell fate
torpedo
receptor
windbeutel
nudel
pipe
dorsal signal
oocyte
nucleus
K10
cornichon
gurken
cappuccino
spire
(no signal on ventral side)
Toll, dorsal
embryonic
pattern
windbeutel
nudel
pipe

signal, which is possibly restricted to the dorsal side of the egg, activation of the receptor kinase might only occur on the dorsal side of the follicle cell epithelium and there, possibly via a second messenger system, serve to implement dorsal follicle cell fates. *torpedo* activity is also required for the establishment of dorsal embryonic cell fates. In *Drosophila*, dorsoventral cell fates of the embryo at the blastoderm stage depend on the concentration of the protein product of the *dorsal* locus present in the zygotic nuclei (Nüsslein-Volhard et al., 1980; Steward et al., 1988; Roth et al., 1989; Steward, 1989). The RNA product of the *dorsal* locus is produced in the nurse cells during oogenesis, transported into the oocyte, and uniformly distributed in the egg. After fertilization the RNA is translated and the protein assumes a high concentration in the zygotic nuclei on the ventral side of the early embryo, whereas it remains mainly cytoplasmic at the dorsal side of the embryo. Eleven additional maternal-effect genes have been identified in *Drosophila*, which are required for the establishment of this dorsoventral gradient of *dorsal* protein (Anderson and Nüsslein-Volhard, 1984; Anderson et al., 1985; Schüpbach and Wieschaus, 1989; Roth et al., 1989; Steward, 1989). With the exception of *cactus*, the genes in this group are thought to act in a cascade, which leads to the differential translocation of the *dorsal* protein from the cytoplasm to the nucleus. In double mutants between *torpedo* and various members of this group of genes (in particular, between *torpedo* and either *dorsal, Toll, windbeutel, pipe*, or *nudel*), the embryos show a dorsalized phenotype, whereas the eggshell remains ventralized (Schüpbach, 1987; Hawkins and Schüpbach, unpublished). This indicates that the ventralization of the embryo produced by mutations at *torpedo* requires the normal function of the *Toll-dorsal* cascade of genes. With respect to embryonic pattern, *torpedo* acts as a negative regulator of this cascade, whereas the effect of *torpedo* on eggshell patterning is independent of this group of genes. When tested in germline mosaics, the three genes *windbeutel* (Table I; see also Stein et al., 1991), *nudel*, and *pipe* (Stein et al., 1991) were found to be somatic dependent, and therefore appear to operate in the same cell type as *torpedo* itself. It is possible that one of these three genes is a target for negative regulation by the *torpedo/DER* kinase activity. In the absence of *torpedo* activation (on the ventral side of the egg chamber), the three genes might be highly active in the follicle cells and produce a new signal that is transmitted to the germline and eventually controls the *Toll-dorsal* cascade. On the dorsal side, where *torpedo* is presumably activated, this activity would interfere with the production of the new signal, and thus prevent the *Toll-dorsal* cascade from becoming active (for further discussion, see also Manseau and Schüpbach, 1989; Stein et al., 1991). This negative effect of *torpedo/DER* activity on the three somatically required genes could be rather indirect and might result as a secondary effect of regulation of transcriptional activity in the follicle cells (i.e., establishment of dorsal follicle cell fate) by the receptor kinase. Alternatively, one of the gene products might constitute a more direct target of the receptor kinase activity. In hypomorphic *torpedo* mutations, we have observed that the degree of ventralization of eggshell and embryo is not always absolutely identical. Some eggs with fused dorsal appendages may occasionally give rise to an almost normal embryo, whereas eggs with two dorsal appendages may sometimes give rise to partially ventralized embryos. This may indicate that the effect of *torpedo* on the embryo is not simply a secondary reflection of the primary effect of *torpedo* on dorsal follicle cell fate. Rather, it suggests that *torpedo* may have a more direct effect on embryonic patterning and control a process that is at least partially independent from its effect on follicle cell patterning. One way this separation might occur is if *torpedo* had multiple substrates in follicle cells, some of which would control follicle cell behavior and others' embryonic patterning. Such a dual effect of *torpedo* would not be unexpected, given that its vertebrate homologues have been shown to interact with multiple substrate molecules (for a recent review, see Ullrich and Schlessinger, 1990).

FUTURE DIRECTIONS

Timing of the Dorsoventral Signaling Processes

It is presently not known when these various signaling processes take place in oogenesis. The first morphological abnormalities in *fs(1)K10, spire, cappuccino, cornichon, torpedo*, and *gurken* become apparent in egg chambers at stage 10B, when in normal egg chambers the difference in shape between the follicle cells on the dorsal and ventral sides becomes distinct. The establishment of dorsoventral asymmetries in the follicle cells has therefore to occur before this stage. The mutant alleles at *gurken*, when in combination with chromosomal deficiencies for the region 29CD, cause an arrest of oogenesis at stage 8, before the oocyte nucleus moves to its acentric, dorsal position. It is therefore possible that the dorsoventral signaling pathway blocked by mutations at *gurken* occurs at or before stage 8. (Alternatively, it is possible that *gurken* participates in additional processes in oogenesis, and that the arrest of the egg chambers at stage 8 is due to a block in a process not directly related to the dorsoventral signaling cascade.) Unfortunately, the complete loss-of-function phenotype of *torpedo* in the egg chamber is not known. Since null alleles at the locus cause zygotic embryonic lethality, it has not been possible to study the effect of such alleles on follicle cells in adult ovaries. *fs(1)K10*, on the other hand, is expressed relatively early in oogenesis, and its protein product is detectable in the oocyte nucleus from stage 8 onward (Haenlin et al., 1987; Proust et al., 1988). However, since *fs(1)K10* acts as a negative regulator of the germline signal, this observation is not necessarily an indication of when in normal development the signal is produced and active. The presence of *fs(1)K10* at certain stages of oogenesis may be necessary to prevent the signal from being produced at that particular time in development.

Further analysis, and ultimately the molecular identification of the signal that binds to *torpedo*, will be required to determine when in oogenesis the dorsoventral signal is first produced. This will also determine whether or not the signal is produced in the oocyte nucleus, and how it is normally distributed within the egg chamber.

Initial Establishment of a Dorsoventral Asymmetry

If the signal is indeed restricted to the dorsal side of the oocyte, a further question will be to determine the primary mechanisms that create this initial asymmetry. Could there be a random mechanism that shifts the oocyte nucleus to one side of the egg chamber and thus establishes the primary dorsoventral asymmetry? Or are there yet-unknown early cues that determine the location of the future dorsal side of the egg and result in the shift of the nucleus toward that side?

Another unresolved problem is the question of why an initial dorsoventral difference in the germline is signaled to the overlying follicle cells and then at a later stage transmitted back to the germline. An ad hoc explanation could be that the initial signal that is produced in the germline may be very crude. Once transmitted to the follicle cell epithelium, the dorsoventral differences might be elaborated and become more graded, ultimately leading to the complex spatial pattern observed in the mature follicle cell epithelium. Such a finer, possibly graded, pattern might then be retransmitted to the oocyte at a later stage, when the oocyte has considerably increased in size. Analysis of the three somatically active genes *windbeutel, nudel*, and *pipe* may eventually establish whether these genes could be part of a gene network operating in the follicle cells that can establish a more refined pattern from a coarse initial signal.

ACKNOWLEDGMENTS

We would like to thank Eileen Shaddix for technical assistance in the production of germline mosaics. We are grateful to Eric Wieschaus for advice throughout the work, as well as for his comments on the manuscript and his help in preparing the figures. We thank Gordon Gray and Cubit Case for providing fly food media, and our colleagues at Princeton for stimulating discussions. This work was supported by a grant from the National Institutes of Health (GM40558) to T.S.

REFERENCES

Anderson KV, Nüsslein-Volhard C (1984): Information for the dorsal-ventral pattern of the *Drosophila* embryo is stored as maternal mRNA. Nature 311:223–227.

Anderson KV, Jürgens G, Nüsslein-Volhard C (1985): Establishment of dorsal-ventral polarity in the *Drosophila* embryo: Genetic studies on the role of the Toll gene product. Cell 42:779–789.

Ashburner M, Thomson P, Roote J, Lasko PF, Grau Y, El Messal M, Roth S, Simpson P (1990): The genetics of a small autosomal region of *Drosophila melanogaster* containing the structural gene for alcohol dehydrogenase. VII. Characterization of the region around the *snail* and *cactus* loci. Genetics 126:679–694.

Baker NE, Rubin GM (1989): Effect on eye development of dominant mutations in *Drosophila* homologue of the EGF receptor. Nature 340:150–153.

Clifford RJ, Schüpbach T (1989): Coordinately and differentially mutable activities of *torpedo*, the *Drosophila melanogaster* homolog of the vertebrate EGF receptor gene. Genetics 123:771–787.

Driever W, Nüsslein-Volhard C (1988a): A gradient of *bicoid* protein in *Drosophila* embryos. Cell 54:83–93.

Driever W, Nüsslein-Volhard C (1988b): The *bicoid* protein determines position in the *Drosophila* embryo in a concentration-dependent manner. Cell 54:95–104.

Haenlin M, Roos C, Cassab A, Mohier E (1987): Oocyte-specific transcription of *fs(1)K10*: A *Drosophila* gene affecting dorsal-ventral developmental polarity. EMBO J 6:801–807.

Kammermeyer KL, Wadsworth SC (1987): Expression of *Drosophila* epidermal growth factor homologue in mitotic cell populations. Development 100:201–210.

King RC (1970): "Ovarian Development in *Drosophila melanogaster*." New York: Academic Press.

Klingler M, Erdelyi M, Szabad J, Nüsslein-Volhard C (1988): Function of *torso* in determining the terminal anlagen of the *Drosophila* embryo. Nature 335:275–277.

Manseau LJ, Schüpbach T (1989): *Cappuccino* and *spire*: Two unique maternal-effect loci required for both the anteroposterior and dorsoventral patterns of the *Drosophila* embryo. Genes Dev 3:1437–1452.

Nüsslein-Volhard C, Lohs-Schardin M, Sander K, Cremer C (1980): A dorso-ventral shift of embryonic primordia in a new maternal-effect mutant of *Drosophila melanogaster*. Nature 283:474–476.

Nüsslein-Volhard C, Frohnhöfer HG, Lehmann R (1987): Determination of anteroposterior polarity in *Drosophila*. Science 238:1675–1681.

Price JV, Clifford RJ, Schüpbach T (1989): The maternal ventralizing locus *torpedo* is allelic to *faint little ball*, an embryonic lethal, and encodes the *Drosophila* EGF receptor homolog. Cell 56:1085–1092.

Proust E, Deryckere F, Roos C, Haenlin M, Pantesco V, Mohier E (1988): Role of the oocyte nucleus in determination of the dorsoventral polarity of *Drosophila* as revealed by molecular analysis of the K10 gene. Genes Dev 2:891–900.

Roth S, Stein D, Nüsslein-Volhard C (1989): A gradient of nuclear localization of the *dorsal* protein determines dorsoventral pattern in the *Drosophila* embryo. Cell 59:1189–1202.

Schejter ED, Shilo B-Z (1989): The *Drosophila* EGF receptor homolog (DER) is allelic to *faint little ball*, a locus essential for embryonic development. Cell 46:1091–1101.

Schüpbach T (1987): Germ line and soma cooperate during oogenesis to establish the dorsoventral pattern of egg shell and embryo in *Drosophila melanogaster*. Cell 49:699–707.

Schüpbach T, Wieschaus E (1986): Germline autonomy of maternal-effect mutations altering the embryonic body pattern of *Drosophila*. Dev Biol 113:443–448.

Schüpbach T, Wieschaus E (1989): Female sterile mutations on the second chromosome of *Drosophila melanogaster*. I. Maternal effect mutations. Genetics 121:101–117.

Stein D, Roth S, Vogelsang E, Nüsslein-Volhard C (1991). An activity present in the perivitelline space of the *Drosophila* egg determines the dorsoventral axis of the embryo. Cell 65:725–735.

Steward R (1989): Relocalization of the *dorsal* protein from the cytoplasm to the nucleus correlates with its function. Cell 59:1179–1188.

Steward R, Zusman SB, Huang LH, Schedl P (1988): The *dorsal* protein is distributed in a gradient in early *Drosophila* embryos. Cell 55:487–495.
Strecker T, Halsell SR, Fisher W, Lipshitz HD (1989): Reciprocal effects of hyper- and hypoactivity mutations in the *Drosophila* pattern gene *torso*. Science 243:1062–1066.
Szabad J, Erdelyi M, Szidonia J (1987): Characterization of *Fs(2)1*, a germ-line dependent dominant female sterile mutation of *Drosophila*. Acta Biol Hungarica 38:257–266.
Ullrich A, Schlessinger J (1990): Signal transduction by receptors with tyrosine kinase activity. Cell 61:203–212.
Wieschaus E (1979): *fs(1)K10*, a female sterile mutation altering the pattern of both the egg coverings and the resultant embryos in *Drosophila*. In Le Douarin N (ed): "Cell Lineage, Stem Cells, and Cell Determination," INSERM Symposium No. 10. Amsterdam: Elsevier, pp 291–302.
Wieschaus E, Marsh JL, Gehring W (1978): *fs(1)K10*, a germline-dependent female sterile mutation causing abnormal chorion morphology in *Drosophila melanogaster*. Roux Arch Dev Biol 184:75–82.
Zak NB, Wides R, Schejter ED, Raz E, Shilo B-Z (1990): Localization of the DER/*flb* protein in embryos: Implications on the *faint little ball* phenotype. Development 109:865–874.

Cell-Cell Interactions in Early Development, pages 175–201

11. Mesenchyme Cell Interactions in the Sea Urchin Embryo

Charles A. Ettensohn
Department of Biological Sciences, Carnegie Mellon University, Pittsburgh, Pennsylvania 15213

INTRODUCTION

Interactions between embryonic cells play a critical role in the specification of cell fates in all multicellular animals. During the embryogenesis of some organisms such as amphibians, mammals, and sea urchins, cell-cell interactions are especially important in regulating the expression of major tissue types and the organization of the body axis (Davidson, 1986; Gerhart et al., 1989). Even in those organisms in which cell diversification is believed to be primarily under the control of maternally derived determinants segregated to different blastomeres during cleavage (e.g., nematodes and ascidians), cellular interactions are critical for the determination of specific cell types (Sulston and White, 1980; Kimble, 1981; Nishida and Satoh, 1989; Sternberg and Horvitz, 1989). Recent studies have provided evidence that cell-cell signaling also plays a significant role in the specification of cell fates in plants (Siegel and Verbeke, 1989; Schiavone and Racusen, 1990).

The sea urchin embryo has been an important model for studying the specification of cell fates by cellular interactions. The analysis of this problem has been facilitated by several characteristics of the sea urchin embryo, especially its rapid, external development and optical transparency, as well as the availability of simple methods for isolating, recombining, and culturing specific embryonic cell types (e.g., Hörstadius, 1939; Okazaki, 1975a; Ettensohn and McClay, 1987; Khaner and Wilt, 1990). Recent studies have benefited from improvements to the classical lineage map of the sea urchin embryo, resulting from the use of fluorescent lineage tracers (Cameron et al., 1987, 1989, 1990; Wray and Raff, 1990). In addition, a significant advance has been the generation of a large battery of molecular probes (polyclonal antibodies, monoclonal antibodies, and cDNAs) that recognize specific cell lineages in the embryo (reviewed by McClay et al., 1983; Davidson, 1986; Ettensohn and Ingersoll, 1991). As discussed elsewhere, the usefulness of such probes is that they allow a less subjective and more rapid method of assaying the results of embryological experiments, often at the resolution of individual cells (Dawid et al., 1990).

Early experimental work with the sea urchin embryo first demonstrated the ability of isolated blastomeres to form complete larvae and defined the phenomenon of embryonic regulation (Driesch, 1892). Cell isolation and recombination studies (reviewed by Hörstadius, 1939, 1973) demonstrated that the fates of early embryonic cells are profoundly influenced by interactions with neighboring cells, and led to the distinction between prospective potency and prospective fate. More recently, lineage-specific molecular markers have been used to examine the expression of cell fates following blastomere recombinations (Livingston and Wilt, 1990; Khaner and Wilt, 1990) and by dissociated embryonic cells in vitro (Hurley et al., 1989; Stephens et al., 1989, 1990). These studies have confirmed the developmental plasticity of early sea urchin blastomeres, especially those of the animal region of the embryo, and have shown that cellular associations modulate patterns of gene expression. Although such studies have focused on interactions that appear to occur early in development (i.e., during cleavage), it is clear that important regulatory interactions continue to take place at later embryonic stages as well. We have investigated such a system of regulatory interactions associated with the selection of cell fates by mesenchymal lineages in the sea urchin embryo.

MESENCHYME CELL LINEAGES AND FATES

The mesoderm of the sea urchin embryo consists of two populations of cells; the primary mesenchyme cells (PMCs), which produce the larval skeleton, and the secondary mesenchyme cells (SMCs), which differentiate into several cell types including muscle, coelom, and pigment cells, but do not participate in skeletogenesis (Fig. 1). (Many reviews have considered various aspects of mesenchyme morphogenesis; see Gustafson and Wolpert, 1967; Okazaki, 1975b; Harkey, 1983; Solursh, 1986; Decker and Lennarz, 1988; Wilt and Benson, 1988; McClay et al., 1990; Ettensohn, 1991; Ettensohn and Ingersoll, 1991.) During blastula formation, progenitors of these two populations of cells become incorporated into the epithelial wall of the embryo at the vegetal pole. As gastrulation proceeds, cells of the PMC and SMC lineages undergo a transformation from an epithelial to a mesenchymal phenotype and move into the blastocoel cavity, a process known as ingression (Kinnander and Gustafson, 1960; Katow and Solursh, 1980; Fink and McClay, 1985; Anstrom and Raff, 1988; Amemiya, 1989). PMCs and SMCs are highly motile, and migrate within the blastocoel by continuously extending and retracting slender filopodial cell processes (Wolpert and Gustafson, 1961; Katow and Solursh, 1981; Karp and Solursh, 1985).

Despite the similarity in their motile behavior, PMCs and SMCs differ with respect to their lineage, time of ingression, and developmental fate. The PMCs are the sole progeny of the large micromere descendants (the VO*M*k, VA*M*k, and 2 VL*M*k blastomeres after Cameron et al., 1987), four cells that arise

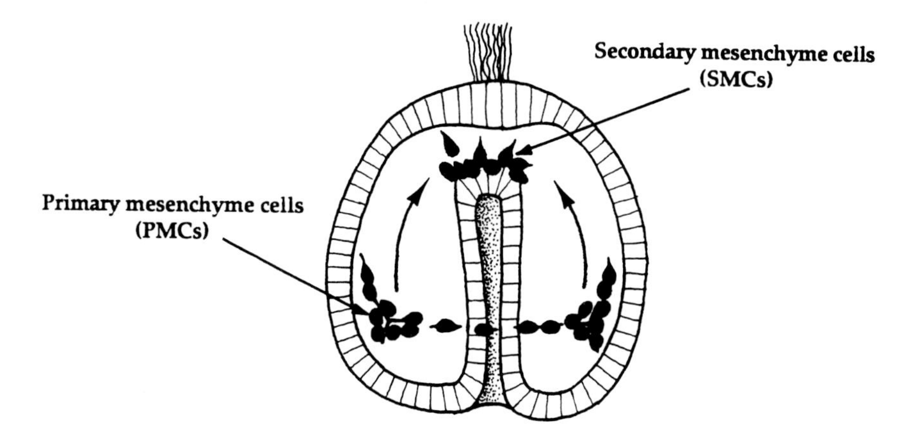

Fig. 1. Diagram of a late gastrula stage embryo, showing primary mesenchyme cells (PMCs) arranged in the subequatorial ring and secondary mesenchyme cells (SMCs) at the start of their ingression into the blastocoel. PMCs interact with SMCs (indicated by arrows) and regulate their choice of fates.

at the vegetal pole of the embryo at the fifth cleavage as a result of the unequal division of the micromeres (Endo, 1966, cited in Harkey, 1983). The ingression of the PMCs marks the beginning of gastrulation. There are an average of 32 or 64 of these cells in each embryo, depending upon the species, and they do not divide following ingression. The substratum on which the PMCs migrate is a thin basal lamina that completely lines the blastocoel (Katow and Solursh, 1979, 1981; Galileo and Morrill, 1985; Spiegel et al., 1989; Amemiya, 1990). During their migratory phase the PMCs gradually arrange themselves in a characteristic ring-like pattern between the vegetal pole and equator of the embryo—the subequatorial PMC ring (see Ettensohn, 1990a). As the ring pattern forms, filopodial processes of the cells contact one another and fuse, resulting in the formation of an extensive syncytial network. Within this syncytium, PMCs synthesize and secrete the larval skeleton, a complex, branched array of crystalline rods (spicules) composed of $CaCO_3$ and $MgCO_3$ deposited in an organic matrix (Wilt and Benson, 1988). The elaborate structure of the skeleton is determined by the arrangement of the PMCs in the blastocoel and by the branching pattern of the filopodial processes they extend (Gustafson and Wolpert, 1967). All PMCs express a skeletogenic fate, and fate mapping studies demonstrate that they are the only cells in the embryo that contribute to the larval skeleton in an undisturbed embryo (Hörstadius, 1973).

The developmental potential of the PMC lineage is restricted very early in embryogenesis, since isolated micromeres cultured in vitro or transplanted to ectopic locations in the embryo give rise only to skeletogenic cells (Hörstadius, 1973; Okazaki, 1975a; Livingston and Wilt, 1990). The asymmetry of the fourth cleavage division is not essential for the specification of the PMC fate (Langelan and Whiteley, 1985). In addition, if micromere formation is induced precociously at the third cleavage, these cells will also form spicules in culture (Kitajima and Okazaki, 1980). In fact, any egg fragment or blastomere that contains cytoplasm from the vegetal region of the egg will give rise to skeletal structures, suggesting that there may exist cytoplasmic determinants stored in the egg that specify skeletogenic differentiation, although such determinants have not yet been characterized.

There is a growing body of information concerning specific changes in gene expression that accompany the development of the PMC lineage. At least nine gene products expressed uniquely or predominantly by PMCs have been identified, including the spicule matrix proteins sm50 and sm30 (Benson et al., 1987; Sucov et al., 1987; Wilt and Benson, 1988; Killian and Wilt, 1989; George, Killian, and Wilt, personal communication), several PMC-specific surface glycoproteins including Meso 1 (Wessel and McClay, 1985), msp130 (McClay et al., 1983; Carson et al., 1985; Anstrom et al., 1987; Leaf et al., 1987; Farach et al., 1987; Harkey et al., 1988; Farach-Carson et al., 1989; Parr et al., 1990), and at least three additional, immunologically related but distinct cell surface proteins (Fuhrman, Knecht, and Ettensohn, in preparation), a putative calcium-binding protein (Iwata and Nakano, 1986), and a collagen transcript (Angerer et al., 1988). In addition, global patterns of protein synthesis during the development of the PMC lineage have been analyzed by two-dimensional gel electrophoresis (Harkey and Whiteley, 1982, 1983; Matsuda et al., 1988), revealing other gene products specifically expressed by the PMC lineage that may play important roles in their morphogenesis. Analysis of patterns of gene expression in the micromere-PMC lineage have revealed that an important transition takes place just before ingression, when many PMC-specific gene products are first expressed.

As the archenteron extends across the blastocoel during gastrulation, another population of mesodermal cells, the secondary mesenchyme, is released from its tip. The secondary mesenchyme is a heterogeneous population of cells that gives rise to portions of the coelomic pouches, muscle cells, pigment cells, and other unidentified derivatives. The SMCs are descendants of eight blastomeres that form the veg_2 tier of the 64-cell stage embryo [the VAM(1- 2)1, VOM(1-2)1, and two pairs of VLM(1-2)l cells after Cameron et al. (1987)]. These eight cells are derived from the four macromeres of the 16-cell stage embryo, the large sister cells of the micromeres. The PMC and SMC lineages are therefore separated from one another very early in development, by the end of the fourth cleavage division. Less is known concerning patterns of gene expression in the SMC lineage, although several molecular markers expressed specifically or predominantly by subpopulations of these cells have been identified (Ishimoda-Takag et al., 1984; Gibson and Burke, 1985, 1988; Cox et al., 1986; D'Alessio et al., 1989, 1990; Saitta et al., 1989; Wessel et al., 1990).

MESENCHYME CELL INTERACTIONS

PMC-SMC Interactions

Interactions between the PMCs and SMCs during gastrulation are important in regulating skeletogenesis and the expression of SMC fates. As noted above, fate mapping studies have shown that the skeletal structures of the larva are synthesized exclusively by PMCs, while SMCs express an array of other fates. Early cell isolation studies demonstrated, however, that the potential for skeletogenic differentiation is not restricted to the large micromere-PMC lineage. Hörstadius examined the development of isolated layers of 64-cell stage embryos and found that in some cases, isolated veg_2 layers gave rise to larvae with rudimentary skeletal elements (reviewed by Hörstadius, 1939, 1973). In undisturbed embryos, the cells of the veg_2 layer do not give rise to skeletal structures but rather contribute to the endoderm of the gut and to the secondary mesenchyme. These studies, then, demonstrated that the progenitors of one or both of these cell types, when isolated early in development, switch to a skeletogenic fate. Direct evidence that SMCs possess skeletogenic potential was provided by a provocative but little-known study by Fukushi (1962), who removed PMCs from embryos of *Glyptocidaris crenularis* and reported that SMCs moved to the region of the PMC ring and acquired a skeletogenic phenotype. Langelan and Whiteley (1985) altered normal cleavage patterns by treating embryos with low concentrations of sodium dodecyl sulfate, and found that skeletogenesis took place even in embryos that lacked both micromeres and early ingressing mesenchyme cells. They suggested that under these conditions, SMCs might alter their fate and compensate for the missing PMCs.

We have investigated the interaction between PMCs and SMCs and its role in regulating skeletogenesis in the developing embryo. Our findings indicate that a subpopulation of SMCs has latent skeletogenic potential, but that an interaction with the PMCs suppresses this potential and directs the SMCs into an alternative developmental pathway. Elimination of the PMCs by microsurgical or fluorescence photoablation methods triggers a transformation of SMCs to the PMC phenotype, a process we have called "SMC conversion."

SMC Conversion

Microsurgical methods allow the rapid removal of all or any fraction of the PMC population from embryos (Ettensohn and McClay, 1988). Elimination of the PMCs triggers a series of profound changes in morphogenetic behavior and gene expression on the part of the SMCs (Figs. 2, 3). Time-lapse video

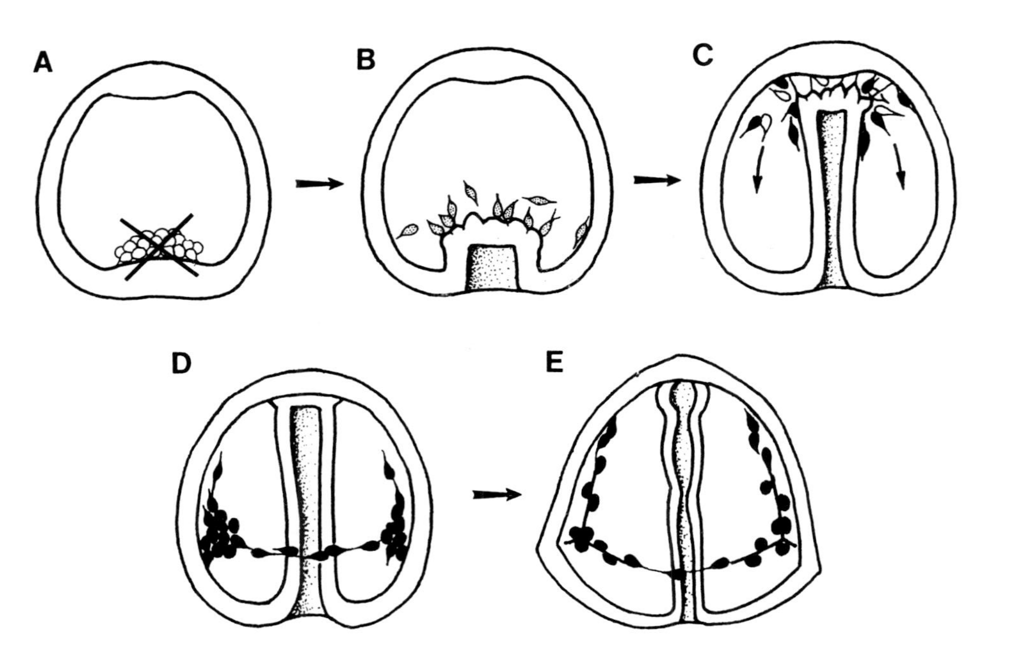
A
B
C
D
E

analysis of PMC-deficient embryos reveals that the converting SMCs are a subpopulation of 65–70 cells that ingress late in gastrulation. These studies show that the ingression of the SMCs takes place over an extended period of gastrulation and that discrete subpopulations of SMCs ingress at different times (Fig. 2). A population of 35–45 early-ingressing SMCs leaves the archenteron during primary invagination (the first phase of archenteron elongation). These cells migrate within the blastocoel for 1–4 h before penetrating the basal lamina and moving into the ectoderm, where they differentiate into pigment cells (see also Young, 1958; Gustafson and Wolpert, 1967; Gibson and Burke, 1985). Fate specification in these cells appears to be independent of any interaction with the PMCs. Large numbers of SMCs leave the gut rudiment during the second phase of gastrulation, as has been observed previously (see Gustafson and Wolpert, 1967). In PMC-deficient embryos, 65–75 of these cells convert to the PMC phenotype. They move in a vegetal direction after leaving the tip of the archenteron and gradually accumulate at the vacant sites on the blastocoel wall normally occupied by the PMCs. The cells initiate skeleton formation at two sites, as the PMCs normally do, and give rise to a correctly patterned larval skeleton (Figs. 2, 3). There are many late-ingressing SMCs that do not express a skeletogenic phenotype after PMC ablation; these cells may contribute to the coelom, the musculature surrounding the foregut, or give rise to additional pigment cells (see Ettensohn and Ingersoll, 1991).

In undisturbed embryos, although PMCs and SMCs exhibit similar motile activity, the directionality of their migration is very different. PMCs move to specific sites on the blastocoel wall and accumulate in a subequatorial ring, while SMCs either penetrate the basal lamina and invade the epithelial wall of the embryo (where they differentiate into pigment cells) or remain scattered within the blastocoel, often in the interior of the cavity rather than on the blastocoel wall. Following PMC removal, however, converting SMCs migrate to those sites on the basal lamina normally occupied by PMCs and arrange themselves in a pattern that closely resembles the subequatorial ring. Because the migration and patterning of PMCs involves specific interactions between these cells and directional cues in the blastocoel (reviewed by Ettensohn, 1991), converted SMCs must acquire the capacity to recognize the same set of guidance cues. These events occur several hours after the removal of the PMCs,

Fig. 2. Diagram illustrating SMC conversion, based on time-lapse videorecordings of PMC-depleted embryos. Elimination of the PMCs (**A**) by microsurgical methods (Ettensohn and McClay, 1988) or fluorescence photoablation (Ettensohn, 1990b) results in the conversion of a subpopulation of SMCs to the PMC fate. One group of SMCs leaves the tip of the archenteron early in gastrulation (**B**); fate specification in these cells is apparently independent of the PMCs. A larger number of SMCs ingress at the late gastrula stage (**C**), and 65–75 of these cells (shown in black) switch to a skeletogenic fate. The converting cells migrate vegetally (arrows), accumulate in a substitute subequatorial ring pattern (**D**), and synthesize skeletal elements (**E**).

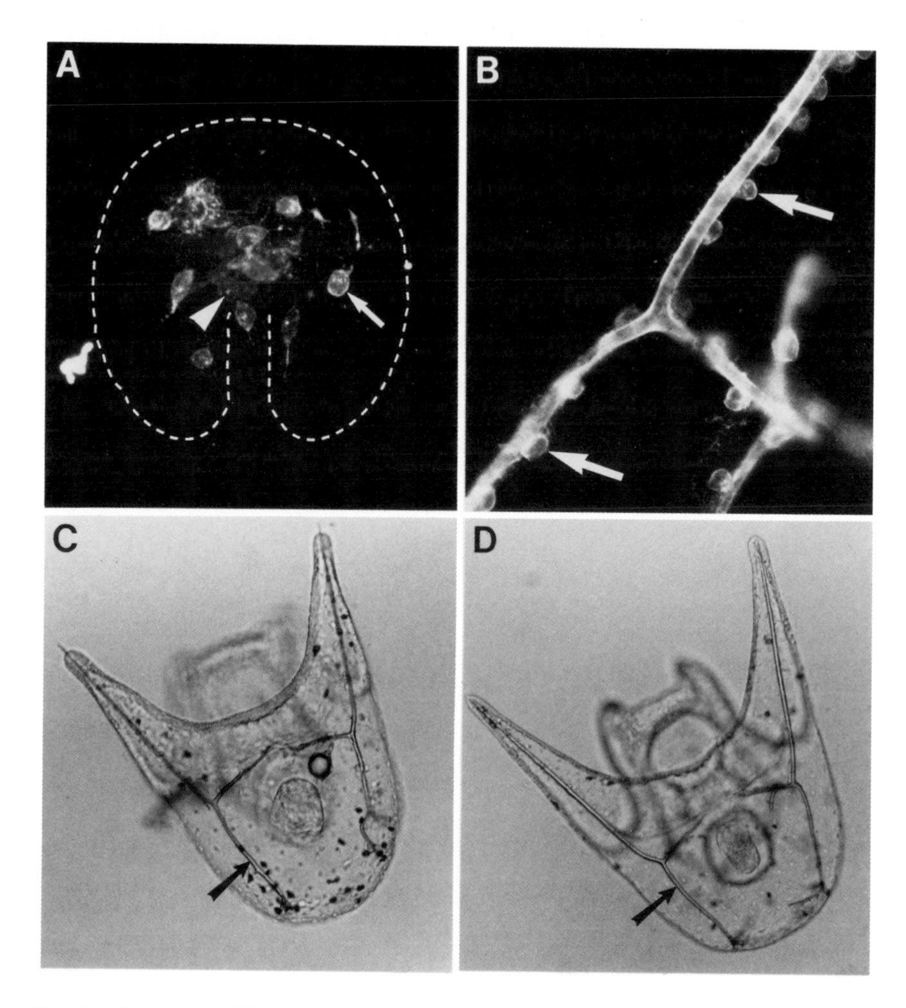

Fig. 3. Expression of PMC-specific cell surface molecules and skeleton formation by converting SMCs. (**A**) Whole mount of a late gastrula stage, PMC-depleted *Lytechinus variegatus* embryo. PMCs were removed microsurgically from this embryo at the early gastrula stage, and after 7 hr (23°C) the embryo was fixed and stained with monoclonal antibody (MAb) 6a9 and a rhodamine-conjugated secondary antibody (Ettensohn and McClay, 1988). MAb 6a9 recognizes a family of PMC-specific cell surface proteins (the msp family) that includes msp130. SMC conversion is just beginning in this embryo, as monitored by the de novo expression of msp proteins by the converting cells. Converting SMCs are concentrated near the tip of the archenteron and are migrating vegetally toward the position of the subequatorial ring. Conversion is asynchronous; in this embryo approximately 20 cells express PMC surface molecules as detected by 6a9 staining, although the number will increase to 65–75 as additional SMCs migrate away from the archenteron [converting cells probably do not divide after ingression (Ettensohn and McClay,

so that while a normal ring pattern and skeleton eventually form, these events are delayed relative to control embryos (Ettensohn and McClay, 1988). The change in SMC migration cannot be explained simply as the availability of sites that would otherwise be obstructed by PMCs, because SMCs from normal donor embryos transplanted into early blastula recipient embryos do not accumulate in a ring configuration (Ettensohn and McClay, 1986). Moreover, SMCs do not move to the subequatorial ring sites when PMCs are prevented from occupying those sites by agglutinating the cells with a lectin, wheat germ agglutinin, or PMC-specific monoclonal antibodies (Ettensohn, unpublished observations). Thus, conversion to a skeletogenic fate is associated with cellular changes on the part of the SMCs that allow these cells to recognize and respond to specific directional cues in the blastocoel.

Conversion of SMCs to the PMC phenotype is associated with the de novo expression of PMC-specific gene products. Several laboratories have isolated monoclonal antibodies (MAbs) that recognize cell surface proteins expressed specifically by the PMCs (McClay et al., 1983; Carson et al., 1985; Leaf et al., 1987). These antibodies have been used to characterize two PMC-specific cell surface glycoproteins, msp130 and Meso 1. We have generated and used a collection of PMC-specific MAbs that recognize these molecules as well as at least three other immunologically related but distinct cell surface proteins expressed exclusively by the PMCs (Fuhrman, Knecht, and Ettensohn, in preparation). All of these proteins are expressed by converted SMCs. The earliest sign of SMC skeletogenesis we have thus far detected is the appearance of immunoreactive granules in the cytoplasm of the SMCs soon after their ingression from the tip of the archenteron (Fig. 3). These granules probably represent the *trans* Golgi network, where PMC surface proteins are concentrated and packaged in preparation for secretion (Wessel and McClay, 1985; Anstrom et al., 1987).

The interaction between PMCs and SMCs takes place in all species that have been examined to date. The species used most extensively has been *Lytechinus variegatus*, because of the optical transparency of its embryos and their special suitability for microsurgical studies. However, PMC-depletion experiments have shown that the PMC-SMC interaction takes place in *L. pictus*

1988)]. The first indication of conversion is the appearance of 1–2 brightly stained granules in the cytoplasm of each SMC (arrowhead). Immunostaining gradually becomes more concentrated on the cell surface (arrow). Epifluorescence optics, ×330. (**B**) Skeletogenesis by converted SMCs. 6a9-stained whole mount showing skeletal elements formed by converted SMCs in a PMC-depleted embryo (pluteus stage). As in control larvae, the spicule rods and all associated skeletogenic cells (arrows) stain intensely with the antibody. Epifluorescence optics, ×1,000. (**C,D**) Bright-field micrographs of PMC-depleted (C, 48 hr postfertilization) and control (D, 36 hr postfertilization) pluteus larvae. The converted SMCs synthesize a skeleton (arrows) that is of the correct size and configuration. ×200.

(Fig. 3) and a sand dollar, *Dendraster excentricus* (Armstrong and McClay, personal communication; see also Langelan and Whiteley, 1985). In addition, recent studies (Ettensohn and Truschel-Peeler, unpublished observations) have demonstrated the expression of a skeletogenic fate by a subpopulation of cells in vitro following dissociation of embryos and removal of the PMCs by "panning" on wheat germ agglutinin-coated dishes (Ettensohn and McClay, 1987). The time course of skeletogenic differentiation and the numbers of cells that express a skeletogenic fate are consistent with the view that the same cells convert to the PMC phenotype in vitro and in vivo. In cell culture, cells of *L. variegatus*, *L. pictus*, as well as a third species too opaque to be used for microsurgical studies, *Arbacia punctulata*, convert to a skeletogenic fate. Because the earlier studies of Fukushi (1962) were carried out on embryos of *Glyptocidaris crenularis*, this brings to at least five the number of species in which the PMC-SMC interaction occurs, and suggests that the interaction is universal at least among the euechinoid subclass of sea urchins.

Quantitative Aspects of the PMC-SMC Interaction

The results of experiments eliminating the endogenous complement of PMCs from embryos raise two questions concerning quantitative aspects of the PMC-SMC interaction. First, what is the threshold at which SMC conversion is suppressed (i.e., is SMC skeletogenesis suppressed when only one-half, or one-quarter, etc., of the normal complement of the PMCs is present)? Second, is the conversion of SMCs to the PMC phenotype an all-or-none response or a graded one (i.e., is the number of SMCs that convert always 65–75, or is the number of converting cells dependent upon the number of PMCs in the blastocoel)?

To address these questions, a double-labeling approach was used (Ettensohn and McClay, 1988). The entire endogenous population of PMCs was removed from host embryos and replaced with varying numbers of PMCs that had been fluorescently labeled with rhodamine B isothiocyanate (Bonhoeffer and Huf, 1980). This labeling technique results in the covalent coupling of the fluorochrome to cell surface proteins and produces an intense fluorescent label that is persistent, not transferred from cell to cell, and does not affect the development of the labeled PMCs (Ettensohn and McClay, 1986). Converted SMCs were later identified by immunostaining whole mounts of embryos with MAb 6a9, an antibody that recognizes PMC-specific cell surface proteins, followed by a fluorescein-conjugated secondary antibody. When such whole mounts were examined by fluorescence microscopy, donor PMCs were labeled with both rhodamine and fluorescein, while SMCs that had switched to a skeletogenic fate were labeled with fluorescein only.

These studies demonstrated that the PMC-SMC interaction is quantitatively graded (Fig. 4). That is, the number of SMCs that convert to a skeletogenic

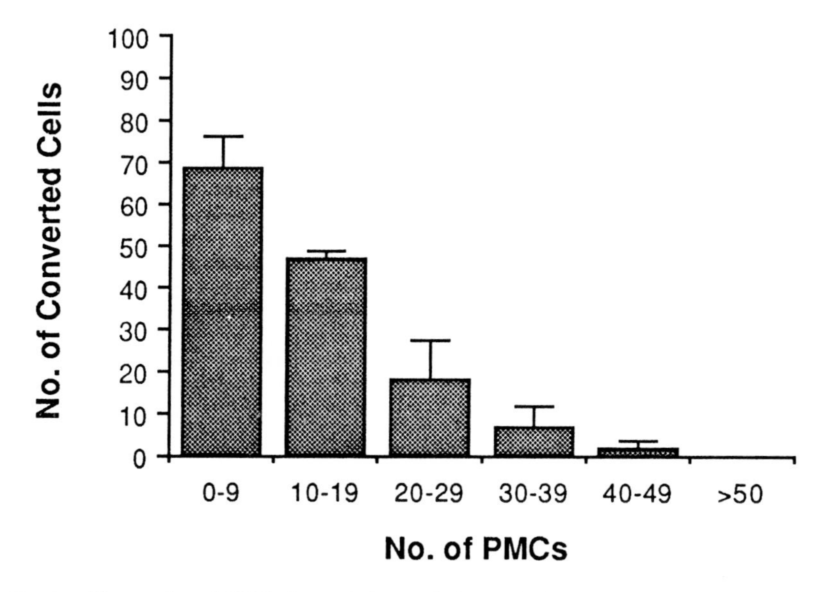

Fig. 4. The number of SMCs that switch to a skeletogenic fate is dependent on the number of PMCs in the blastocoel. This graph summarizes data from double-label experiments (see text). For each of the six classes of embryos, four to eight specimens were scored (35 total embryos). Bars indicate standard deviations (no bar is shown for those embryos with >50 PMCs; of six such embryos examined, none had any converted cells). When fewer than 50 PMCs are present in the blastocoel, SMC conversion takes place. The number of converting cells is inversely proportional to the number of PMCs in the blastocoel. Reprinted from Ettensohn and McClay (1988), with permission from Academic Press.

phenotype is inversely proportional to the number of PMCs in the blastocoel. The threshold is approximately 50 PMCs; this number of cells is sufficient to completely block SMC skeletogenesis. When intermediate numbers of PMCs are present in the blastocoel, intermediate numbers of SMCs convert to the PMC phenotype. Under such conditions, the larval skeleton is produced by a mixed population of PMCs and converted SMCs. These cells exhibit a remarkable ability to cooperate in the formation of a ring pattern and larval skeleton. In most cases, the PMCs distribute themselves evenly throughout the subequatorial ring pattern. In some embryos, however, they become concentrated in part of the pattern, for example, in one of the two ventrolateral clusters. When this occurs, most of the converted SMCs join the other ventrolateral cluster so that a rough balance between the two is maintained. In such embryos, the spicule rudiment associated with the ventrolateral cluster composed predominantly of PMCs is invariably larger than its counterpart, indicating that the PMCs continue their program of spiculogenesis on an intrinsic timetable. Nevertheless, the converted SMCs have a surprising ability to integrate into the partial pattern and to cooperate with PMCs in the construction of a relatively normal larval skeleton. The remaining PMCs do not regulate by dividing under these conditions—instead, their number remains constant (Ettensohn and McClay, 1988).

Models of the PMC-SMC interaction must take into consideration the apparent ability of the SMC population to ''count'' PMCs and to modulate accordingly the numbers of cells that switch fate. This aspect of the PMC-SMC interaction can be accounted for if two simple conditions are true: 1) Individual SMCs vary in the timing of their developmental decisions so that their switch to a skeletogenic fate is not simultaneous; and 2) a feedback mechanism operates such that converted SMCs (as well as PMCs) act to suppress the skeletogenic potential of as-yet unconverted SMCs. This would suggest that when the total number of skeletogenic cells (PMCs and converted SMCs) reaches a threshold level of approximately 50 cells, the transformation of additional SMCs to a skeletogenic fate is blocked. The first condition is known to be true; immunostaining of whole mounts at various times after PMC depletion indicates that the SMCs express PMC-specific cell surface molecules as they leave the tip of the archenteron, a process that may take several hours (see Fig. 3). The asynchrony in SMC conversion may therefore be related directly to the asynchrony of SMC ingression. The second condition has not yet been experimentally tested, although it should now be possible to do so.

Timing of the PMC-SMC Interaction

The timing of the PMC-SMC interaction has been analyzed by eliminating the PMCs at various developmental stages using a fluorescence photoablation technique (Ettensohn, 1990b). Although microsurgical methods can be used to remove PMCs from embryos early in gastrulation, this approach is impractical at later surfaces because the cells disperse within the blastocoel and adhere firmly to the basal stages of overlying epithelial cells. To ablate PMCs at later stages, the endogenous complement of cells was removed immediately after ingression and replaced with an equal number of PMCs that had been covalently labeled with rhodamine B isothiocyanate (Fig. 5A). The embryos were allowed to develop to different stages before the labeled PMCs were selectively ablated by prolonged excitation of the fluorochrome with green light (λmax = 550 nm).

Others have used fluorescence methods to eliminate embryonic cells by microinjecting labeled dextrans into early blastomeres and photoablating the labeled progeny at later development stages (Shankland, 1984; Nishida and Satoh, 1989). In our studies, cells were labeled by covalent coupling of the isothiocyanate derivative of the fluorochrome to amine groups on cell surface proteins. Such an approach may be useful in labeling large populations of cells (for example, in cell recombination experiments carried out in tissue culture), cells that are difficult to microinject, or cells whose embryological precursors are not known.

Photoablation of the PMCs at early stages of gastrulation results in a complete SMC conversion response (approximately 70 cells switching to the PMC phenotype) as monitored by immunoreactivity with MAb 6a9 (Fig. 5B). Surprisingly, photoablation of the PMCs even 8 hr after ingression results in a nearly complete conversion response (>50 SMCs switching to a skeletogenic phenotype). By this stage (late gastrula), the PMCs have completed their phase of most active migration, arranged themselves in a well-formed subequatorial ring pattern with two ventrolateral clusters, and initiated the synthesis of the skeleton. Even by this late stage no irreversible interaction has taken place between the PMCs and SMCs. Photoablation of the PMCs 8–10 hr after ingression results in intermediate numbers of converted cells, while ablation at the prism stage (10–12 hr after PMC ingression) does not result in SMC conversion. The period of sensitivity in these studies corresponds closely to the time of SMC ingression and migration and indicates that the critical signal passes between PMCs and SMCs at that time. The fact that SMCs no longer express a skeletogenic fate after the prism stage suggests that they may become committed to other pathways of differentiation at that time. An alternative interpretation is that by the prism stage, signals that suppress SMC skeletogenesis are stable even in the absence of the PMCs. For example, putative signaling molecules might be present in the blastocoel at sufficiently high levels so that ablation of PMCs does not trigger conversion, or SMCs with skeletogenic potential may migrate into the ectodermal layer and be prevented from converting by interactions with cells other than PMCs.

The data summarized in Figure 5 demonstrate that no critical signal passes between PMCs and SMCs during the early or midgastrula stages. This could indicate that no signal is transmitted by the PMCs during that time or that the SMCs are not competent to respond to the signal until late in gastrulation. To distinguish between these possibilities, PMC-deficient embryos were allowed to develop for progressively longer periods of time before 50–60 rhodamine-labeled PMCs were reintroduced into the blastocoel (Ettensohn, 1990b). These studies showed that injection of PMCs into the blastocoel at the late mesenchyme blastula or early gastrula stage resulted in complete inhibition of SMC skeletogenesis. Reintroduction of PMCs as late as the midgastrula stage, only some 3 hr before the initiation of SMC conversion as assayed by MAb 6a9 immunoreactivity, led to a significant reduction in the numbers of converted cells ($\tau = 25.7$, $n = 11$), although SMC conversion was not completely suppressed. Thus, PMCs need not be present in the blastocoel continuously to block SMC skeletogenesis. A period of 3 hr is sufficient to partially suppress SMC conversion, and complete suppression occurs in 3–6 hr. Results of fluorescence photoablation studies, however, show that PMCs present in the blastocoel for even longer periods of time fail to block SMC conversion if the cells are ablated before the time of SMC ingression. Taken together, then, these observations indicate that inhibitory signals are established by the PMCs

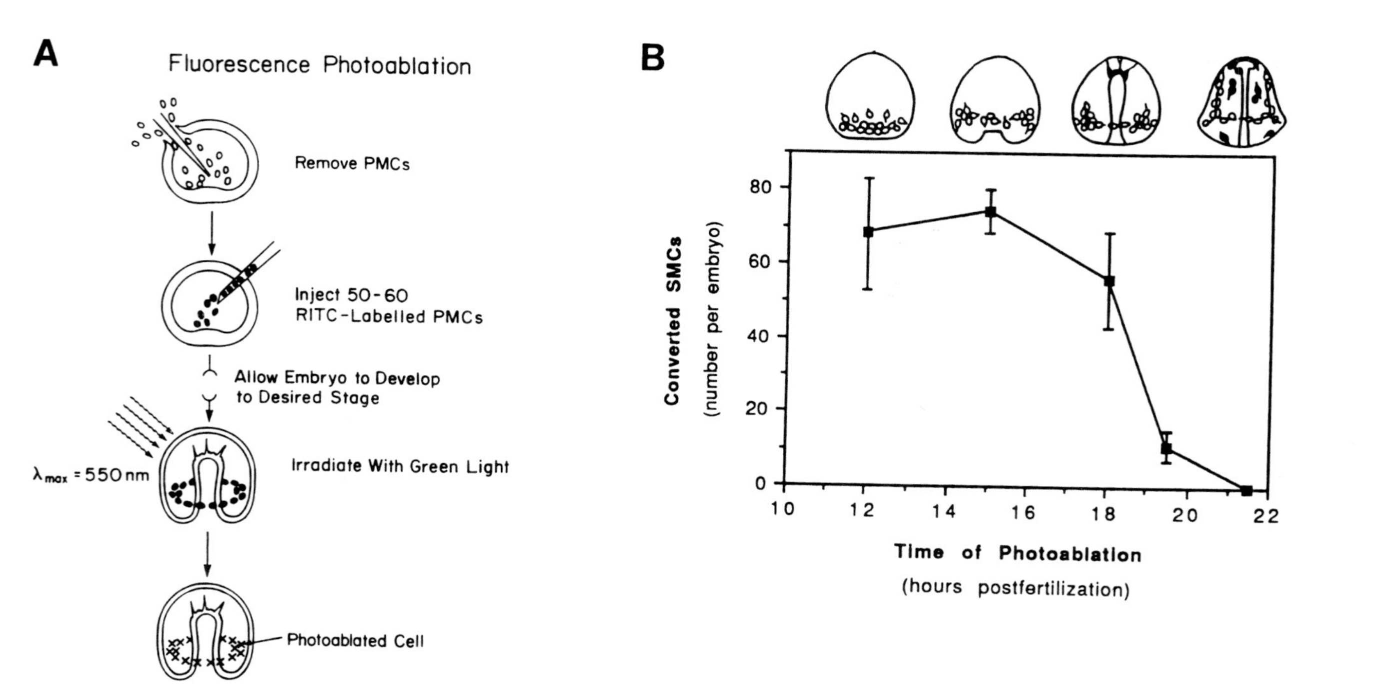

Fig. 5. Timing of the PMC-SMC interaction. (**A**) Summary of the fluorescence photoablation technique. The entire complement of PMCs is removed from an early gastrula stage embryo and replaced with an equal number of rhodamine-labeled PMCs. At different times the labeled cells are photoablated by exposing the embryos for 1–4 min to conventional epifluorescence illumination, using a 100 W mercury arc lamp and standard rhodamine filter set. (**B**) Results of fluorescence photoablation experiments. PMCs were photoablated at various developmental stages and the numbers of converting SMCs scored by immunostaining with MAb 6a9. The developmental stages of the embryos at the time of photoablation are illustrated by the diagrams at the top of the graph. Bars indicate standard errors (95% confidence limits on the mean). For each developmental stage, 5–11 embryos were scored. Reprinted from Ettensohn (1990b), with permission from the American Association for the Advancement of Science.

prior to SMC ingression, but SMCs are not sensitive to those signals until they leave the tip of the archenteron and begin to migrate. In addition, the conversion of SMCs to a skeletogenic fate even when PMCs are ablated late in gastrulation demonstrates that the signal is transient and decays rapidly upon elimination of the PMCs.

MECHANISM OF THE PMC-SMC INTERACTION

A central problem is the mechanism by which PMCs interact with SMCs to regulate their fates. Several models for this interaction can be envisioned (Fig. 6). In a formal sense, each of these mechanisms could involve either a negative interaction (the suppression of SMC skeletogenic potential by the PMCs) or a positive one (the induction of an alternative fate). In fact, there may be no meaningful distinction between these two alternatives at the molecular level; for example, if the activation of one genetic program invariably involves the repression of a second set (Jamrich et al., 1987; Symes et al., 1988; Dawid et al., 1990). In the absence of a complete molecular understanding of the phenomenon, it is therefore more appropriate to simply view the PMC-SMC interaction as one that regulates the choice of cell fate by the SMCs.

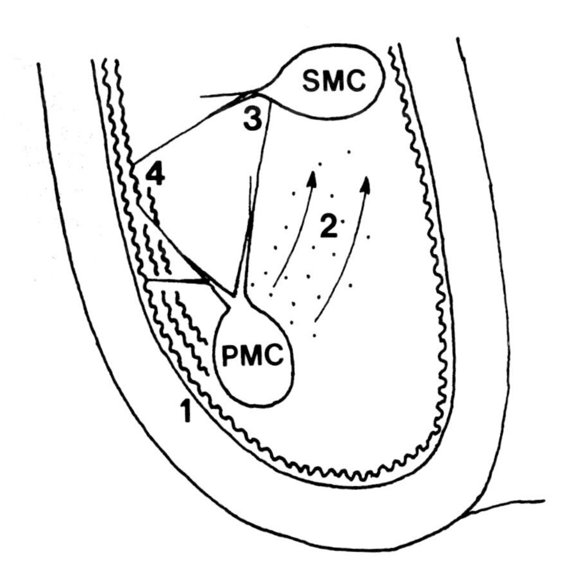

Fig. 6. Possible mechanisms of the PMC-SMC interaction. The PMCs might regulate SMC fates by occupying special sites on the blastocoel wall that promote skeletogenesis (1). Alternatively, the signaling mechanism might involve the release of a diffusible signaling molecule into the blastocoel by the PMCs (2), direct contact between the PMCs and SMCs (possibly mediated by the exploratory filopodia of the cells) (3), or the secretion of extracellular matrix components into the basal lamina/blastocoel matrix by PMCs (4).

One possibility is that there may exist special sites on the basal lamina (presumably those at the position of the subequatorial PMC ring) that promote skeletogenic differentiation. Such sites are occupied by PMCs during normal development, but would be made available to SMCs following elimination of the PMCs. Indirect evidence against this model comes from the observation that SMC conversion, as monitored by the expression of PMC-specific cell surface proteins, begins well before the cells become localized in the substitute ring pattern (Ettensohn and McClay, 1988) (Fig. 3). Nevertheless, filopodia extended by the SMCs might still contact such sites on the blastocoel wall before the bodies of the cells move to those positions. More direct evidence comes from the observation that when PMCs are prevented from migrating and occupying their normal sites on the basal lamina by microinjecting a lectin, wheat germ agglutinin, or PMC-specific monoclonal antibodies into the blastocoel, SMC skeletogenesis is completely inhibited (Ettensohn, unpublished observations). The PMC-SMC interaction cannot, therefore, be explained by a model that supposes the PMCs block access to special sites on the basal lamina that promote skeletogenesis.

A second simple hypothesis is that PMCs secrete a soluble signaling molecule into the blastocoel that regulates SMC fate. Examples of small molecules that regulate cell differentiation have been described (e.g., Strickland et al., 1980; Urist et al., 1983; Ekblom, 1984; Lyons et al., 1990). The best-characterized factors of this kind in developing embryos are fibroblast growth factor (FGF)- and transforming growth factor-β (TGF-β)-like molecules that are responsible for the induction of mesoderm in amphibian embryos (reviewed by Slack et al., 1989; Smith et al., 1989; Dawid et al., 1990). These molecules act as diffusible morphogens that appear to have an effective range of several cell diameters in the embryo (Grunz and Tacke, 1986; Gurdon, 1989). We have attempted to test whether a similar mechanism mediates the PMC-SMC interaction by microinjecting blastocoel fluid from control embryos into PMC-deficient embryos at various developmental stages (Ettensohn, unpublished observations). Such manipulations do not result in a reduction in the numbers of converted SMCs and therefore argue against (but do not disprove) a mechanism based on a diffusible factor.

The first indication in the embryo of SMC skeletogenesis, the expression of PMC-specific cell surface molecules, correlates with the time of SMC ingression and the initiation of migration. In addition, results of fluorescence photoablation studies indicate that the SMCs receive a critical stimulus from the PMCs late in gastrulation. These temporal correlations raise the possibility that delamination of SMCs from the endodermal epithelium and filopodial exploration by these cells are involved in the signaling process. Direct observations of filopodial motility in living embryos show that these cell processes can be sufficient in length to span the blastocoel cavity (Gustafson and Wolpert, 1967). The PMCs also extend filopodia, and by the late gastrula stage produce an extensive network of filopodial processes that covers a large region of the blastocoel wall. These observations raise the possibility that direct cell-cell contact between PMCs and SMCs mediates the interaction between these populations of cells. Recent studies on the *Notch* and *delta* gene products suggest that these transmembrane proteins may act as receptor and ligand at the cell surface, either as adhesive molecules or as part of a signal-transduction system (or both) (Fehon et al., 1990; Hoppe and Greenspan, 1990). Other genetically defined molecules that regulate cell fate decisions in *Drosophila* (Tomlinson, 1988; Rubin, 1989) and *C. elegans* (Greenwald et al., 1983; Yochem et al., 1988; Austin and Kimble, 1989), some of which are closely related to the *Notch* gene product, may act by similar mechanisms.

A fourth possibility is that the PMCs mediate cell-cell interactions by synthesizing and secreting constituents of the extracellular matrix (ECM). The blastocoel is a complex environment composed of fibrous and granular ECM components that fill the cavity and are incorporated into the thin embryonic basal lamina (see review by Ettensohn and Ingersoll, 1991). The PMCs condition this microenvironment by secreting ECM components, including collagen, into the blastocoel (Angerer et al., 1988; D'Alessio et al., 1990; Benson et al., 1990). Collagens (and possibly other ECM molecules) are therefore candidates for molecular mediators of the PMC-SMC interaction that should be tested. A large body of evidence has shown that the ECM regulates gene expression in embryonic cells in vitro (e.g., Maxwell and Forbes, 1987; Perris et al., 1988; Morrison-Graham et al., 1990). Genetic evidence supporting such a model comes from work with the *glp-1* gene in *C. elegans*, a gene involved in interactions between the distal tip cell and germline syncytium in the developing gonad (Austin et al., 1989). At least two suppressors of *glp-1* have been isolated and found to encode collagens, suggesting that the ECM plays a role in regulating this cell-cell interaction (Kramer et al., 1988; Maine and Kimble, 1989).

PMC-SMC INTERACTIONS, REGULATION, AND EQUIVALENCE GROUPS

The ability of SMCs to assume the phenotype and function of PMCs that have been removed microsurgically or ablated by fluorescence methods is a striking example of regulative development—the formation of structures from alternative progenitors following the ablation of the normal progenitors of those structures. The ability of embryos to compensate for missing parts has intrigued developmental biologists since the earliest days of experimental embryology (Driesch, 1892; McClendon, 1910; Spemann, 1938). In fact, the term "regulation" was originally used to describe the development of whole embryos or larvae from isolated blastomeres or fragments of embryos. Embryonic regulation of this kind may represent a complex combination of morphogenetic changes in tissue and embryo shape, as well as changes in the presumptive fates of embryonic cells. Here we have analyzed an example of regulation at the cellular level, and have shown that ablation of the cells that will normally give rise to the larval skeletal structures leads to their replacement by cells of a distinctly different lineage which acquire the differentiated function of the missing cells. The conversion of SMCs to a skeletogenic phenotype may explain certain long-standing observations concerning the regulative properties of the sea urchin embryo, such as the appearance of skeletal elements in isolated veg_2 layers (Hörstadius, 1939, 1973).

In the larva of *Caenorhabditis elegans*, an organism with a well-described and nearly invariant fate map, cell lineage analysis and cell ablation experiments have revealed examples of cellular interactions that regulate cell fates. These studies have led to the identification of "equivalence groups": populations of cells, usually closely related with respect to their developmental history (e.g., sibling cells of a specific lineage or cells descended from bilaterally symmetrical lineages), with similar or overlapping developmental potentials (Sulston and White, 1980; Kimble, 1981; Greenwald, 1989; Sternberg, 1990). Several such groups of cells have been identified in *C. elegans*, both by virtue of natural variations in cell lineages and by the demonstration of cell-type conversion after cell ablations. Interactions between cells in an equivalence group constitute a kind of "competition" for a preferred, or primary, fate (usually defined by the characteristic pattern of divisions a cell will subsequently undergo) that is ultimately expressed by only one or a limited subset of cells in the group. Removal of a cell expressing this primary fate leads to a regulative event and the conversion of a nearby cell in the equivalence group to the primary fate. (In some cases there exists a more complex hierarchy of possible fates, e.g., Sternberg and Horvitz, 1989.) Cell populations that exhibit characteristics of equivalence groups have been identified in the developing nervous systems of leech (Weisblat and Blair, 1984; Martindale and Shankland, 1990), insect (Kuwada and Goodman, 1985; Doe and Goodman, 1985), and ascidian (Nishida and Satoh, 1989) embryos.

Although the PMCs and SMCs appear to compete with one another for a primary (skeletogenic) fate, there are other reasons to conclude that these cells do not constitute an equivalence group. First, although SMCs can convert to a skeletogenic phenotype, it does not appear that the converse transformation can occur, and therefore the two cell populations do not have equivalent developmental potential. Statements concerning the developmental potential of any embryonic cell population should be made cautiously, since this characteristic can only be operationally defined (Harrison, 1933; Trinkaus, 1956). Nevertheless, cell isolation and transplantation experiments carried out with micromeres (Hörstadius, 1973; Okazaki, 1975a; Livingston and Wilt, 1990) and PMCs (Ettensohn and McClay, 1986; Ettensohn, 1990a) under a variety of experimental conditions indicate that this lineage is rigidly specified to adopt a skeletogenic fate. The stability of the PMC skeletogenic potential means that, in contrast to many examples of equivalence groups, there is little or no natural variability in the lineages that will give rise to skeletal structures. In addition, although PMCs and SMCs show many similarities in their morphogenetic behavior, the two cell populations share no obvious lineage homologies of the kind usually shown by cells comprising equivalence groups.

EVOLUTIONARY CONSIDERATIONS

One of the most intriguing questions raised by examples of cellular interactions such as those that take place between PMCs and SMCs or between cells in equivalence groups is why the embryos of multicellular animals should possess such apparently redundant systems. Why should an embryo be constructed with multiple cell lineages that share a common developmental potential but which then, through a process of cellular interactions, restrict the expression of that fate to cells of only one lineage? The answers to these questions lie at the interface between evolutionary and developmental biology and will emerge from an understanding of how evolution has acted to create and modify developmental programs.

Normal lineage patterns and the phenotypes of developmental mutations in *C. elegans* suggest that cell lineages may evolve by a process of cell duplication followed by a modification of one copy of the duplicated lineage. It has been proposed that such duplicated lineages may evolve progressively into autonomously determined cell populations with unique fates, but that an intermediate stage in such a process may be characterized by interactions between multipotential cells that restrict their fates (Sternberg and Horvitz, 1984). This view implies that the evolutionary progression of developmental programs within any individual taxon is from a ''regulative'' to a more ''mosaic'' pattern, with an ever-increasing supply of autonomously specified cell lineages.

Heterochrony, a shift in the timing of a developmental program, is another important mechanism of evolutionary change (Raff and Kaufman, 1983; Ambros and Horvitz, 1984). In sea urchins, informative comparisons can be made between the euechinoids (Subclass Euechinoidea), those species used most commonly in the laboratory, and the cidaroid urchins (Subclass Perischoechinoidea), which are believed to more closely resemble the ancestral stock from which modern sea urchins arose (Smith, 1984; Wray and McClay, 1988; Parks et al., 1989). The development of one cidaroid, *Eucidaris tribuloides*, has been examined in detail (Schroeder, 1981; Wray and McClay, 1988; Urben et al., 1988). This species exhibits variable numbers of micromeres and relatively few (approximately 16) skeletogenic cells. Moreover, *Eucidaris* does not exhibit two distinct populations of mesenchymal cells that ingress at different times; instead, mesenchyme cells (including presumptive skeleton-forming cells) are released from the tip of the archenteron in a single burst at the midgastrula stage. The appearance of a specialized, early-ingressing population of skeletogenic cells may therefore represent a relatively recent modification to this ancestral developmental program. The appearance of such a cell population might have been accompanied by cellular interactions that limit the expression of that same fate by other cell lineages.

The above arguments essentially treat regulative cell interactions as temporary (on a historical time scale) by-products of the way new cell lineages arise during evolution. An alternative possibility, however, is that redundant systems of cell fate specification have long-term usefulness to those organisms that exhibit them and therefore are selectively favored. Indirect support for this view comes from a comparison of modes of embryogenesis among different taxa, which demonstrates that regulative interactions are a prominent feature of the development of all higher metazoans (Davidson, 1990). In the sea urchin, there is natural variability in certain species in the numbers of micromeres, although most such embryos give rise to normal pluteus larvae (Schroeder, 1981; Langelan and Whiteley, 1985). There is also considerable variability in the numbers of PMCs found in embryos of a single species; even those derived from the eggs of a single female (Ettensohn, 1990a). The existence of regulative cell interactions might serve to buffer the developing embryo from variabilities in the division, movements, or differentiation of essential cell lineages that would otherwise have critically deleterious consequences to the embryo. [Note that such a view treates the embryo (organism) as the unit of selection, while others have argued that it is more appropriate to view the cell lineage as the unit of selection (Buss, 1987).]

Comparisons between different classes of echinoderms reveal that evolution has modified the program of larval skeletogenesis in this phylum in a variety of ways (see Korschelt and Heider, 1895; Schroeder, 1981). Ophiuroids (brittle stars) do not form micromeres, but nevertheless release an early mesenchyme that produces an elaborate larval skeleton closely resembling that of the echinoids. Asteroids (starfish) lack both an early ingressing mesenchyme and a larval skeleton. In such species, both components of the PMC-SMC system must therefore be lacking. Holothuroids and crinoids (see cucumbers and sea lilies, respectively) do not form micromeres or an early ingressing mesenchyme, but construct simple skeletal elements consisting of fenestrated plates or calcareous "wheels." These significant differences are observed among those echinoderms that develop via an intermediate larval form (indirect development); still other variations are observed in species that exhibit a direct mode of development (Raff, 1987). Comparative studies of the nature of mesenchyme cell interactions in these organisms, especially in developmentally "primitive" sea urchins such as *Eucidaris*, may provide a glimpse of how such systems evolve.

FUTURE DIRECTIONS

Present studies have defined the role of cell-cell interactions in regulating skeletogenesis in the sea urchin embryo and have identified important features of the quantitative nature and timing of the PMC-SMC interaction. Several central questions need now to be addressed: 1) It will be important to better characterize the subpopulation of converting SMCs. The secondary mesenchyme is a heterogeneous population of cells, and the normal fate of the converting cells has not yet been clearly established. Do these cells adopt a nonskeletogenic fate in normal larvae (e.g., giving rise to muscle cells, coleomic cells, or pigment cells) or do they remain in an undifferentiated state? One interesting possibility is that the converting SMCs are a population of cells that normally participates in the formation of adult skeletal structures following metamorphosis. The PMCs and larval skeletal structures are not incorporated into the juvenile sea urchin that forms during metamorphosis, and other cells, of as yet undefined lineage, deposit the calcified structures (test, teeth, spines, etc.) of the adult. Substantial molecular homology has been found between skeletogenic cells in the larva and those in the adult (Wilt and Benson, 1988; Drager et al., 1989; Parks et al., 1989). 2) A central question concerns the nature of the signal transmitted by the PMCs. Do these cells act directly on the SMCs, or is the signal transmitted indirectly via an intermediate population of cells? Is the interaction mediated by direct cell-cell contact or by the synthesis and secretion of extracellular molecules by the PMCs? Components known to be secreted by the PMCs, such as collagen, are candidates that can now be tested. 3) The response of the converting SMCs now can be analyzed in more detail, given the growing availability of molecular probes specific for the micromere-PMC lineage. It will be possible to determine whether converting SMCs undergo the same spectrum of changes in gene expression normally exhibited by skeletogenic cells of the PMC lineage. This information may have implications with regard to the kinds of possible signal transduction mechanisms that operate to modulate patterns of gene expression in these cells.

ACKNOWLEDGMENTS

This work was supported by National Institutes of Health grant HD24690, a Basil O'Connor Starter Scholar Award from the March of Dimes Foundation, and a National Science Foundation Presidential Young Investigator Award.

REFERENCES

Ambros V, Horvitz HR (1984): Heterochronic mutants of the nematode *Caenorhabditis elegans*. Science 226:409–416.

Amemiya S (1989): Electron microscopic studies on primary mesenchyme cell ingression and gastrulation in relation to vegetal pole cell behavior in sea urchin embryos. Exp Cell Res 183:453–462.
Amemiya S (1990): Development of the basal lamina and its role in migration and pattern formation of primary mesenchyme cells in sea urchin embryos. Dev Growth Diff 31:131–145.
Angerer LM, Chambers SA, Yang Q, Venkatesan M, Angerer RC, Simpson RT (1988): Expression of a collagen gene in mesenchyme lineages of the *Strongylocentrotus* embryo. Genes Dev 2:239–246.
Anstrom JA, Chin JE, Leaf DS, Parks AL, Raff RA (1987): Localization and expression of msp 130, a primary mesenchyme lineage-specific cell surface protein of the sea urchin embryo. Development 101:255–265.
Anstrom JA, Raff RA (1988): Sea urchin primary mesenchyme cells: Relation of cell polarity to the epithelial-mesenchymal transformation. Dev Biol 130:57–66.
Austin J, Kimble J (1989): Transcript analysis of *glp-1* and *lin-12*, homologous genes required for cell interactions during development of *C. elegans*. Cell 58:565–571.
Austin J, Maine E, Kimble J (1989): Genetics of intercellular signalling in *C. elegans*. Development 107(Suppl):53–57.
Benson S, Smith L, Wilt F, Shaw R (1990): The synthesis and secretion of collagen by cultured sea urchin micromeres. Exp Cell Res 188:141–146.
Benson SC, Sucov HM, Stephens L, Davidson EH, Wilt FH (1987): A lineage specific gene encoding a major matrix protein of the sea urchin embryo spicule. I. Authentication of the cloned gene and its developmental expression. Dev Biol 120:499–506.
Bonhoeffer F, Huf J (1980): Recognition of cell types by axonal growth cones *in vitro*. Nature 288:162–164.
Buss L (1987): "The Evolution of Individuality." Princeton: Princeton University Press.
Cameron RA, Fraser SE, Britten RJ, Davidson EH (1989): The oral–aboral axis of a sea urchin is specified by first cleavage. Development 106:641–647.
Cameron RA, Fraser SE, Britten RJ, Davidson EH (1990): Segregation of oral from aboral ectoderm precursors is completed at fifth cleavage in the embryogenesis of *Strongylocentrotus purpuratus*. Dev Biol 137:77–85.
Cameron RA, Hough-Evans BR, Britten RJ, Davidson EH (1987): Lineage and fate of each blastomere of the eight-cell sea urchin embryo. Genes Dev 1:75–84.
Carson DD, Farach MC, Earles DS, Decker GL, Lennarz WJ (1985): A monoclonal antibody inhibits calcium accumulation and skeleton formation in cultured embryonic cells of the sea urchin. Cell 41:639–648.
Cox KH, Angerer LM, Lee JJ, Davidson EH, Angerer RC (1986): Cell lineage-specific programs of expression of multiple actin genes during sea urchin embryogenesis. J Mol Biol 188:159–172.
D'Alessio M, Ramirez F, Suzuki HR, Solursh M, Gambino R (1989): Structure and developmental expression of a sea urchin fibrillar collagen gene. Proc Natl Acad Sci USA 86:9303–9307.
D'Alessio MD, Ramirez F, Suzuki HR, Solursh M, Gambino R (1990): Cloning of a fibrillar collagen gene expressed in the mesenchymal cells of the developing sea urchin embryo. J Biol Chem 265:7050–7054.
Davidson EH (1986): "Gene Activity in Early Development," 3rd ed. Orlando, FL: Academic Press, pp 213–246, 493–504.
Davidson EH (1990): How embryos work: Comparative view of diverse modes of cell fate specification. Development 108:365–389.
Dawid IB, Sargent TD, Rosa F (1990): The role of growth factors in embryonic induction in amphibians. In Nilsen-Hamilton M (ed): "Growth Factors and Development." New York: Academic Press, Curr Top Dev Biol 24:261–288.
Decker GL, Lennarz WJ (1988): Skeletogenesis in the sea urchin embryo. Development 103:231–247.

Doe CQ, Goodman CS (1985): Early events in insect neurogenesis. II. The role of cell interactions and cell lineage in the determination of neuronal precursor cells. Dev Biol 111:206–219.

Drager BJ, Markey MA, Iwata M, Whiteley AH (1989): The expression of embryonic primary mesenchyme genes of the sea urchin, *Strongylocentrotus purpuratus,* in the adult skeletogenic tissues of this and other species of echinoderms. Dev Biol 133:14–23.

Driesch H (1892): Entwicklungsmechanische Studien. I. Der Werth der beiden esten Furchungszellen in der Echinodermenentwicklung. Experimentelle Erzeugen von Theil—und Doppelbildung. Zeit fur Wiss Zool 53:160–178.

Ekblom P (1984): Basement membrane proteins and growth factors in kidney differentiation. In Trelstad RL (ed): ''The Role of Extracellular Matrix in Development.'' New York: Alan R. Liss, 42nd Symp Soc Dev Biol, pp 173–206.

Ettensohn CA (1990a): The regulation of primary mesenchyme cell patterning. Dev Biol 140:261–271.

Ettensohn CA (1990b): Cell interactions in the sea urchin embryo studied by fluorescence photoablation. Science 248:1115–1118.

Ettensohn CA (1991): Primary mesenchyme cell migration in the sea urchin embryo. In Keller RE (ed): ''Gastrulation: Movement, Patterns and Molecules.'' New York: Plenum Press (in press).

Ettensohn CA, Ingersoll EP (1991): Morphogenesis of the sea urchin embryo. In Rossomando EF, Alexander S (eds): ''Morphogenesis: Analysis of the Development of Biological Structures.'' New York: Marcel Dekker (in press).

Ettensohn CA, McClay DR (1986): The regulation of primary mesenchyme cell migration in the sea urchin embryo: Transplantations of cells and latex beads. Dev Biol 117:380–391.

Ettensohn CA, McClay DR (1987): A new method for isolating primary mesenchyme cells of the sea urchin embryo: Panning on wheat germ agglutinin-coated dishes. Exp Cell Res 168:431–438.

Ettensohn CA, McClay DR (1988): Cell lineage conversion in the sea urchin embryo. Dev Biol 125:396–409.

Farach MC, Valdizan M, Park HR, Decker GL, Lennarz WJ (1987): Developmental expression of a cell surface protein involved in calcium uptake and skeleton formation in sea urchin embryos. Dev Biol 122:320–331.

Farach-Carson MC, Carson DD, Collier JL, Lennarz WJ, Park HR, Wright GC (1989): A calcium-binding, asparagine-linked oligosaccharide is involved in skeleton formation in the sea urchin embryo. J Cell Biol 109:1289–1299.

Fehon RG, Kooh PJ, Rebay I, Regan CL,Xu T, Muskavitch MA, Artavanis-Tsakonas S (1990): Molecular interactions between the protein products of the neurogenic loci *Notch* and *Delta*, two EGF-homologous genes in *Drosophila*. Cell 61:523–534.

Fink RD, McClay DR (1985): Three cell recognition changes accompany the ingression of sea urchin primary mesenchyme cells. Dev Biol 107:66–74.

Fukushi T (1962): The fates of isolated blastoderm cells of sea urchin blastulae and gastrulae inserted into the blastocoel. Bull Marine Biol Stat Asamushi 11:21–30.

Galileo DS, Morrill JB (1985): Patterns of cells and extracellular matrix material of the sea urchin *Lytechinus variegatus* (Echinodermata: Echinoidea) embryo, from hatched blastula to late gastrula. J Morphol 185:387–402.

Gerhart J, Danilchik M, Doniach T, Roberts S, Rowning B, Stewart R (1989): Cortical rotation of the Xenopus egg: Consequences for the anteroposterior pattern of embryonic dorsal development. Development 107(Suppl):37–51.

Gibson AW, Burke RD (1985): The origin of pigment cells in the sea urchin *Strongylocentrotus purpuratus*. Dev Biol 107:414–419.

Gibson AW, Burke RD (1988): Localization and characterization of an integral membrane protein antigen expressed by pigment cells in embryos of the sea urchin *Strongylocentrotus purpuratus*. Dev Growth Diff 30:283–292.

Greenwald I (1989): Cell-cell interactions that specify certain cell fates in *C. elegans* develpoment. Trends Genet 5:237–241.

Greenwald IS, Sternberg PW, Horvitz HR (1983): The *lin-12* locus specifies cell fates in *Caenorhabditis elegans*. Cell 34:435–444.

Grunz H, Tacke L (1986): The inducing capacity of the presumptive endoderm of *Xenopus laevis* studied by transfilter experiments. Roux Arch Dev Biol 195:467–473.

Gurdon JB (1989): The localization of an inductive response. Development 105:27–33.

Gustafson T, Wolpert L (1967): Cellular movement and contact in sea urchin morphogenesis. Biol Rev 42:441–498.

Harkey MA (1983): Determination and differentiation of micromeres in the sea urchin embryo. In Jeffery WR, Raff RA (eds): "Time, Space, and Pattern in Embryonic Development." New York: Alan R. Liss, pp 131–155.

Harkey MA, Whiteley AH (1982): Cell-specific regulation of protein synthesis in the sea urchin gastrula: A two-dimensional electrophoretic analysis. Dev Biol 93:453–462.

Harkey MA, Whiteley AH (1983): The program of protein synthesis during the development of the micromere-primary mesenchyme cell line in the sea urchin. Dev Biol 100:12–28.

Harkey MA, Whiteley HR, Whiteley AH (1988): Coordinate accumulation of primary mesenchyme-specific transcripts during skeletogenesis in the sea urchin embryo.Dev Biol 125:381–395.

Harrison RG (1933): Some difficulties in the determination problem. Am Nat 67:306–321.

Hoppe PE, Greenspan RJ (1990): The *Notch* locus of *Drosophila* is required in epidermal cells for epidermal development. Development 109:875–885.

Hörstadius S (1939): The mechanism of sea urchin development, studied by operative methods. Biol Rev Camb Phil Soc 14:132–179.

Hörstadius S (1973): "Experimental Embryology of Echinoderms." Oxford: Clarendon Press.

Hurley DL, Angerer LM, Angerer RC (1989): Altered expression of spatially regulated embryonic genes in the progeny of separated sea urchin blastomeres. Development 106:567–579.

Ishimoda-Takag T, Chino I, Sato H (1984): Evidence for the involvement of muscle tropomyosin in the contractile elements of the coelom-esophagus complex of sea urchin embryos. Dev Biol 105:365–376.

Iwata M, Nakano E (1986): A large calcium-binding protein associated with the larval spicules of the sea urchin embryo. Cell Differ 19:229–236.

Jamrich M, Sargent TD, Dawid IB (1987): Cell type-specific expression of epidermal cytokeratin genes during gastrulation of *Xenopus laevis*. Genes Dev 1:124–132.

Karp GC, Solursh M (1985): Dynamic activity of the filopodia of sea urchin embryonic cells and their role in directed migration of the primary mesenchyme *in vitro*. Dev Biol 112:276–283.

Katow H, Solursh M (1979): Ultrastructure of blastocoel material in blastulae and gastrulae of the sea urchin, *Lytechinus pictus*. J Exp Zool 210:561–567.

Katow H, Solursh M (1980): Ultrastructure of primary mesenchyme cell ingression in the sea urchin, *Lytechinus pictus*. J Exp Zool 213;231–246.

Katow H, Solursh M (1981): Ultrastructural and time-lapse studies of primary mesenchyme cell behavior in normal and sulfate deprived sea urchin embryos. Exp Cell Res 136:233–245.

Khaner O, Wilt F (1990): The influence of cell interactions and tissue mass on differentiation of sea urchin mesomeres. Development 109:625–634.

Killian CE, Wilt FH (1989): The accumulation and translation of a spicule matrix protein mRNA during sea urchin embryo development. Dev Biol 133:148–156.

Kimble J (1981): Alterations in cell lineage following laser ablation of cells in the somatic gonad of *Caenorhabditis elegans*. Dev Biol 87:286–300.

Kinnander H, Gustafson T (1960): Further studies on the cellular basis of gastrulation in the sea urchin larva. Exp Cell Res 19:278–290.

Kitajima T, Okazaki K (1980): Spicule formation *in vitro* by the descendants of precocious micromeres formed at the 8-cell stage of the sea urchin embryo. Dev Growth Diff 22:265–279.

Korschelt E, Heider K (1895): "Textbook of the Embryology of Invertebrates." London: Swan Sonnenscheins.

Kramer JM, Johnson JJ, Edgar RS, Basch C, Roberts S (1988): The *sqt-1* gene of *C. elegans* encodes a collagen critical for organismal morphogenesis. Cell 55:555–565.

Kuwada J, Goodman CS (1985): Neuronal determination during embryonic development of the grasshopper nervous system. Dev Biol 110:114–126.

Langelan RE, Whiteley AH (1985): Unequal cleavage and the differentiation of echinoid primary mesenchyme. Dev Biol 109:464–475.

Leaf DS, Anstrom JA, Chin JE, Harkey MA, Showman RM, Raff RA (1987): Antibodies to a fusion protein identify a cDNA clone encoding msp 130, a primary mesenchyme-specific cell surface protein of the sea urchin embryo. Dev Biol 121:29–40.

Livingston BT, Wilt FH (1990): Range and stability of cell fate determination in isolated sea urchin blastomeres. Development 108:403–410.

Lyons KM, Pelton RW, Hogan BLM (1990): Organogenesis and pattern formation in the mouse: RNA distribution patterns suggest a role for bone morphogenetic protein-2A (BMP-2A). Development 109:833–844.

Maine E, Kimble J (1989): Identification of genes that interact with *glp-1*, a gene required for inductive interactions in *Caenorhabditis elegans*. Development 106:133–143.

Martindale MQ, Shankland M (1990): Neuronal competition determines the spatial pattern of neuropeptide expression by identified neurons of the leech. Dev Biol 139:210–226.

Matsuda R, Kitajima T, Ohinata H, Katoh Y, Higashinakagawa T (1988): Micromere differentiation in the sea urchin embryo: Two-dimensional gel electrophoretic analysis of newly synthesized proteins. Dev Growth Diff 30:25–33.

Maxwell GD, Forbes EM (1987): Exogenous basement membrane-like matrix stimulates adrenergic development in avian neural crest cultures. Development 101:767–776.

McClay DR, Alliegro MC, Hardin JD (1990): Cell interactions as epigenetic signals in morphogenesis of the sea urchin embryo. In Stocum D (ed): "Cellular and Molecular Biology of Pattern Formation." (In press).

McClay DR, Cannon GW, Wessel GM, Fink RD, Marchase RB (1983): Patterns of antigenic expression in early sea urchin development. In Jeffery WR, Raff RA (eds): "Time, Space, and Pattern in Embryonic Development." New York: Alan R. Liss, pp 157–169.

McClendon JF (1910): The development of isolated blastomeres of the frog's egg. Am J Anat 10:425–430.

Morrison-Graham K, West-Johnsrud L, Weston JA (1990): Extracellular matrix from normal but not *Steel* mutant mice enhances melanogenesis in cultured mouse neural crest cells. Dev Biol 139:299–307.

Nishida H, Satoh N (1989): Determination and regulation in the pigment cell lineage of the ascidian embryo. Dev Biol 132:355–367.

Okazaki K (1975a): Spicule formation by isolated micromeres of the sea urchin embryo. Am Zool 15:567–581.

Okazaki K (1975b): Normal development to metamorphosis. In Czihak G (ed): "The Sea Urchin Embryo: Biochemistry and Morphogenesis." New York: Springer-Verlag, pp 177–232.

Parks AL, Bisgrove BW, Wray GA, Raff RA (1989): Direct development in the sea urchin *Phyllacanthus parvispinus* (Cidaroidea): Phylogenetic history and functional modification. Biol Bull 177:96–109.

Parr BA, Parks AL, Raff RA (1990): Promoter structure and protein sequence of msp130, a lipid-anchored sea urchin glycoprotein. J Biol Chem 265:1408–1413.

Perris R, von Boxberg Y, Löfberg J (1988): Local embryonic matrices determine region-specific phenotypes in neural crest cells. Science 241:86–89.

Raff RA (1987): Constraint, flexibility, and phylogenetic history in the evolution of direct development in sea urchins. Dev Biol 119:6–19.

Raff RA, Kaufman TC (1983): "Embryos, Genes and Evolution." New York: Macmillan Press.

Rubin GM (1989): Development of the *Drosophila* retina: Inductive events studied at single cell resolution. Cell 57:519–520.
Saitta B, Buttice G, Gambino R (1989): Isolation of a putative collagen-like gene from the sea urchin *Paracentrotus lividis*. Biochem Biophys Res Comm 158:633–639.
Schiavone FM, Racusen RH (1990): Microsurgery reveals regional capabilities for pattern reestablishment in somatic carrot embryos. Dev Biol 141:211–219.
Schroeder TE (1981): Development of a "primitive" sea urchin (*Eucidaris tribuloides*): Irregularities in the hyaline layer, micromeres, and primary mesenchyme. Biol Bull 161:141–151.
Shankland M (1984): Positional determination of supernumerary blast cell death in the leech embryo. Nature 307:541–543.
Siegel BA, Verbeke JA (1989): Diffusible factors essential for epidermal cell redifferentiation in *Catharanthus roseus*. Science 244:580–582.
Slack JMW, Darlington BG, Gillespie LL, Godsave SF, Isaacs HV, Paterno GD (1989): The role of fibroblast growth factor in early *Xenopus* development. Development 107(Suppl):141–148.
Smith A (1984): "Echinoid Paleobiology." London: Allen and Unwin.
Smith JC, Cooke J, Green JBA, Howes G, Symes K (1989): Inducing factors and the control of mesodermal pattern in *Xenopus laevis*. Development 107(Suppl):149–159.
Solursh M (1986): Migration of sea urchin primary mesenchyme cells. In Browder L (ed): "Developmental Biology: A Comprehensive Synthesis," Vol 2. New York: Plenum Press, pp 391–431.
Spemann H (1938): "Embryonic Development and Induction." New Haven: Yale University Press.
Spiegel E, Howard L, Spiegel M (1989): Extracellular matrix of sea urchin and other marine invertebrates. J Morphol 199:71–92.
Stephens LE, Kitajima T, Wilt F (1989): Autonomous expression of tissue specific genes in dissociated sea urchin embryos. Development 107:299–308.
Stephens LE, Shiflet GW, Wilt FH (1990): Gene expression, DNA synthesis and protein synthesis in cells from dissociated sea urchin embryos. Dev Growth Diff 32:103–110.
Sternberg PW (1990): Genetic control of cell type and pattern formation in *Caenorhabditis elegans*. Adv Genet 27:63–116.
Sternberg PW, Horvitz HR (1984): The genetic control of cell lineage during nematode development. Ann Rev Genet 18:489–524.
Sternberg PW, Horvitz HR (1989): The combined action of two intercellular signaling pathways specifies three cell fates during vulval induction in *C. elegans*. Cell 58:679–693.
Strickland S, Smith KK, Marotti KR (1980): Hormonal induction of differentiation in teratocarcinoma stem cells: Generation of parietal endoderm by retinoic acid and dibutyryl cAMP. Cell 21:347–355.
Sucov HM, Benson S, Robinson JJ, Britten RJ, Wilt F, Davidson EH (1987): A lineage-specific gene encoding a major matrix protein of the sea urchin embryo spicule. II. Structure of the gene and derived sequence of the protein. Dev Biol 120:507–519.
Sulston JE, White JG (1980): Regulation and cell autonomy during postembryonic development of *Caenorhabditis elegans*. Dev Biol 78:577–597.
Symes K, Yaqoob M, Smith JC (1988): Mesoderm induction in *Xenopus laevis*: Responding cells must be in contact for mesoderm formation but suppression of epidermal differentiation can occur in single cells. Development 104:609–618.
Tomlinson A (1988): Cellular interactions in the developing *Drosophila* eye. Development 104:183–193.
Trinkaus JP (1956): The differentiation of tissue cells. Am Nat 90:273–189.
Urben S, Nislow C, Spiegel M (1988): The origin of skeleton forming cells in the sea urchin embryo. Roux Arch Dev Biol 197:447–456.
Urist MR, DeLange RJ, Finerman GA (1983): Bone cell differentiation and growth factors. Science 220:680–686.

Weisblat DA, Blair SS (1984): Developmental interdeterminancy in embryos of the leech *Helobdella triserialis*. Dev Biol 101:326–335.

Wessel GM, McClay DR (1985): Sequential expression of germ layer specific molecules in the sea urchin embryo. Dev Biol 111:451–463.

Wessel GM, Zhang W, Klein WH (1990): Myosin heavy chain accumulates in dissimilar cell types of the macromere lineage in the sea urchin embryo. Dev Biol 140:447–454.

Wilt FH, Benson SC (1988): Development of the endoskeletal spicule of the sea urchin embryo. In Varner JE (ed): "Self-Assembling Architecture." New York: Alan R. Liss, 46th Symp Soc Dev Biol, pp 203–227.

Wolpert L, Gustafson T (1961): Studies on the cellular basis of morphogenesis of the sea urchin embryo. Development of the skeletal pattern. Exp Cell Res 25:311–325.

Wray GA, McClay DR (1988): The origin of spicule-forming cells in a "primitive" sea urchin (*Eucidaris tribuloides*), which appears to lack primary mesenchyme cells. Development 103:305–315.

Wray GA, Raff RA (1990): Novel origins of lineage founder cells in the direct-developing sea urchin *Heliocidaris erythrogramma*. Dev Biol 141:41–54.

Yochem J, Weston K, Greenwald I (1988): The *Caenorhabditis elegans lin-12* gene encodes a transmembrane protein with overall similarity to *Drosophila Notch*. Nature 335:547–550.

Young RS (1958): Development of pigment in the larvae of the sea urchin *Lytechinus variegatus*. Biol Bull 114:394–403.

Cell-Cell Interactions in Early Development, pages 203–225

12. Lineage Specification During Early Embryonic Development of the Zebrafish

Charles B. Kimmel, Donald A. Kane, and Robert K. Ho
Institute of Neuroscience, University of Oregon, Eugene, Oregon 97403

INTRODUCTION

As embryos develop, distinctive lineages emerge that produce functionally related subsets of body cells. A major problem faced by developmental biologists is to try to understand the events that are associated with such lineage segregations, including the instructive events that ultimately cause the segregations to occur. This problem provides the theme for this chapter. Here we review studies with the embryonic zebrafish *Brachydanio rerio*, including the cell lineage and recent experimental work involving cell transplantation and mutational analysis that yield information about the regulative capacities of the embryonic cells. These findings suggest that cells by the early gastrula stage are expressing programs of development that are position-specific and that normally restrict their fates, but which seem to be reversible. This interpretation contrasts with classical views about vertebrate development in which the instructions that direct a cell to follow a particular course of development are assumed to also restrict the cell's ability to follow other ones (e.g., Weiss, 1939; Waddington, 1956).

SPECIFICATION AND CELL LINEAGE ANALYSIS

Specification refers to the processes by which distinctive cells or cell lineages are normally established (or "individuated," Waddington, 1956) during development (see Davidson, 1990). The word is often used interchangeably with commitment, but this has confounded our understanding. *Commitment* denotes autonomous and heritable restriction of cell *potential* to a particular developmental fate (Stent, 1985). *Differentiation*, or specialization of the cell to the final phenotype, presumably follows sometimes after commitment. It is clear that in embryos of higher animals, cells do indeed lose developmental potential, or the ability to regulate, during development. Specifications could commit cells, but they do not necessarily do so (Davidson, 1990). Specifications might be reversible, but commitments are not. Later we argue why it can be important that initial lineage specifications not restrict cell potential.

We need to examine normal development in order to see features that might distinguish different cells present at the same stage, and thus provide evidence whether specifications have occurred. Regularly observed differences among the cells which correlate with their positions in the embryo or with their ancestries, or expression of some biochemical or morphological marker, can suggest that the cells have been specified. Thus in *Xenopus* at the onset of zygotic transcription during the midblastula transition (Newport and Kirshner, 1982), animal hemisphere cells newly express a particular cytokeratin gene (Jonas et al., 1985), and vegetal hemisphere cells newly express a particular homeobox gene (Rosa, 1989). These observations provide very powerful and direct evidence that animal and vegetal cells are separately specified by this stage, and the observations are given meaning because the differences can easily be related to the fate map: Animal cells will normally give rise to ectoderm, and vegetal cells form endoderm and mesoderm (e.g., Dale and Slack, 1987; Moody, 1987).

Currently, cell lineage analysis provides the best means available to learn the fates of embryonic cells. The same analysis also reveals patterns of divisions, movements, and cellular associations that, like selective patterns of gene expression, can provide evidence about specifications. Among vertebrates, the embryos of certain small teleosts, such as the zebrafish, have important advantageous features for the study of embryonic cell lineages. Zebrafish can be studied genetically (Streisinger et al., 1981; Kimmel, 1989), and their embryos are accessible for observation and manipulation at any stage, such that, for example, single cells can be specifically labeled by intracellular injection with tracer dyes. The embryos are small enough to fit under a high-power objective of a compound microscope, and transparent enough so that even cells located very deeply in the living, intact embryo are visible. The embryos develop rapidly, hatching into swimming larvae by three days. Indeed, within a single day after fertilization the primary organs of the body have formed and the embryo contains many types of postmitotic differentiated cells that can be distinguished from one another by morphology and position (Kimmel and Warga, 1987a). If one examines the lineages of these early cells, in particular, problems associated with doing lineage analyses in a vertebrate are very considerably reduced. Lineage analysis is ongoing, and the work is beginning to yield some insights that appear relevant to the problem of specification.

THE BLASTULA ARISES BY A REGULAR, BUT VARIABLE, CELL LINEAGE

Development from the zygote through the first day of embryogenesis in zebrafish can be subdivided into the periods of cleavage, blastula, gastrula, and segmentation (Fig. 1). We consider a series of lineage segregations that occur during about 3 hours starting in the blastula (Fig. 2). The blastula stage begins at the seventh cleavage, at 2.3 hours after fertilization (h), by a stereotyped but, as we shall see, not an invariant set of events. These regular feature of early development are coupled only very loosely to the later developmental history of the descendants of the early cells, the blastomeres, that are present during cleavage. Lineage analysis suggests that blastomeres may still be unspecified as the early blastula forms.

An initial animal–vegetal axial polarity of the zygote is evidenced by cytoplasmic streaming, which begins within 20 minutes after fertilization, towards the animal pole, the site where the sperm enters and polar bodies emerge. This streaming produces the blastodisc, a region of the cell that consists of largely yolk-free cytoplasm and to which the divisions that generate the embryonic cells will be confined. Streaming continues to add cytoplasm to the margin of the blastodisc during cleavage, although it progressively diminishes in amount.

The early cleavages are partial; the vegetal yolk-rich region of the zygote is left entirely uncleaved. In the midblastula (see below) there will occur abruptly a complete segregation between the blastomeres and this uncleaved yolk-rich region, present thereafter as a single giant multinucleate syncytial cell, the yolk cell.

After the first cleavage, at 40 min, nine subsequent cleavages occur very synchronously at 15-min intervals. The polarities of the early five or six cleavages are often fixed with reference to the animal–vegetal axis and one another (Fig. 3). Cleavages 1–5 are produced by vertically oriented furrows, with odd- and even-numbered divisions alternating at right angles. Cleavages 1 and 2 produce single furrows through the animal pole, and cleavages 3 and 4 produce pairs of furrows; at the 16-cell stage the blastomeres lie in one plane (or "tier"), in a 4×4 array. The furrows undercut the blastomeres such that, as revealed by dye injections (Kimmel and Law, 1985a), the four central blastomeres at this stage (stippled in Fig. 3E) are partitioned completely from the others. The surrounding outer margin of so-called "marginal" blastomeres that borders the yolk cell still possesses cytoplasmic bridges to the yolk cell. Commonly, cleavage 5 consists of a set of three vertical furrows paralleling those of 1 and 3, and producing a single 4×8 array of cells (Fig. 3F), and cleavage 6 is horizontal (in the plane of the page in Fig. 3F), producing essentially another 4×8 tier of cells.

However, often one or more obliquely oriented furrows occur during these cleavages, producing different patterns of cells (e.g., see Kimmel and Law, 1985a) that still can produce fully formed normal embryos. Changes in pattern can be observed as early as the eight-cell stage, and their frequency increases progressively. For example, during cleavages 5 through 10 there is much variability, but the patterns are not random; a tendency remains for alternation between vertical and horizontal furrows that is reminiscent of the odd–even alternation observed during cleavages 1 through 5 (Kimmel and Law, 1985b).

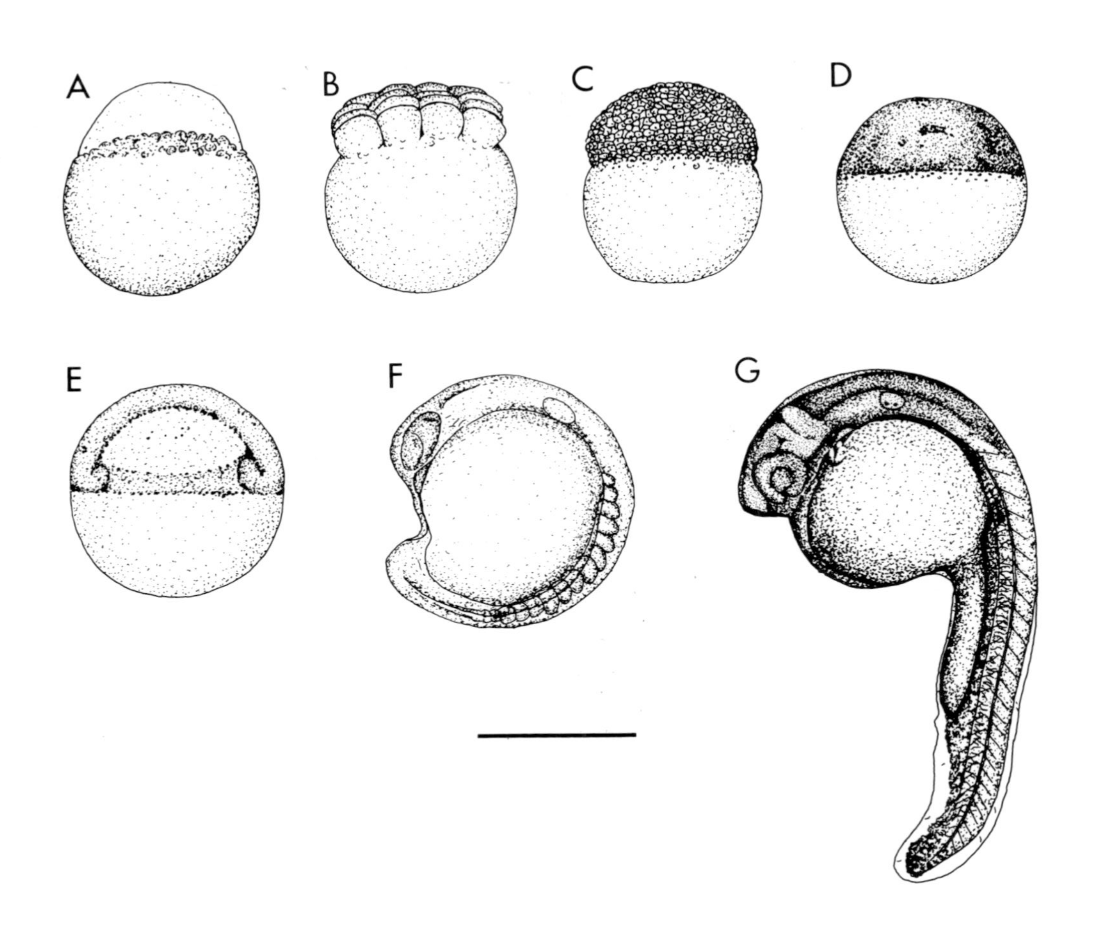
A
B
C
D
E
F
G

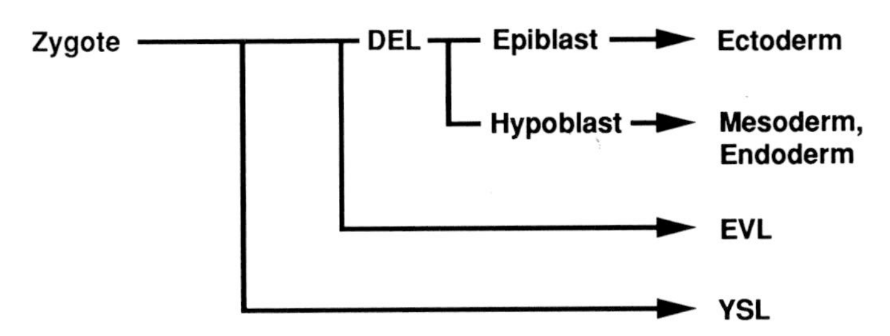

Fig. 2. Lineage segregations occur sequentially in subsets of blastula and gastrula cells. Segregations are observed as normally occurring clonal restrictions to individual fates, as discussed in the text. Epiblast cells, after gastrulation, are tissue-restricted to one or another ectodermal fate, e.g., epidermis or nervous system. Hypoblast cells are similarly tissue-restricted to either mesodermal or endodermal fates (see Fig. 9). Note that we have not detected a hierarchy, but can understand the sequence as a series of restrictions occurring separately in a single unspecified stem lineage.

Even in regularly cleaving embryos, the fates of the descendants of the blastomeres are always highly variable, or indeterminate (Kimmel and Warga, 1987a). Clones founded by early blastomeres include cells of many types, scattered among other cells over very large areas of the embryo. Examining the clonal progeny in each of 10 embryos of a particular cell at the 64-cell stage, i.e., one formed by an identical set of six cleavages, revealed 10 very different sets of fates. Evidently early lineage does not establish cell fate in any significant way (Kimmel and Warga, 1988).

Fig. 1. Development of the zebrafish during the first day of embryogenesis. Stages are named as in Warga and Kimmel (1990). The drawings, made by Seth Kimmel, are all oriented as left-side views and scaled identically. The animal pole is to the top in the early stages and cells deriving from the animal pole form anterior structures later. **A:** Zygote, the one-cell stage at 0.5 hours after fertilization at 28.5°C (h). The lower part of the cell is yolk-rich, the upper part is the blastodisk, from which the cellular blastoderm arises. **B:** Cleavage, the 16-cell stage at 1.5 h. Cleavages are confined to the blastodisk; the pattern is shown from an animal pole view in Figure 3. **C:** Midblastula, the 1k-cell stage at 3 h. The 10th cleavage, which produces this stage, is the last synchronous one. The nuclei of the yolk cell lie in a band of yolk-free cytoplasm just beyond the blastoderm margin. **D:** Late blastula, the sphere stage at 4 h. Expansion has produced the shape change, and at this time the blastoderm occupies the upper one third of the volume. After this stage it thins to a cup shape (see Fig. 7). **E:** Gastrula, the shield stage at 6 h. The blastoderm has thinned, spread by epiboly to cover half of the yolk cell. Involution has produced the thickened margin of the blastoderm (the germ ring), and convergent extension have made the germ ring thicker dorsally (to the right in the drawing), this region being known as the embryonic shield. **F:** Segmentation, the 14-somite stage at 16 h. Epiboly is completed and the tail bud is growing out. The optic vesicle and otic vesicle are apparent, and somites are forming at a rate of two per hour. **G:** End of segmentation, 30-somite stage at 24 h. Many cell types have differentiated; for example, the embryo is motile, requiring function of both nerve and muscle, as well as neuromuscular connections.

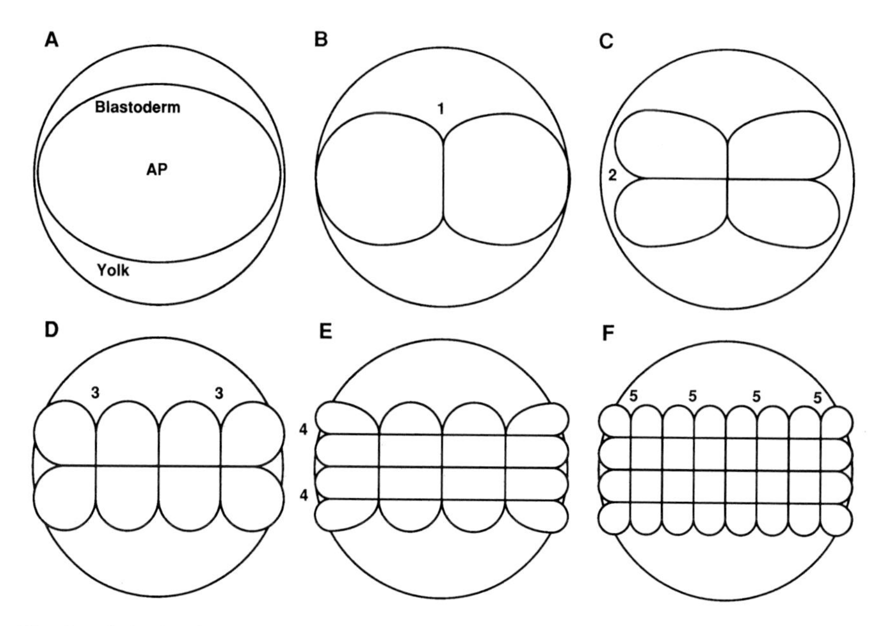

Fig. 3. Animal polar views of the planes' first five cleavages. The outer circle is the yolk and the inner ellipse, present initially, is the blastodisk (compare with the side view in Fig. 1A). The first and other odd-numbered cleavages cut along the short axis of the ellipse. The second and other even-numbered cleavages cut along the long axis. The plane of bilateral symmetry of the embryo is established along the second cleavage plane in many but not all cases.

A major factor contributing to the indeterminacy of the fates of the blastomeres is that the plane of bilateral symmetry of the embryo is not strictly aligned with an early cleavage plane (Kimmel and Law, 1985c). This differs from the situation in a number of other kinds of embryos. For example, in ascidians, which are primitive chordates, the first cleavage plane and the plane of bilateral symmetry are invariably one and the same. On the other hand, the relationship is not random in zebrafish (Kimmel and Warga, 1987a). Two axes must be specified in order to establish the plane of bilateral symmetry. One, the anterior–posterior axis, is normally present in the egg, for the animal pole of the egg will form, as exactly as can be ascertained experimentally, at the anterior end of the embryo (the equivalent of the embryo's nose). Establishment of the other axis, the dorsal–ventral axis, has not been studied in the zebrafish, but may occur beginning at the one-cell stage as in *Xenopus* (discussed in Kimmel, 1989), and it is frequently aligned with the second cleavage plane (Kimmel and Warga, 1987a; unpublished data). Thus, as for orientations of the cleavages themselves, there is a preferred relationship, but not an invariant one. Individual blastomeres must thus necessarily be multipotential.

SEGREGATION OF THE YOLK CELL LINEAGE IN THE MIDBLASTULA

The first strictly obeyed lineage segregation known in the zebrafish (Fig. 2) is that of the yolk cell lineage, and occurs in the blastula during, or just preceding, a "midblastula transition" that seems identical in its features and control (D.A.K., unpublished results) to the midblastula transition that has been described for *Xenopus*. At this stage, two separate lineage *compartments* (see below) arise. By the definition established for the *Drosophila* integument (Garcia-Belido et al., 1973) a lineage compartment is a "polyclone"; a spatially defined territory that includes all of the clonal descendants of a group of cells, and excludes all other cells. Here, beginning in the midblastula, one of the compartments is the multinucleate syncytial yolk cell, and the other is the blastoderm, the complete set of mononucleate cells (Kimmel and Law, 1985b).

Remarkably, one can watch the compartmentation occur. It was described for teleost embryos more than 100 years ago (Agassiz and Whitman, 1884; Wilson, 1889). After cleavage 9, at the 512-cell stage, the marginal blastomeres, those present in a ring at the blastoderm rim, have the appearance of single cells (Fig. 4A), although in fact they are syncytial, remaining connected to the yolk cell by cytoplasmic bridges left from the incomplete early cleavages as desribed above. At cleavage 10 (occurring at 3 h) the nuclei of these cells undergo mitosis as usual, but cytokinesis does not occur. Instead, in a process that seems to begin even before the nuclear division, the cell membranes of adjacent marginal blastomeres partly disappear (Fig. 4B). Consequently, as interphase nuclei are reformed after cleavage 10, those nuclei that were derived from marginal blastomeres now share a common pool of cytoplasm, the so-called yolk syncytial layer (YSL) of the yolk cell (Fig. 4C).

After the YSL forms, the yolk cell and blastoderm lineages are entirely separate, at least during the next several cleavages, as required by the compartment hypothesis (Kimmel and Law, 1985b). Furthermore, no cryptic compartmentation occurred before the YSL forms, for through cleavage 9 the marginal blastomeres contribute variably to both the YSL and non-YSL lineages.

The YSL formation, and the subsequent lineage compartmentation, indicates that by the midblastula stage the yolk cell and the blastoderm have been separately specified. There are a number of other features that mark the divergence in their development: the YSL nuclei continue to divide synchronously (Wilson, 1889) with a distinctive shorter rhythm than the asynchronously dividing blastoderm cells (D.A. Kane, unpublished observations), in a manner reminiscent of mitotic "domains" that accompanies specification of fate map boundaries in *Drosophila* (Foe, 1989).

As is also observed for *Drosophila* (Weir and Lo, 1982), the lineage compartmental boundary is a "communication" compartmental boundary, as

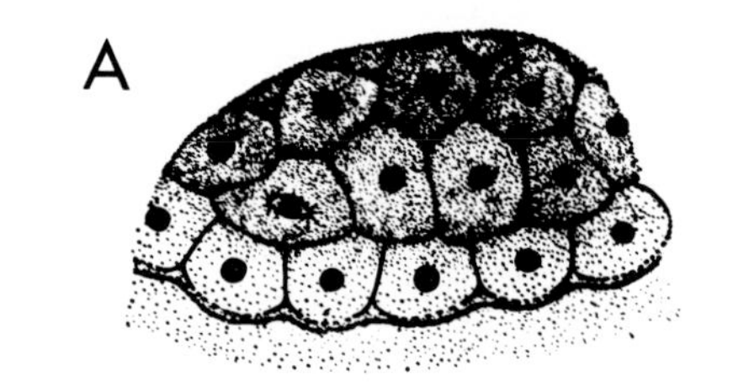
A

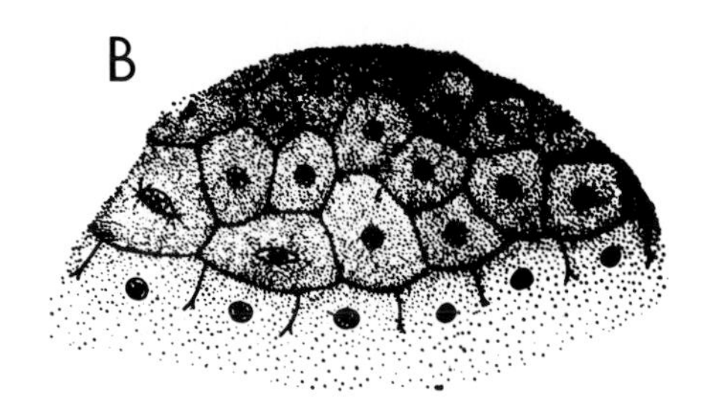
B

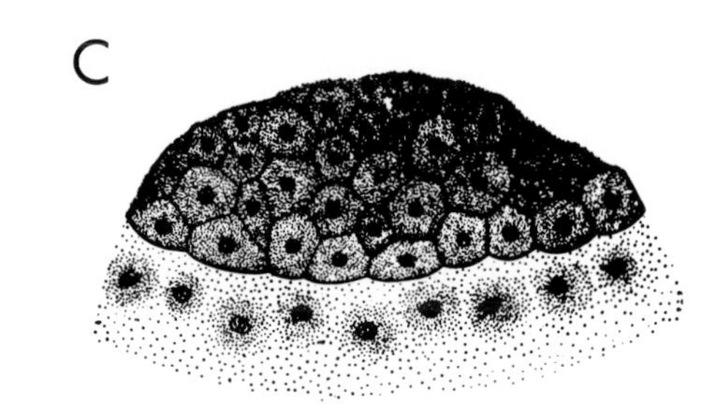
C

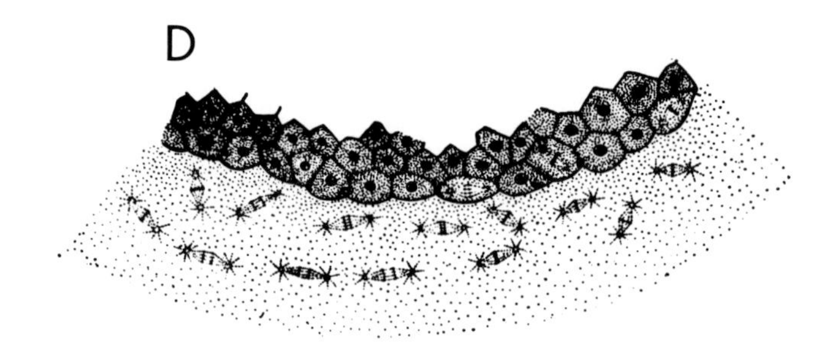
D

revealed by the pattern of transfer of the dye Lucifer yellow, which can pass from cell to cell though gap junctions. At the time the YSL forms, all of the blastomeres are extensively dye-coupled to one another and to the yolk cell. A little over 2 hours later, at the onset of gastrulation, the blastoderm cells are still dye-coupled to one another, but their coupling to the yolk cell is lost (Kimmel and Law, 1985b).

The yolk cell also begins to have specialized developmental functions. In *Fundulus*, and presumably in all teleosts, the yolk cell appears to play a crucial morphogenetic role beginning in the late blastula, literally pulling the blastoderm around itself (in the fashion of putting on a stocking) in the cell movement known as blastoderm epiboly (Trinkaus, 1984). Epiboly depends, at least in part, on actin-based cytoskeletal activity in the YSL (Betchaku and Trinkaus, 1978). The yolk cell clearly may also play some nutritive role in the utilization of the yolk itself during embryogenesis, but what this role is, and what eventually becomes of the yolk cell, is unknown. It makes no known contribution to the definitive embryonic tissues.

ENVELOPING LAYER LINEAGE SEGREGATION AT EARLY EPIBOLY

After cleavage 6, two populations of blastomeres are present, a monolayer of cells that form the blastoderm surface, termed enveloping layer (EVL) cells, and deep layer (DEL) cells, underlying the surface, eventually in a multilayer. Once DEL cells appear, they or their descendants do not normally ever reenter the EVL, but remain interior to it. However, the early EVL cells are not so restricted. They can divide in the plane of the EVL, so as to generate two EVL daughters, or they can divide such that only one of the daughters remains in the EVL and the other daughter, produced just beneath this layer, enters the DEL. Which of these two types of divisions a particular EVL cell will enter cannot be predicted with certainty, but their frequencies depend upon the number of the particular cleavage (Kimmel and Law, 1985c).

By the late blastula stage, at 4 h, divisions in which EVL cells generate DEL daughters no longer occur. Afterwards, EVL cells are clonally restricted

---

Fig. 4. Formation of the YSL. (**A**) At the 256-cell stage, before the YSL begins to form, the marginal blastomeres have fairly distinct lower borders. (**B**) These borders become indistinct as the cells at the 512-cell stage enter mitosis for the 10th cleavage. (**C**) The YSL is present after this cleavage; the cell membranes that were originally present between the marginal blastomeres have disappeared. (**D**) YSL nuclei continue to divide synchronously for several cycles of nuclear mitosis, and with a different rhythm than the cell cycles of the blastoderm. (Note that most blastoderm nuclei are in interphase.) The drawings are of the sea bass embryo, from Wilson (1889), but he did not report at which particular cleavages the YSL forms; those described here are for the zebrafish.

to their outer epithelial layer, which then goes on to develop directly into an embryonic "periderm," a specialized outer squamous epithelium covering the embryo. The later fate of the periderm has not been established, but it is reportedly discarded in *Salmo* (Bouvet, 1976).

As for the yolk cell lineage restriction, the EVL lineage restriction is accompanied by dramatic morphogenetic changes. At the beginning of the midblastula period the blastoderm is perched high upon the yolk cell, rather separately from it, which gives the embryo something of a dumbbell appearance (Fig. 1C). During the hour in which the EVL restriction arises in the midblastula, the blastoderm flattens upon the yolk cell, producing a spherically shaped embryo (Fig. 1D) exactly at the time that EVL lineage restriction is complete. It seems likely that this change in shape could be accompanied by an increase in tension at the surface of the EVL, a change that itself might be responsible for the lineage restriction, for increased tension would influence the plane of mitosis and therefore the plane of division of EVL cells, such that both resulting daughter cells remain in this layer.

As early as epiboly begins in the late blastula, EVL cells behave distinctly from the underlying DEL cells, and this provides evidence that they have been separately specified. EVL cells become very flattened in shape and are tightly joined to one another (and also joined to the yolk cell; Betchaku and Trinkaus, 1978) as a epithelial monolayer, whereas DEL cells are only loosely associated with one another (or with the yolk cell or EVL) until well into gastrulation. EVL cells divide more slowly than DEL cells in the late blastula and gastrula, as revealed in lineage-tracing experiments (D.A. Kane, in preparation), again recalling, as for the yolk cell restriction, *Drosophila* mitotic domains. Whereas DEL cells begin morphogenetic rearrangements in the late blastula (see below), EVL cells do not (Warga and Kimmel, 1990), even though they are not completely passive during epiboly (Keller and Trinkaus, 1987; Kageyama, 1982). As a consequence, clones descended from EVL blastomeres are more compactly arranged than are clones derived from DEL cells, and are located in the embryo nearer the site where they originated in the blastula (Fig. 5B, and see Kimmel et al., 1990).

The EVL lineage restriction shares several features with the yolk cell restriction that occurred an hour earlier in development. Both lineages may be extraembryonic. The restrictions form geographically separate lineage compartments, are associated with pronounced morphogenetic changes in the embryo, and the restricted cells show distinctive mitotic and morphogenetic behavior. Both lineages thus seem to arise by specifications occurring in separate subsets of cells before gastrulation.

SPECIFICATION OF THE EMBRYONIC SOMATIC LINEAGES

Lineages that form the tissues of the body proper seem not to be separately specified during blastula stages, for single DEL cells at these stages give rise

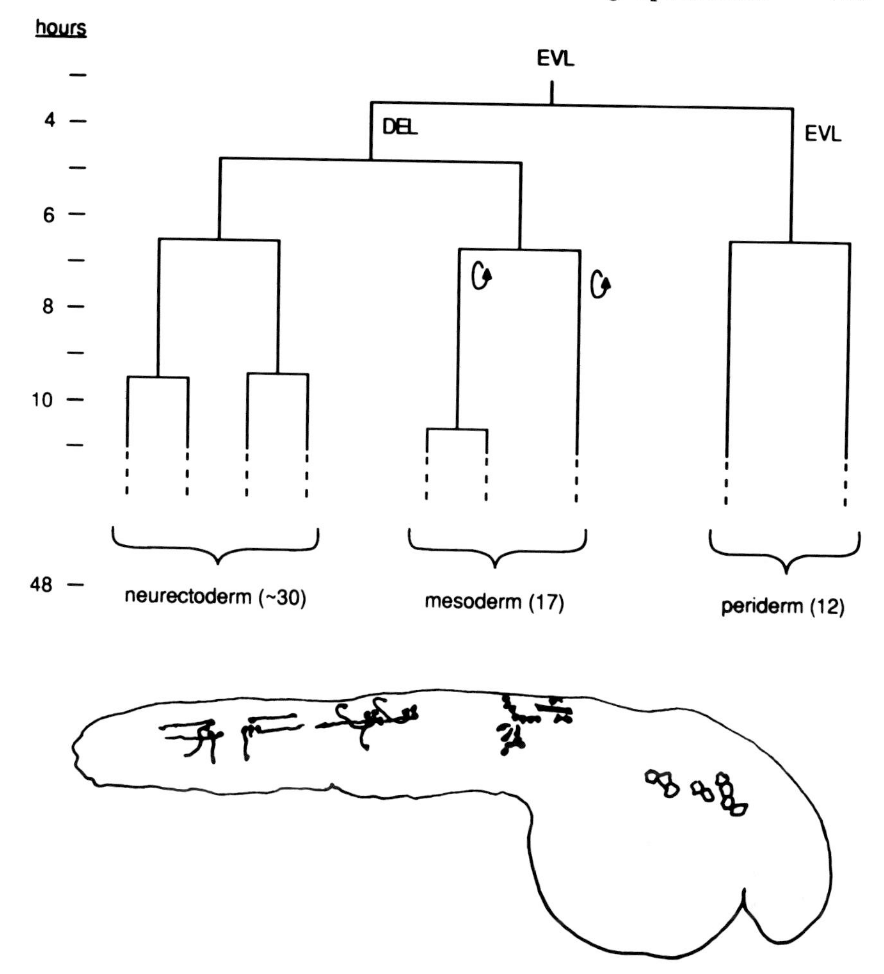

Fig. 5. Segregation of restricted EVL and DEL cell lineages in the blastula. **Upper:** Lineage diagram beginning with an EVL cell that was injected with lineage tracer dye, and showing numbers, in parentheses, and fates of its EVL and DEL descendants. Vertical lines are cells and horizontal lines are divisions. Time after fertilization is shown on the left. The curved arrows indicate times of involution. **Lower:** A view of the same clone in the live embryo at 24 h. The three clusters of labeled cells are of three different tissue types, corresponding to the descriptions shown just above the cells (from Kimmel et al., 1990).

to scattered clones containing diverse cell types. DEL cells scatter by so-called radial cell intercalations (Keller, 1980) that begin at about 4.3 h, an hour before gastrulation. The cell rearrangement accompanies or is part of epiboly, and in concert with changes in the shape of the yolk cell accomplishes flattening and expansion of the blastoderm into the form it has at the time gastrulation begins, a thinner cup-shaped structure inverted over the yolk cell (Fig. 6). The move-

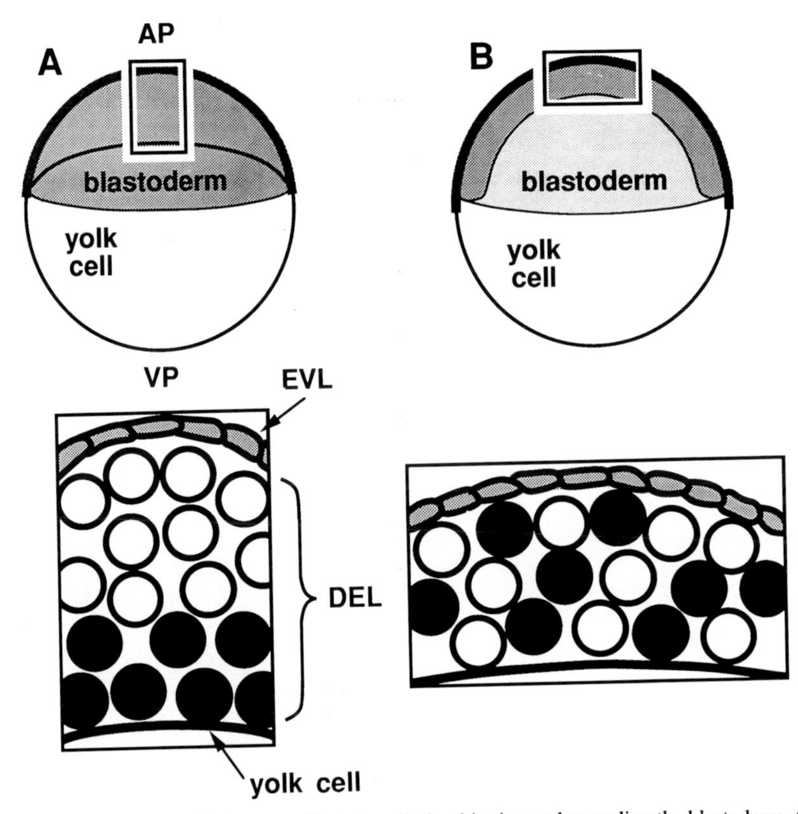

Fig. 6. Radial intercalations contribute to epiboly, thinning and spreading the blastoderm. Cutaway views of the midblastula (**A**) and through the onset of gastrulation (**B**). Deep DEL cells move outward (radially), intercalating among more superficial DEL cells but not among EVL cells (from Warga and Kimmel, 1990).

ments are directed in the sense that deeply lying cells move preferentially outward along radii to produce the thinning, and intercalate to produce the expansion of the cell sheet. As revealed by watching the intermixing in double-label experiments in which the deeper cells were labeled differently from the shallower ones (Warga and Kimmel, 1990), one has no sense that separate regions of the DEL are behaving differently from one another; rather, the DEL might be singly specified as a unified cell population in the blastula, at least with respect to the early movements. This interpretation fits with the fact that their lineages are unrestricted, even as to germ layer (Kimmel et al., 1990). Additionally, examining DEL cell cycles yields no evidence for distinctive mitotic domains within the DEL in the late blastula (D.A.K., in preparation).

Tissue-restricted DEL cell lineages are observed as gastrulation starts at 5.2 h, which suggests that separate specifications are beginning to arise within the DEL. DEL cells frequently divide once during the hour that comprises the late blastula period, a division often skipped in the EVL (Fig. 5A). DEL cells arising from this division, the 12th zygotic one, usually generate clones of cells that are all of the same tissue type. These lineages, termed ''tissue-restricted'' by Nishida (1987) and ''histospecific'' by Davidson (1990), may contain distinctive *cell* types, e.g., several distinct functional classes of neurons are often represented in a nerve tissue-restricted clone.

One might argue for a trivial explanation for the tissue-restricted lineages, proposing, for example, that there were not enough cell divisions or cell movements within a clone originating in the early gastrula for it to come to occupy more than a single tissue (see Kimmel and Warga, 1986). However, DEL cells divide at least 2–3 times during gastrulation (albeit these divisions occur at a slower rate than earlier; see Fig. 5A). Furthermore, the gastrulation movements of involution (Fig. 7) and particularly convergent extension (Fig. 8) separate clonally related cells as did the radial intercalations earlier, but now without scattering them into the rudiments of separate tissues. This more orderly behavior allows construction of a fate map at the stage gastrulation begins (Fig. 9).

Examining the morphogenetic movements of gastrula cells also provides evidence that separate specifications have occurred within the DEL by this stage, as contrasted with the blastula (see above). Involution occurs as DEL cells located initially near the margin move first toward the margin, and then turn underward to move away from the margin closer to the underlying yolk cell surface (Fig. 7). Involution is rigorously correlated with cell fate, for the upper layer or epiblast is the rudiment of the ectoderm, and the lower layer or hypoblast is the rudiment of the mesoderm and endoderm (Warga and Kimmel, 1990). Cells in single clones labeled at gastrulation onset generally either involute or not, in an all-or-nothing fashion.

Similarly, during gastrulation DEL cells undergo convergent extension, accumulating in and lengthening the embryonic axis (Fig. 8). The axis forms on the dorsal side of the gastrula, which becomes recognizable by 6 h as converging cells accumulate there within the germ ring. The accumulation forms the so-called embryonic shield (shown within the box in Fig. 8). Convergence is a directed movement (i.e., cells move toward the dorsal midline) but, unlike the directed radial intercalations observed in the late blastula, different early gastrula cells undergo these mediolateral intercalations (Keller and Danilchick, 1988; Keller and Tibbetts, 1989) to greater or lesser extents, apparently as a function of their blastoderm positions. Thus, dorsal cells appear to more actively undergo intercalations than ventral cells, as evidenced by a dorsally located clone becoming much more dispersed along the axis than a ventral one at the same stage (R.M. Warga and C.B.K., unpublished work in progress). This activity of cells produces the dramatic extension of the axis that occurs on the dorsal side of the gastrula at the expense of more ventral cells. Mediolateral

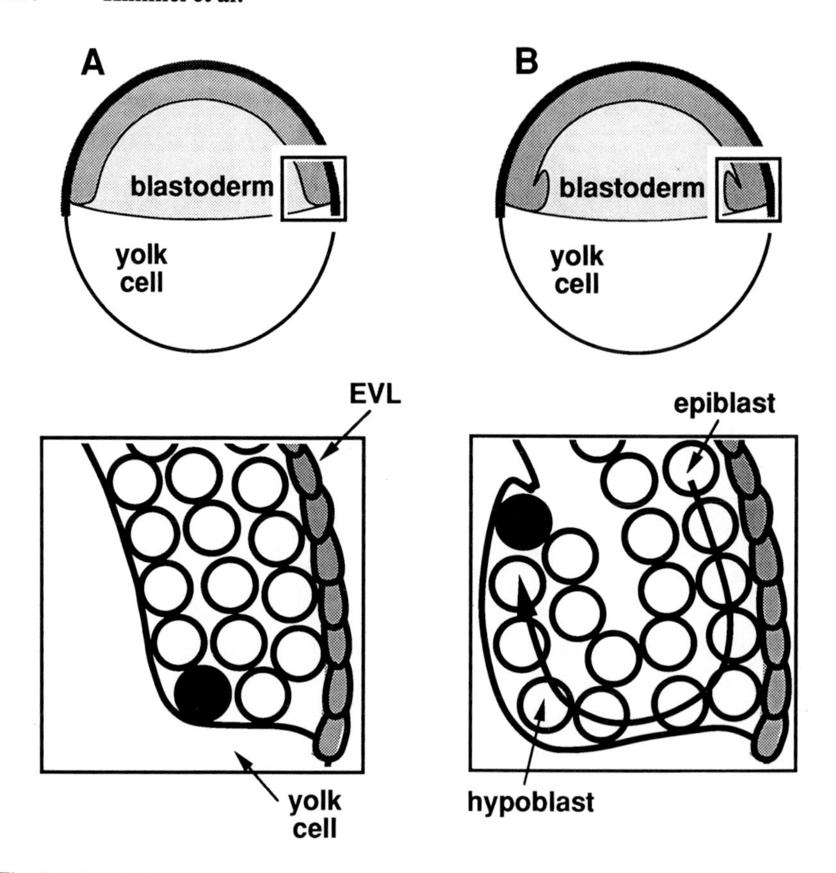

Fig. 7. Involution generates the hypoblast. Cut-away side views at gastrulation onset (**A**) and when the germring has formed, 20 min later (**B**). A DEL cell first at the blastoderm margin (black) is at the front of the wave of involuting cells. EVL cells do not involute (from Warga and Kimmel, 1990).

intercalations occur within both the epiblast and hypoblast, but whether there are quantitative differences between these layers has not been studied.

The tissue-restricted cell lineages and region-specific cell movements of the early gastrula argue that DEL cells are specified in the early gastrula, even though they may be only broadly specified, not to the level of individual differentiated cell types at specific locations. Convergent extension scatters cells of a single clone along the anterior–posterior axis such that it is unlikely that any precise anterior–posterior positional specifications occur at the single-cell level during early gastrulation (Kimmel and Warga, 1987b). Also, as pointed out above (see also Kimmel and Warga, 1986), tissue-restricted clones may have several cell types that are normally characteristic of one kind of tissue; in

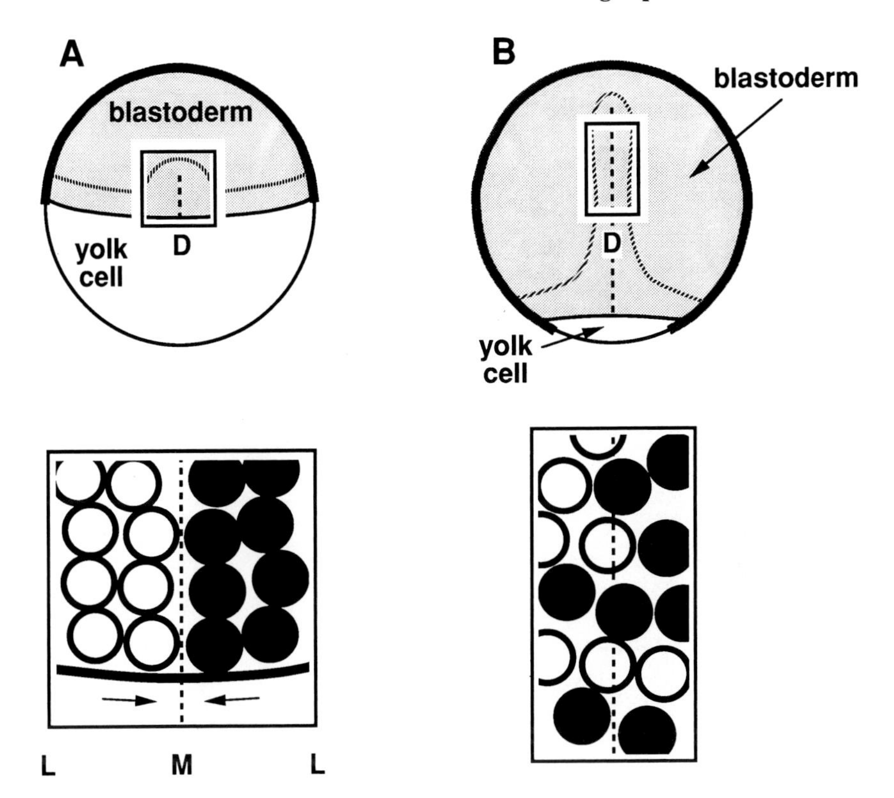

Fig. 8. Mediolateral intercalations produce convergent extension. Dorsal (D) views of early (**A**) and late (**B**) gastrula. DEL cells converge from lateral (L) to medial (M) positions, lengthening the embryonic axis (dashed line) (from Warga and Kimmel, 1990).

at least some lineages that generate the more complex tissues, the specification of exact cell fate may occur much later. More work needs to be done to learn to what extent specifications of somatic lineages occur in a progressive, hierarchical fashion.

THE *spt-1* MUTATION DIFFERENTIALLY AFFECTS GASTRULATION MOVEMENTS

We imagine that the early specifications produce subsets of cells at particular blastoderm positions whose distinctive behaviors are underlain by the differential expression of zygotic genes. That embryonic cells differentially transcribe genes in a position-specific manner as early as midblastula transition is well known in *Xenopus* (e.g., Rosa, 1989). Mutational analysis, for which zebrafish have some advantages (Streisinger et al., 1981), provides a

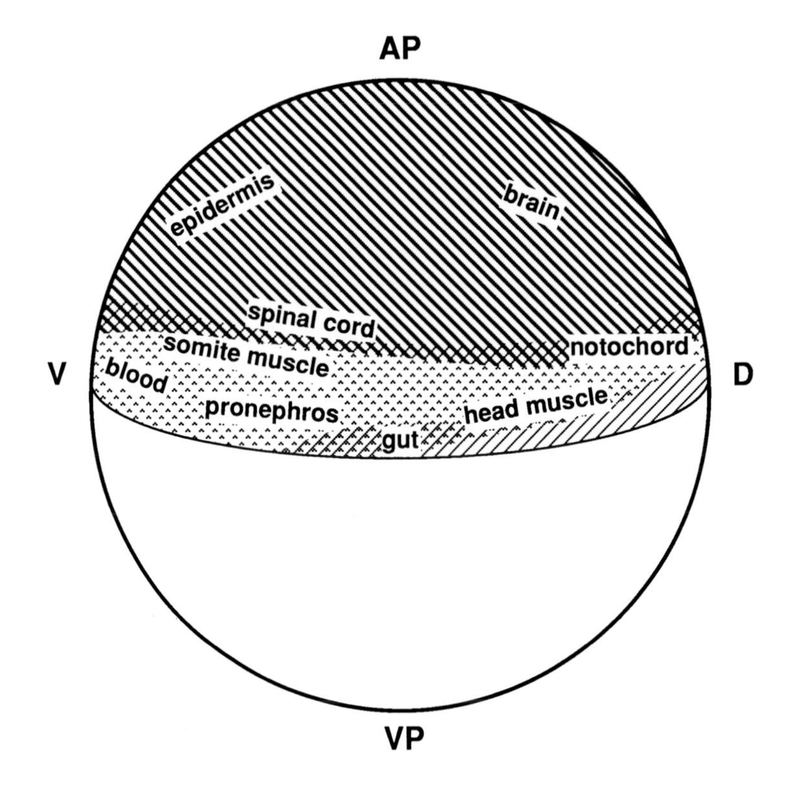

Fig. 9. Fate map for the zebrafish at the beginning of gastrulation. The blastoderm covers the upper half of the yolk cell. Shadings of different textures show overlapped regions, from the animal pole to the margin, where DEL cells give rise to ectoderm, mesoderm, and endoderm (from Kimmel et al., 1990).

functional approach toward learning about the genes and the roles they play in early embryos, and systematic screens for zygotic lethal mutations that disrupt patterning of the embryos are underway. ''Spadetail,'' *spt-1(b104)*, is the first mutation discovered that acts during gastrulation; it appears to directly produce a very discrete change in the movements of a single population of gastrula cells. However, as we shall see, this single initial change has severe consequences for later development.

In the motile embryo at 24 h there are two very prominent aspects of the *spt-1* phenotype; somitic mesoderm is severely depleted and disrupted in the trunk of the body, and there is an extra mass of cells accumulated in a ball at the end of the tail. At 10 h, when somites normally begin to form, one can observe the early changes that lead to both abnormalities; at this stage in the mutants there is a deficiency of paraxial mesoderm from which the somites

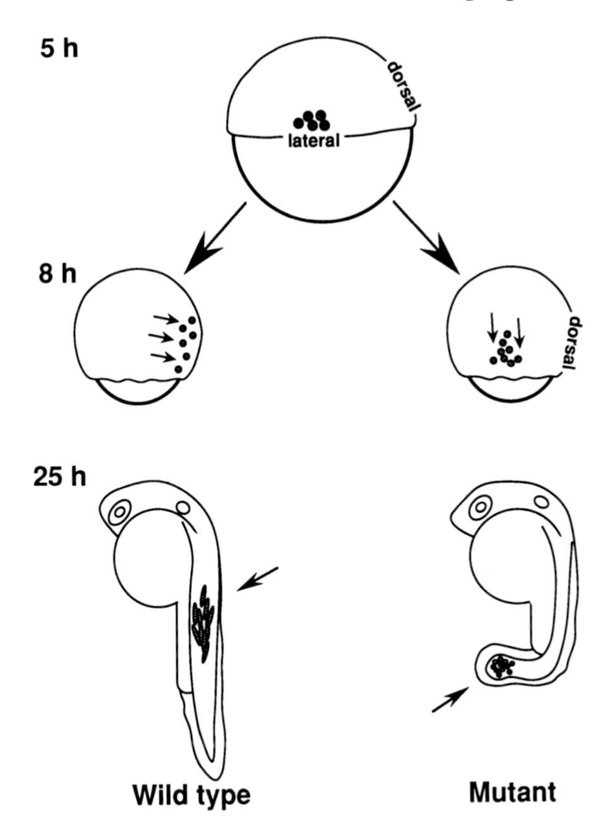

Fig. 10. The "spadetail" mutation, *spt-1*, blocks dorsalward convergence movements of prospective trunk somitic cells. DEL cells, labeled with lineage tracer, and present near the lateral margin of the blastoderm just before gastrulation begins (5 h, top) move dorsally and form trunk muscle in the wild type (left), but move posteriorly and form tail mesenchyme and other mesodermal cell types in the mutant (right).

normally arise. At the same stage the newly formed tail bud is larger than normal. In fact, cells begin to be abnormally located even earlier, as subtle changes in the shape of the gastrula can be detected as early as 7–8 h (Kimmel et al., 1989; D.A.K., unpublished results).

These observations suggested an explanation for the mutant phenotype, namely, that cells which should have formed somites in the trunk instead are present in the tail rudiment by the end of gastrulation (Fig. 10). Trunk somites normally arise from involuting cells present along the lateral margin of the early gastrula, which converge dorsally to take up their dorsal position along the axis by the time epiboly ends (Kimmel and Warga, 1987b). By labeling these cells in mutants with lineage tracer dye and following their movements

(Kimmel et al., 1989), we observed that the lateral marginal cells in fact do not converge correctly, but move in an abnormal posterior direction to enter the tail bud. Developing in the tail they can express a variety of mesodermal fates, including myotomal (somite-derived) tail muscle, for *spt-1* mutants make tail somites fairly normally in spite of not making them, at least on schedule, in the trunk. Other fates of the misplaced cells include mesenchyme and notochord. These are mesodermal fates, as is muscle, but, especially for notochord, are inappropriate considering the fate map (Fig. 9). Notochord invariably arises from dorsal cells of the wild-type gastrula, but these *spt-1* tail notochord cells come from lateral positions. Finally, cells remaining in the terminal swelling after formation of the last tail somite generally expressed no differentiated fate at all and died *en masse* in the late embryo. Cell death has also been noticed in the corresponding location in the wild type, and it is likely that cell death, magnified as it is in the mutant, is a normal fate for cells remaining at the end of the segmental plate after segmentation is completed. Making cells in excess, and then eliminating the extra ones, is a way to assure that enough cells are present to form the complete series of somites, as previously proposed for leech (Shanklin, 1984) and chick embryos (Bellairs, 1986).

The *spt-1* mutation thus may directly interfere with morphogenesis, and as a consequence, cells change their fates in accordance with their new positions. Only a single gastrulation movement, convergence, is defective; the cells appear to undergo normal epiboly, involution, and extension movements. Disruption of convergence can be interpreted as a specific loss of function if we assume that the pathway to the tail is a default one, rather than a gain of a new function. Loss of function is in keeping with the fact that the mutation produces a recessive phenotype. We argued above that during gastrulation wild-type cells have already become specified to one or another tissue-level fate. The gastrula stage is the time when morphogenesis is disturbed in the mutant, and, since the displaced mutant cells developed regulatively in the tail to generate such fates as notochord, they must have been *uncomitted* during gastrulation. We thus believe that these mutant cells, initially specified to make muscle, must have been *re*specified after entering the tail bud to make notochord. This interpretation requires that specification be reversible and separable from commitment. An interesting but unlikely (see below) alternative explanation is that *spt-1* acts directly to change specification and commitment in the early gastrula, and as a consequence the cells migrate abnormally.

Disruption of the convergence movement appears to be specific for the lateral mesoderm. Prospective neuroectodermal cells that form the trunk spinal cord converge to the dorsal side of the gastrula from remote distances in wild types and mutants alike, and make their correct fates. A very discrete case where neuroectodermal development occurs correctly is the set of primary sensory neurons known as Rohon-Beard cells. Rohon-beard progenitors fate map to the ventral half of the blastoderm (Kimmel et al., 1990); they need to converge dorsally for very long distances in order to take up their definitive locations in the dorsal spinal cord. However, in *spt-1* embryos, Rohon- Beard cells are made in the correct numbers and locations (J.S. Eisen, unpublished observations).

Disruption of convergence in only a subset of cells that normally undergo this movement suggests that the *spt-1* mutation acts in this subset specifically, rather than perturbing general navigation cues that lead cells to the dorsal trunk. To learn in which cells the mutation exerts its direct effects, Ho and Kane (1991) used cell transplantation to produce genetically mosaic embryos. Mixtures of wild-type and mutant cells, distinctively labeled with lineage tracer dyes of two colors, were transplanted together into one blastoderm location in unlabeled wild-type or mutant hosts at or before the beginning of gastrulation, and the subsequent development of the cells determined. If placed into the lateral margin, wild-type and mutant donor cells segregated from one another, irrespective of the host genotype. The mutant cells went to the tail and made such derivatives as mesenchyme, as they would have starting from that position in the mutant. The wild-type cells converged dorsally to form trunk paraxial mesoderm muscle. This segregation of wild-type and mutant cells was only observed if the mixtures were transplanted into the lateral margin, from which paraxial mesoderm arises. For example, if placed at the dorsal margin the transplanted cells remained mixed together and made notochord, and if placed laterally but further from the margin they converged together, remained mixed together, and made nervous system. Thus, *spt-1* acts specifically and autonomously in lateral marginal cells of the gastrula.

In these experiments, observing the time when the transplanted wild-type and mutant cells begin to segregate provides a way to learn the stage of development when the wild-type (*spt-1*+) gene function is required for normal morphogenesis. Such analysis, utilizing time-lapse video (D.A.K. and R.K.H., unpublished), has revealed that mutant cells move incorrectly as they are involuting in the early gastrula, only a few hours after midblastula transition, when the gene might first be transcribed.

These studies are important ones for our understanding of cell specification, for they show that separate populations of cells differentially require *spt-1*+ function in order to gastrulate correctly. Possibly the gastrulating cells differentially *express* the gene as well. Learning if this so, which will require molecular identification of *spt-1*+, will provide direct evidence pertinent to the idea that a genetically based specification of gastrulating lateral marginal cells is crucial for correct development. The early defect that *spt-1* produces, whatever its molecular nature, has severe consequences with respect to later development. Presumably because of the early dislocation of cells, several tissues begin to develop abnormally after the gastrula stage, and eventually the mutants die as young larvae.

EARLY GASTRULA CELLS ARE UNCOMMITTED

Because diverse phenotypes are expressed by *spt-1* cells that enter the tail, not just a single one, it seems unlikely that the direct action of the mutation is to change the specification and commitment of early lateral marginal cells. The transplantation studies carried out in the analysis of *spt-1* also provide insight about whether cells are committed to their normal fates in the zebrafish early gastrula. The experiments were done by taking cells from *any* dorsoventral blastoderm location from the donors, and putting them into *any* location in the hosts. No attempt was made to transplant the cells orthotopically; indeed this would have been impossible, for dorsoventral polarity of the blastoderm is invisible until the embryonic shield forms at 6 h, an hour or so after transplantation. Yet, how the transplanted cells developed in the mosaic embryos was completely predictable from their genotypes and their starting positions in the hosts. Furthermore, in a set of controls for the mosaic analysis in which cells from two wild-type donors were put together into the same location in wild-type host, the transplanted cells invariably both developed the same way, in accordance with their new positions in the hosts. They undoubtedly had come from separate donor locations, in the vast majority of cases, and therefore to develop as they did in the hosts, the cells would have had to switch from their old to a new developmental pathway. Thus, by this test, the cells appear uncommitted at the beginning of gastrulation, in accord with the experimental findings in other teleost embryos made years ago (Oppenheimer, 1936, 1938; Luther, 1936). Further experiments, in which single early gastrula cells were deliberately repositioned (e.g., between the animal pole and blastoderm margin) also showed that early gastrula cells can switch fates, confirming that they are not committed at this stage (R.K.H., unpublished observations).

SEPARABILITY OF SPECIFICATION AND COMMITMENT

Most of this review has dwelt on observations suggesting that as gastrulation begins, or in the case of the yolk cell even as early as the midblastula transition, separate subpopulations of cells in the zebrafish are developing along distinctive pathways. Yet transplantation experiments show that early gastrula cells are not committed to their normal fates. Although the current evidence is incomplete, the experiments we have reviewed suggest that specification and commitment are processes that are temporally separable; cells are specified before they are committed. Evidence is accumulating from other organisms as well, including frogs (Snape et al., 1987; Wylie et al., 1987) and sea urchins (Chapter 11, Ettensohn, this volume), for developmental plasticity in specified lineages. Thus it may be conceptually useful to consider specification and commitment as generally separable and distinct events. One must not assume, as is sometimes done, that acquisition of a specific behavior or marker phenotype by a developing cell means the cell has undergone commitment.

Why should early specification be reversible? There are at least two classes of explanations, not mutually exclusive: 1) For whatever reason, a specific cell may require time to consolidate the committed state (see e.g., Roussant, 1977). Imagine that a single inductive interaction mediates a developmental transition in which a molecular change can be experimentally detected nearly immediately. Thus the responding cell is specified following the interaction, but it may not express a restriction in potential until, e.g., a cell division occurs (Brown, 1984). By this kind of argument a specification might function only as a preliminary step on a course that normally leads to commitment; the specified "state" might have no independent role. 2) On the other hand, specification and commitment might well be separable for reasons that are functionally important, in terms of the normal course of development. We should like to argue for this point of view. In particular, cells might be specified at a certain stage in part so they may carry out position-dependent functions soon afterward, irrespective of their later course of development.

This view predicts that at least some of the molecules uniquely expressed by specified cells will have an early functional role, not one that will only become important after commitment and differentiation. Thus the cytokeratins synthesized by animal cap cells in the frog midblastula might function in the cell-specific behaviors of these cells themselves (e.g., during epibolic spreading), and not in their descendants that undergo definitive cytodifferentiation at a much later stage. We have argued above that the different early patterns of cell division and movements observed in distinct subsets of early cells provide evidence for early specifications, and clearly how each of the types plays out its role is important for the orchestration of the development of the whole embryo. We have just seen, in the analysis of *spt-1*, the profound effects that can result when perhaps one component of morphogenesis is changed in one subset of gastrula cells.

Given that early specifications could be functionally significant, why might a cell opt, during normal development, for a developmental pathway different from the one for which it was initially specified? In other words, why is it important that the early decisions not irrevocably determine final fate? We suggest that this is because of morphogenesis itself. The most dramatic movements of cells during embryogenesis occur in the gastrula, after initial specifications have occurred, but before cells occupy their final destinations. It is only after these movements are completed that it is critical that cell commitment be in accord with cell position. The morphogenetic cell movements of epiboly, convergence, and extension that occur during gastrulation are all accompanied by cell intercalation events (Keller, 1986; Warga and Kimmel, 1990). The rearrangements that occur, although highly regulated, appear to be imprecise with respect to the movements and neighbor changes of individual cells. Cells that have begun to develop along one pathway might be dislocated, during these rearrangements, into positions where that pathway is inappropriate. Even though cells must express genes in accordance with their positions (and their functions) before and during gastrulation, it also could be crucial that they remain potent to develop in accordance with a range of possible environments to which the cell-scattering morphogenetic movements might bring them.

ACKNOWLEDGMENTS

We thank Nigel Holder, John Postlethwait, Rachel Warga, Tom Schilling, and Jay Mittenthal for contributing to the ideas presented here. Our work was supported by NSF grant BNS-8708638 and NIH grant HD22486.

REFERENCES

Agassiz A, Whitman CO (1884): On the development of some pelagic fish eggs—preliminary notice. Daedulus 20:23–75.

Bellairs R (1986): The tail bud and cessation of segmentation in the chick embryo. In Bellairs R, Ede DA, Lash JW (eds): ''Somites in Developing Embryos.'' New York: Plenum, pp 161–178.

Betchaku T, Trinkaus JP (1978): Contact relations, surface activity, and cortical microfilaments of marginal cells of the enveloping layer and of the yolk syncytial and yolk cytoplasmic layers of *Fundulus* before and during epiboly. Exp Zool 206:381–425.

Bouvet J (1976): Enveloping layer and periderm of the trout embryo. (*Salmo trutta fario L.*). Cell Tiss Res 170:367–382.

Brown DD (1984): The role of stable complexes that repress and activate eukaryotic genes. Cell 37:359–365.

Dale L, Slack JMW (1987): Fate map for the 32-cell stage of *Xenopus laevis*. Development 99:527–551.

Davidson EH (1990): How embryos work: A comparative view of diverse modes of cell fate specification. Development 108:235–389.

Foe VE (1989): Mitotic domains reveal early commitment of cells in *Drosophila* embryos. Development 107:1–22.

Garcia-Bellido A, Ripoll P, Morata G (1973): Developmental compartmentalization of the wing disc of *Drosophila*. Nature New Biol 245:251–253.

Ho RK, Kane DA (1991): The zebrafish *spt-1* mutation acts autonomously in specific mesodermal precursors. Nature 334:728–730.

Jonas E, Sargent TD, Dawid IB (1985): Epidermal keratin gene expressed in embryos of *Xenopus laevis*. Proc Natl Acad Sci USA 82:5413–5417.

Kageyama T (1982): Cellular basis of epiboly of the enveloping layer in the embryo of the Medaka, *Oryzyas latipes*. II. Evidence for cell rearrangement. J Exp Zool 219:241–256.

Keller RE (1980): The cellular basis of epiboly: An SEM study of deep cell rearrangement during gastrulation in *Xenopus laevis*. J Embryol Exp Morphol 60:201–234.

Keller RE (1986): The cellular basis of amphibian gastrulation. In Browder L (ed): ''Developmental Biology: A Comprehensive Synthesis,'' Vol 2. New York: Plenum, pp 241–327.

Keller RE, Danilchik M (1988): Regional expression, pattern and timing of convergence and extension during gastrulation of *Xenopus laevis*. Development 103:193–209.

Keller RE, Tibbetts P (1989): Mediolateral cell intercalation in the dorsal, axial mesoderm of *Xenopus laevis*. Dev Biol 131:539–549.

Keller RE, Trinkaus JP (1987): Rearrangement of enveloping layer cells without disruption of the epithelial permeability barrier as a factor in *Fundulus* epiboly. Dev Biol 120:12–24.

Kimmel CB (1989): Genetics and early development of zebrafish. Trends Genetics 5:283–288.
Kimmel CB, Kane DA, Walker C, Warga RM, Rothman MB (1989): A mutation that changes cell movement and cell fate in the zebrafish embryo. Nature 337:358–362.
Kimmel CB, Law RD (1985a): Cell lineage of zebrafish blastomeres. I. Cleavage pattern and cytoplasmic bridges between cells. Dev Biol 108:78–85.
Kimmel CB, Law RD (1985b): Cell lineage of zebrafish blastomeres. II. Formation of the yolk syncytial layer. Dev Biol 108:86–93.
Kimmel CB, Law RD (1985c): Cell lineage of zebrafish blastomeres. III. Clonal analyses of the blastula and gastrula stages. Dev Biol 108:94–101.
Kimmel CB, Warga RM (1986): Tissue-specific cell lineages originate in the gastrula of the zebrafish. Science 231:365–368.
Kimmel CB, Warga RM (1987a): Indeterminate cell lineage of the zebrafish embryo. Dev Biol 124:269–280.
Kimmel CB, Warga RM (1987b): Cell lineages generating axial muscle in the zebrafish embryo. Nature 327:234–237.
Kimmel CB, Warga RM (1988): Cell lineage and developmental potential of cell in the zebrafish embryo. Trends Genetics 4:68–74.
Kimmel CB, Warga RM, Schilling TF (1990): Origin and organization of the zebrafish fate map. Development 108:581–594.
Luther W (1936): Potenzprufungen an isolierten teilstucken der forellenkeimscheibe. Arch Entw Mech Org 135:359–383.
Moody SA (1987): Fates of the blastomeres of the 16-cell stage *Xenopus* embryo. Dev Biol 119:560–578.
Newport J, Kirshner M (1982): A major developmental transition in early *Xenopus* embryos. I. Characterization and timing of cellular changes at the midblastula stage. Cell 30:675–686.
Nishida H (1987): Cell lineage analysis in ascidian embryos by intracellular injection of a tracer enzyme. III. Up to the tissue restricted stage. Dev Biol 121:526–541.
Oppenheimer JM (1936): Processes of localization in developing *Fundulus*. J Exp Zool 73:405–444.
Oppenheimer JM (1938): Potencies for differentiation in the teleostean germ ring. J Exp Zool 79:185–212.
Rosa FM (1989): *Mix.1*, a homeobox mRNA inducible by mesoderm inducers, is expressed mostly in the presumptive endodermal cells of *Xenopus* embryos. Cell 57:965–974.
Roussant J (1977): Cell commitment in early rodent development. In Johnson MH (ed): "Development in Mammals," Vol 2, Amsterdam: Elsevier-North Holland, pp 119–150.
Shanklin M (1984): Positional determination of supernumerary blast cell death in the leech embryo. Nature 307:541–543.
Snape A, Wylie CC, Smith JC, Heaseman J (1987): Changes in states of commitment of single animal pole blastomeres of *Xenopus laevis*. Dev Biol 119:503–510.
Stent GS (1985): The role of cell lineage in development. Phil Trans Roy Soc Lond B 312:3–19.
Streisinger G, Walker C, Dower N, Knauber D, Singer F (1981): Production of clones of homozygous diploid zebra fish (*Brachydanio rerio*). Nature 291:293–296.
Trinkaus JP (1984): Mechanisms of *Fundulus* epiboly—a current view. Am Zool 24:673–688.
Waddington CH (1956): "Principles of Embryology." London: Allen & Unwin, 510 pp.
Warga RM, Kimmel CB (1990): Cell movements during epiboly and gastrulation in zebrafish. Development 108:569–580.
Weir MP, Low CW (1982): Gap junctional communication compartments in the *Drosophila* wing disk. Proc Natl Acad Sci USA 79:3232–3235.
Weiss P (1939): "Principles of Development." New York: Holt, 601 pp.
Wilson HV (1889): The embryology of the sea bass. Bull US Fish Comm 9:209–278.
Wylie CC, Snape A, Heaseman J, Smith JC (1987): Vegetal pole cells and commitment to form endoderm in *Xenopus laevis*. Dev Biol 119:496–502.

Cell-Cell Interactions in Early Development, pages 227–239

13. Neuronal Development in the *Drosophila* Compound Eye: Role of the *rap* Gene

Tadmiri Venkatesh and Jon Karpilow
Institutes of Neuroscience (T.V.) and Molecular Biology (J.K.), University of Oregon, Eugene, Oregon 97403

INTRODUCTION

Nervous systems of multicellular organisms are comprised of a variety of cell types with complex patterns of connectivity. The developmental mechanisms that lead to the determination of these cellular phenotypes are not well understood. In recent years, studies in both vertebrates as well as invertebrates have shown that cellular interactions and environmental cues play an important role in the determination of cell fate and pattern formation. The *Drosophila* compound eye is a highly precise neural pattern that is readily accessible to both classical genetic and molecular analysis and, thus, provides an excellent system to study the cellular and molecular mechanisms underlying pattern formation. In this chapter, we present a brief description of the organization and development of the compound eye followed by some of our studies on the role of *rap* (retina aberrant in pattern) gene in photoreceptor pattern formation.

STRUCTURE AND DEVELOPMENT OF THE *DROSOPHILA* COMPOUND EYE

The eye comprises some 800 identical modules called ommatidia, in a highly regular array (Fig. 1A). Each adult ommatidium contains eight photoreceptor neurons (R cells) and 14 accessory cells. Some of these accessory cells serve to optically isolate each ommatidium and secrete the lens. The eight photoreceptor cells are organized in a distinctive pattern; six outer R cells (R1–6) surround two central cells (R7 and R8). Each R cell has a radially oriented photosensitive organelle, the rhabdomere. The rhabdomeres are formed by the multiple infolding of the plasma membrane of the R cells and contain the photosensitive pigment opsin characteristic of each cell type. Cells R1–6 share the same action spectrum and form synapses in the first optic ganglion (the lamina), whereas R7 and R8 each have unique action spectra and each synapses at a different level of the second optic ganglion, the medulla. Thus, the retina

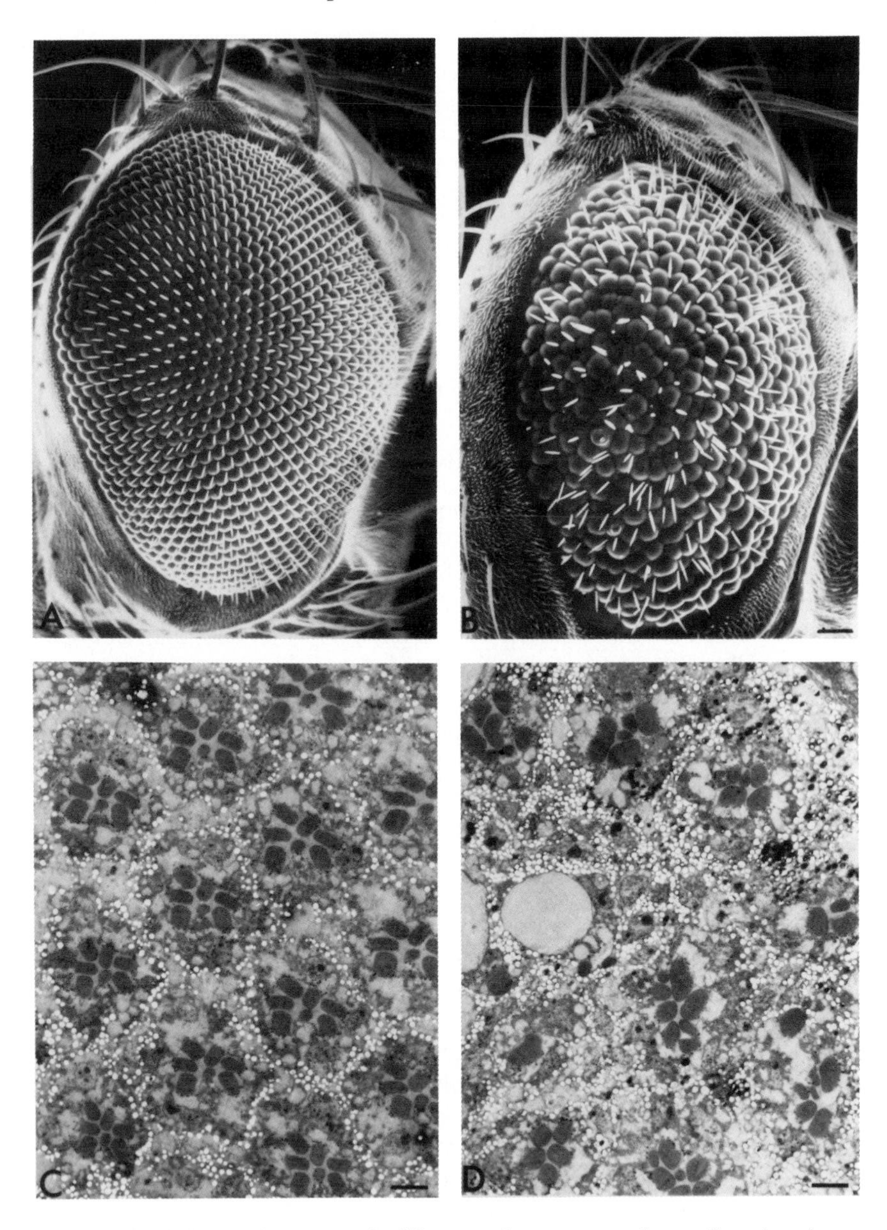

Fig. 1. Morphology and anatomy of wild-type and *rap* compound eyes. Scanning electron micrographs of (**A**) wild-type (×185) and (**B**) rap^3 (×210) compound eyes. The wild-type eye has modular structure with a smooth hexagonal array, whereas the *rap* mutants show unevenly sized ommatidia with aberrant spacing. Tangential sections of (**C**) wild-type and (**D**) *rap*^{3} (×300) compound eyes as viewed under transmission electron microscope. Note the regular trapezoidal arrangement of R cells in each ommatidium. In *rap* eyes (D), the ommatidia are irregular and are comprised of a variable number of R cells.

contains three distinct neuronal types (R1–6, R7, and R8), each with different synaptic specificity. The axonal projections of the R cells are such that the visual field is topographically reconstructed in the optic ganglia. The curvature of the eye and the orientation of the rhabdomeres in each ommatidium necessitate a precise, interlacing pattern of R cell axonal projections into the optic ganglia. This consitutes a spectacular example of neuronal organization and synaptic specificity (Trujillo-Cenoz and Melamed, 1966; Braintenberg, 1967; Kirschfeld, 1967; Harris et al., 1976).

The adult compound eye develops from a monolayer of undifferentiated epithelial cells called the eye imaginal disc. During the third larval instar, pattern formation begins as a wave of morphogenesis moves from posterior to anterior across the disc. This wave is marked by a depression on the disc surface (the morphogenetic furrow), behind which small and precisely distributed cell clusters differentiate into photoreceptor neurons. Anatomical studies employing neuron-specific antibodies have shown that the R cells differentiate in a sequential manner. Photoreceptor cell R8 expresses neural antigens first and is thereafter joined in a pairwise fashion by cells R2 and R5, and R3 and R4. Subsequently, R1 and R6 differentiate, with R7 joining last to form the mature eight-cell cluster (Tomlinson and Ready, 1987a). These events involve the precise movement of cell nuclei along the disc epithelium and are thought to require specific cell-cell contacts (Tomlinson, 1985).

Mosaic analysis has shown that there are no strict lineage relationships between R cells within an ommatidium (Ready et al., 1976; Lawrence and Green, 1979). This finding has led to the view that cell-cell interactions and environmental cues play an important role in the determination of R cell fates. Studies of mutations that affect retinal development support this inference by revealing the presence of genes that are involved in specific inductive interactions between developing R cells, and have provided clues to the nature of molecular strategies involved (for a review, see Tomlinson et al., 1988; Rubin, 1989; Ready, 1989; Banerjee and Zipursky, 1990). While these and other studies have provided insights to the nature of cell-cell interactions and their importance in differentiation, the mechanisms that initiate ommatidial patterning events are poorly understood. Anatomical and immunocytological studies have shown that pattern formation begins with the differentiation and precise positioning of photoreceptor R8. Thus, it is likely that one or more R8-specific functions are required for these processes. By identifying genes that play a role in these early developmental events, it may be possible to determine the mechanism by which pattern formation is initiated.

In this chapter, we describe our studies on the genetic locus *rap*, whose functions are critical for neural pattern formation in the compound eye. A detailed genetic and phenotypic characterization of *rap* has been presented by Karpilow et al. (1989). The *rap* locus exhibits many of the attributes we would expect of a gene involved in ommatidial pattern initiation. Specifically, we present evidence showing that *rap* gene function is required during the stage of development when ommatidial assembly is known to occur. Moreover, immunocytological analysis of eye imaginal discs shows that a mutation in the *rap* gene results in a disruption of patterning events directly behind the morphogenetic furrow. These initial findings prompted a mosaic analysis of the *rap* gene. The results of these studies show that *rap*$^+$ gene function is required only in photoreceptor cell R8 for proper ommatidial assembly. Together, the data presented here detail the significance of *rap* gene function in eye pattern formation and contribute to the growing body of evidence that shows that photoreceptor cell R8 plays a central role in ommatidial development.

rap MUTANTS EXHIBIT A VARIABLE NUMBER OF R CELLS IN EACH OMMATIDIUM

The relatively smooth texture of the wild-type compound eye results from the precise and ordered distribution of 800 hexagonally shaped ommatidia. These ommatidia are arranged in rows and show a high degree of homology in both their size and shape (Fig. 1A). In contrast, the configuration of ommatidial units found in the *rap* compound eye is extremely disordered. The *rap* mutant ommatidia are aberrant in their size, shape, and alignment with respect to neighboring units. These gross changes in structure and pattern are accompanied by abnormalities in the distribution of mechanosensory bristles and a slight reduction in the total surface area of the eye. As a result of these events, the exterior eye morphology of the *rap* mutant is extremely rough (Fig. 1B).

The underlying basis for the change in ommatidial size and shape was revealed when mutant eyes were sectioned and examined by transmission electron microscopy. Whereas tangential sections of wild-type compound eyes show each ommatidium to contain the stereotypic number and position of photoreceptor cells (Fig. 1C), the number of R cells per ommatidia in *rap* mutants varies considerably (Fig. 1D). Furthermore, the R cell position and rhabdomere arrangement in *rap* mutants is altered. This aspect of the *rap* phenotype makes it difficult to assign R cell identities to the photoreceptor cells without the aid of cell-specific markers.

MUTATION IN THE *rap* GENE AFFECTS EARLY CELL RECRUITMENT EVENTS

The various stages of R cell differentiation and cluster formation can be visualized in a single disc by staining with monoclonal antibodies (MAb) that recognize specific antigens in the eye disc (Zipursky et al., 1984; Venkatesh et al., 1985). We used neuron-specific monoclonal antibody MAb22C10 to investigate pattern formation in eye-antennal discs taken from third-instar *rap* mutant larvae. The staining patterns observed with MAb22C10 show that the process by which cells are sequestered into clusters behind the furrow is abnor-

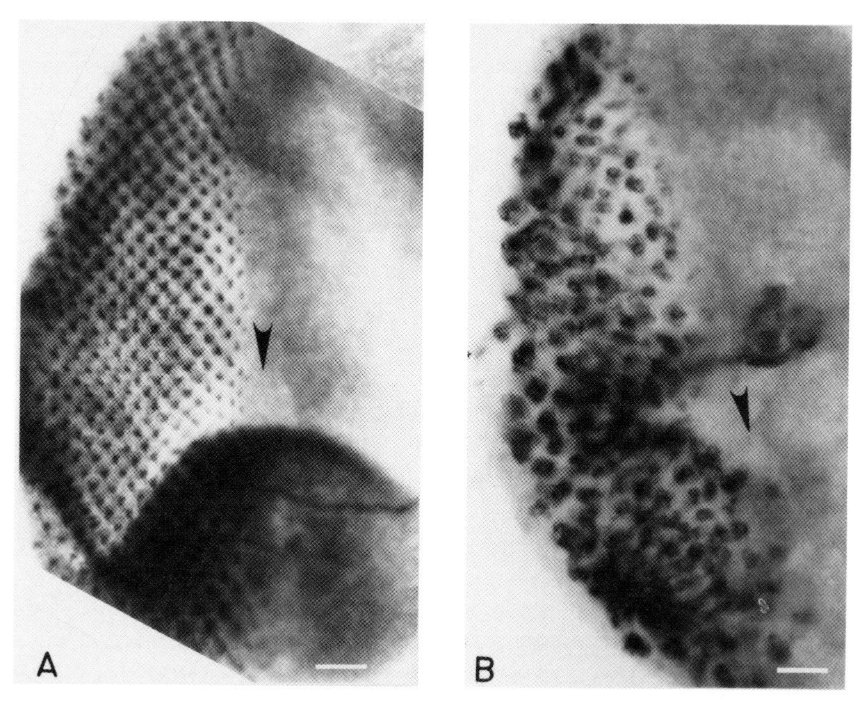

Fig. 2. Developing R cell clusters stained with neuron specific antibody MAb22C10. (**A**) Wild-type disc showing an ordered array of R cell clusters (arrowhead) with mature eight-cell clusters at the posterior region of the disc. (**B**) *rap* mutant disc showing abnormal patterning resulting in variably sized R cell clusters distributed in the region posterior to the furrow (arrowhead).

mal. Typically in wild-type discs, clusters differentiate in a spatially symmetrical fashion along both the posterior–anterior and lateral axes. Five-cell preclusters are observed a few rows behind the furrow and more mature clusters, containing seven and eight cells, are tightly packed together in more posterior regions of the disc (Fig. 2A). As shown in Figure 2B, many of the normal attributes of disc patterning are lost in the *rap* mutant. We have observed that the arrangement of photoreceptor preclusters in *rap* discs is extremely disorganized. Clusters are no longer distributed in an ordered array, and the progressive increase in cluster size that is observed in wild-type tissues is absent in *rap* mutant. In addition, the level at which photoreceptor neurons differentiate appears to have been altered. Typically, the cell bodies of the developing neurons observed in wild-type tissue are apically positioned within the developing disc. This is not the case in the *rap* mutant. Using camera lucida reconstructions of the MAb22C10-stained disc, we observed that the R cells of *rap* eye imaginal discs differentiate at various levels within the tissue (Karpilow et al., 1989).

Examination of regions close to the furrow suggests that the pattern abnormalities observed in *rap* mutants occur early during ommatidial assembly and may result from a disruption in the events involved in the differentiation of R8. Single cells can be observed behind the morphogenetic furrow, yet their distribution and orientation is unlike that found in the wild-type disc.In wild-type tissue, the staining pattern of cells directly behind the furrow is concentrated at single points, thus making it difficult to observe the neuronal soma (Fig. 2A). In contrast, MAb22C10 staining of *rap* eye imaginal discs clearly outlines the cell body of single cells, suggesting that the orientation of these cells or the distribution of antigen recognized by MAb22C10 is altered (Fig. 2B). The disruption in early patterning events is propagated to other regions within the disc. Just posterior to the furrow, two- and three-cell clusters are frequently found adjacent to groups containing larger numbers of differentiated R cells. Conversely, in regions where one typically finds eight-cell clusters in the wild-type disc, small clusters containing few cells are visible. These groups are often intermixed with clusters containing more than eight photoreceptor cells, suggesting that *rap* mutants lack the ability to restrict cell number to the normal wild-type complement.

rap GENE FUNCTION IS REQUIRED DURING THE THIRD-INSTAR STAGE OF LARVAL DEVELOPMENT

To identify the time during development when *rap* gene function was necessary for normal eye pattern formation, we isolated a temperature-sensitive allele (rap^{R22ts}) of the *rap* mutation. When grown at permissive temperature (17°C), rap^{R22ts} flies show normal external eye morphology (Fig. 3A). In contrast, when rap^{R22ts} flies are grown at 29°C, their eyes show a rough phenotype (Fig. 3B). Two kinds of temperature shift experiments were performed with rap^{R22ts} flies. In temperature upshift experiments, rap^{R22ts} flies were grown at 17°C and shifted to 29°C at the specific times indicated (Fig. 4). In the complementary downshift experiments, rap^{R22ts} flies were grown at 29°C and shifted to 17°C (Fig. 4). The results show that when rap^{R22ts} flies are shifted to the higher temperature at any time prior to pupation, normal eye development is forfeited and the adult is rough-eyed. In contrast, rap^{R22ts} flies shifted from 17°C to 29°C after larvae have entered pupation exhibit a normal (wild-type) exterior. In the downshift experiment (from 29°C to 17°C), it was observed that flies transferred to 17°C prior to the third-instar larval stage exhibited a wild-type exterior morphology typical of rap^{R22ts} revertants maintained at low temperature over the full course of their development. This was not the case when flies were shifted to the permissive temperature during later stages. Flies downshifted during pupation always showed a rough-eye phenotype over the entire surface of the eye. These two experiments show that the third-instar stage of larval development is the period in which *rap* function is necessary

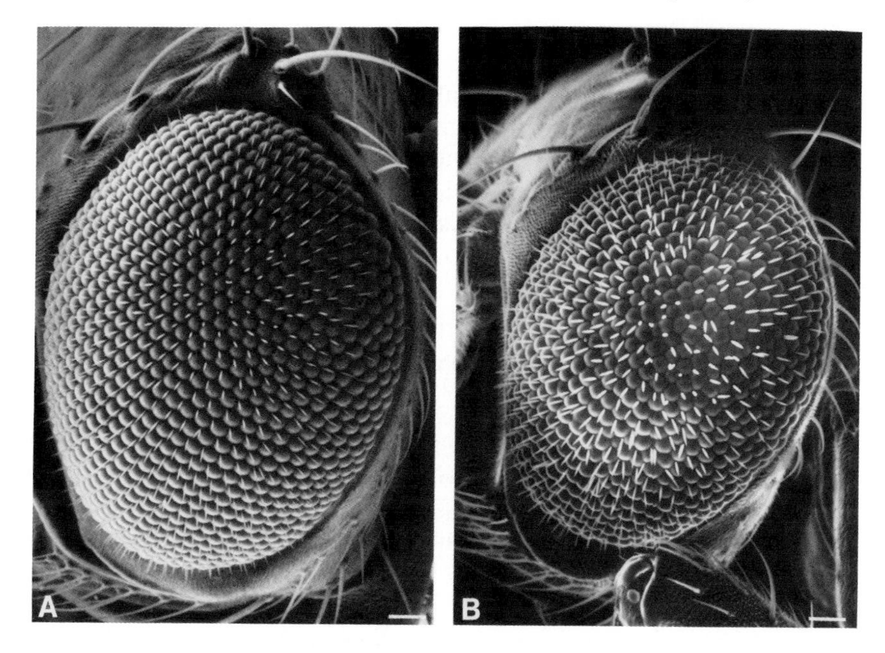

Fig. 3. Phenotype of the temperature-sensitive allele rap^{R22ts}. Scanning electron micrograph of a compound eye of rap^{R22ts} reared at (**A**) 17°C (×209) and (**B**) at 29°C (×185).

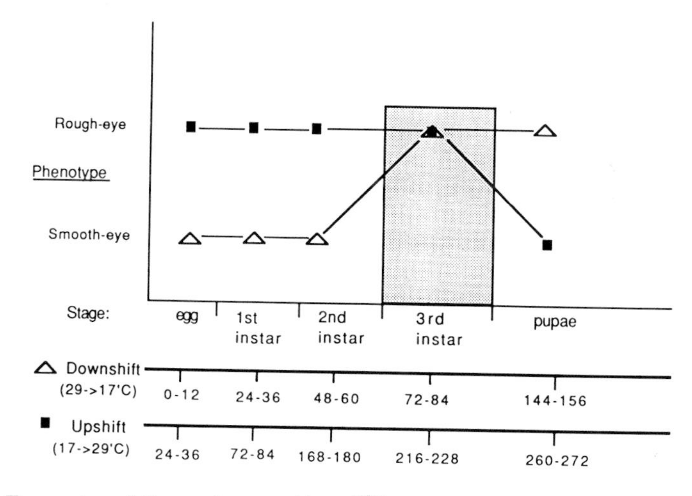

Fig. 4. Temperature shift experiments with rap^{R22ts}, showing that *rap* gene function is specifically required only during third larval instar for normal ommatidial patterning. Upshift, 17°C→29°C. Downshift, 29°C→17°C. Eggs were collected for 12 hr and then shifted at appropriate times. Points on the curve represent stages in development when shift occurred. Shaded area represents the time during development when *rap* gene function is required for normal ommatidial development.

for normal eye morphology. When third-instar larvae of *rap*^{R22ts} flies were downshifted, we commonly found a disruption in eye pattern that was confined to the posterior region of the eye, while a wild-type morphology was exhibited more anteriorly (Fig. 5A,B). This distribution of roughness is consistent with the observation that pattern formation progresses anteriorly across the eye imaginal disc and emphasizes the point that normal *rap* function in one quadrant of the eye fails to rescue the mutant phenotype in regions where *rap* function is absent.

rap GENE FUNCTION IS REQUIRED IN PHOTORECEPTOR CELL R8 FOR PROPER OMMATIDIAL PATTERNING

The immunocytological studies described above suggested that *rap* function is involved in the early patterning events of the eye imaginal disc. To determine whether *rap* expression is required in all cells or is instead necessary in a discrete subclass of photoreceptor, mosaic flies were constructed using the *rap* mutant. Mosaic eyes that contain patches of genotypically mutant cells bordered by wild-type tissue frequently contain ommatidia of mixed genotype. Mutant and wild-type cells can be distinguished by placing a recessive, closely linked, cell autonomous marker on the *rap* chromosome. In our studies, *white(w)*, the gene that is required for the formation of pigment granules, was used to identify cells carrying the *rap* mutation. Using the scheme described in Figure 6 (see below), *rap, w* mosaics were generated. Our analysis then focused on scoring phenotypically normal ommatidia that contained both *rap*$^+$ and *rap*$^-$ cells. Phenotypically abnormal ommatidia were excluded from this study because of the difficulty in determining R cell identity in such clusters.

Fifty-four phenotypically normal ommatidia containing both *rap*$^+$ and *rap*$^-$ R cells were scored from eight mosaic eyes (Fig. 6B). The data, summarized in Table I, show that no correlation exists between the wild-type ommatidia pattern and the presence of the *rap*$^-$ allele in any of the photoreceptor cells R1–7. In contrast, in all of the 54 ommatidia examined, R8 was found to be *rap*$^+$. The likelihood that a distribution thus skewed would arise by chance is less than 0.1% (P, 0.001). Thus these results indicate that the *rap* expression is required only in R8 for normal ommatidial pattern formation.

In addition to demonstrating the R8-specific requirement for *rap* expression, mosaic studies have led to a second interesting observation. As expected, our mosaic flies contained four major classes of ommatidia: 1) phenotypically wild-type ommatidia containing only wild-type (*rap*$^+$) R cells; 2) wild-type containing *rap*$^+$ and *rap*$^-$ R cells; 3) mutant ommatidia containing *rap*$^+$ and *rap*$^-$ R cell types; and 4) mutant ommatidia containing only mutant (*rap*$^-$) cell types. Unexpectedly, a fifth class of ommatidia was observed. In five of the eight mosaic eyes examined, we found cases where phenotypically mutant ommatidia contained only wild-type (*rap*$^+$) R cells. The number of *rap*$^+$ R cells

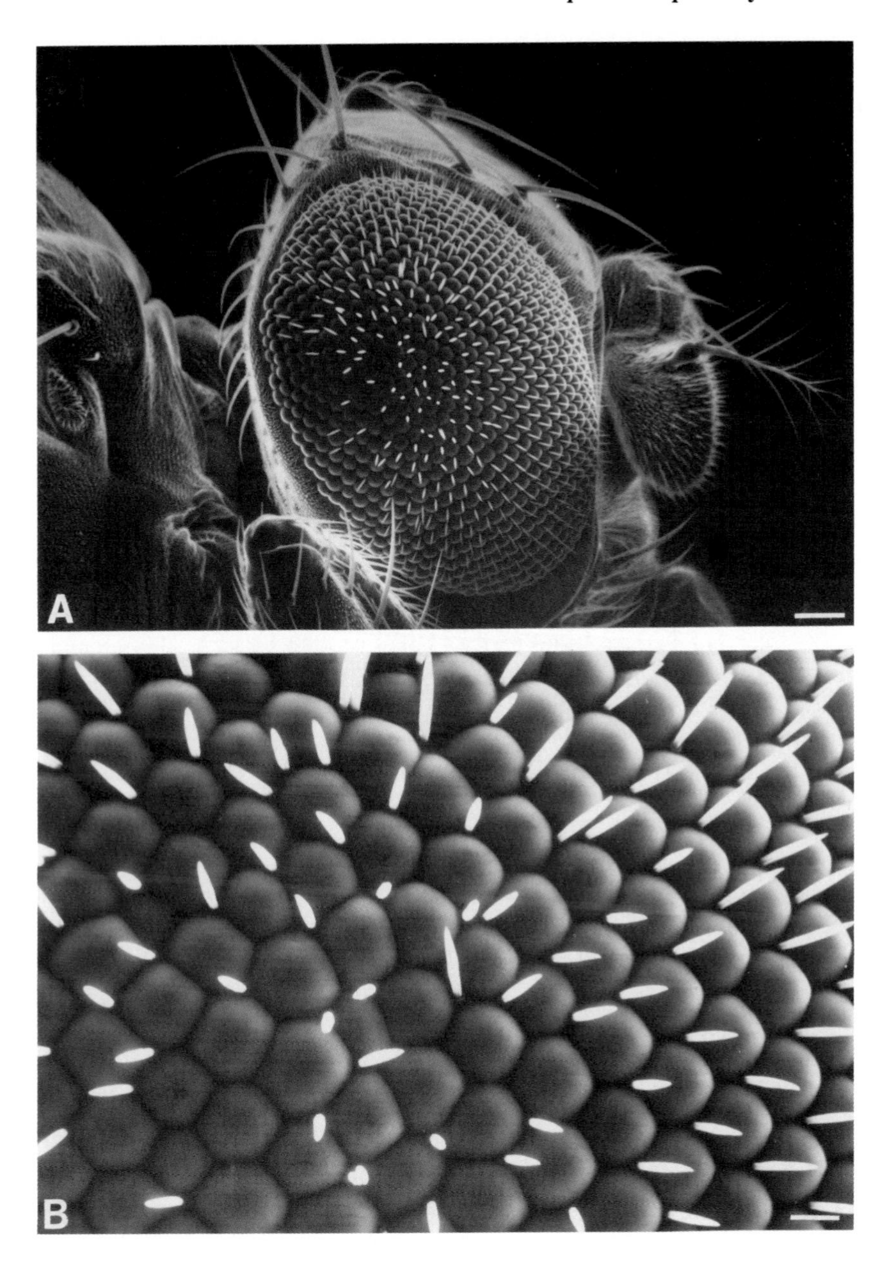

Fig. 5. (**A**) Scanning electron micrograph of a mosaic compound eye ($\times$ 155) induced by shifting rap^{R22ts} flies from nonpermissive temperature to permissive temperature (29°C→17°C), during third instar. (**B**) A high-magnification ($\times$ 760) view of the mosaic eye showing mutant-wild-type border region. The eye is oriented with the posterior on the left.

A

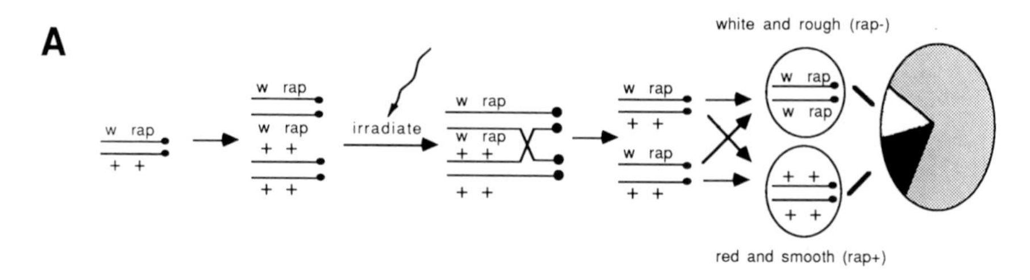

B

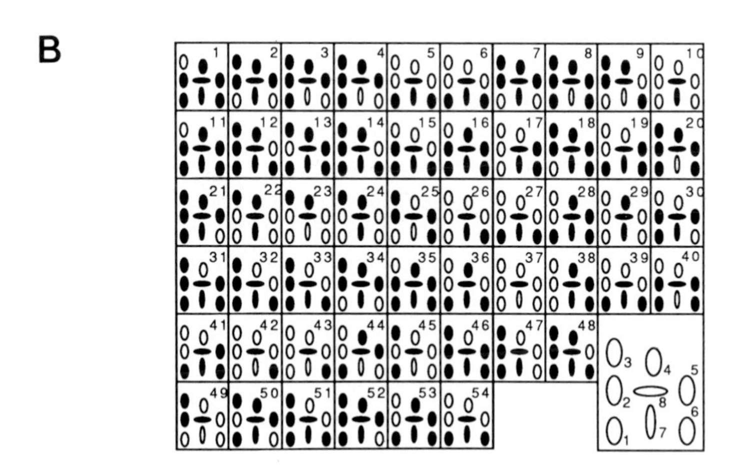

Fig. 6. Mosaic analysis of *rap*. (**A**) Schematic showing how *rap* mosaic flies were generated using gamma irradiation. (**B**) Summary of data obtained from 54 ommatidia having normal R cell pattern and mixed cell genotype. Open circles indicate rap^- genotype. Closed cricles represent the rap^+ genotype.

in these clusters varied greatly with examples of three-, four-, six-, and eight-cell clusters being observed. Furthermore, although all of these clusters were localized to regions that were in close proximity to the border separating mutant and wild-type tissue, their position was not limited to either side.

ROLE OF *rap* FUNCTION IN OMMATIDIAL ASSEMBLY

Our data indicate that the *rap* gene product plays an essential role in eye pattern formation. The timing of *rap* function overlaps the interval in which ommatidial assembly is initiated in the eye imaginal disc and thus presents a correlation between *rap* function and the emergence of ommatidial pattern. With temperature downshift experiments we have generated rap^{R22ts} flies with half-mutant/half-wild-type eyes. In these eyes the rough phenotype is confined to the posterior regions and the border that separates mutant and wild-type tissue is

TABLE I.
Summary of Results From Mosaic Analysis

R cell	rap^-	Frequency (%)
R1	18	33
R2	24	44
R3	26	48
R4	25	46
R5	26	48
R6	19	35
R7	16	30
R8	0	0

aligned in a dorsoventral manner. These features argue that *rap* function is closely associated with the patterning events taking place behind the morphogenetic furrow and thus make it unlikely that the *rap* gene product is a diffusible substance that is distributed throughout the disc prior to morphogenesis.

Genetic mosaic studies further support the notion that the *rap* gene plays an important role in pattern formation. Results presented in this chapter demonstrate that rap^+ function is required only in photoreceptor cell R8 for normal ommatidial pattern formation. Although rap^+ function is not required in cells R1–7, our studies do not rule out the possibility that a low level of functional gene product is made in these cells. This concern can be addressed by using an allele for which the lack of functional gene product has been established. Furthermore, although our analysis does not take into consideration the role of accessory cells, we feel such contributions are unlikely. The pigment and mechanosensory cells that make up the majority of the accessory cells in the compound eye do not differentiate until pupation. Because rap^+ function has been shown to be required during the third-instar stage of larval development, contributions made by accessory cells seem improbable.

In addition to the mosaic and developmental data, immunocytological studies of *rap* eye discs support a role for *rap* gene function in eye pattern formation. Antibody staining of discs from third-instar larvae show that a disruption of the *rap* gene function leads to abnormal ommatidial distribution and photoreceptor cell recruitment. These events suggest the *rap* gene function has a central role in pattern formation and provide clues as to its time of action during ommatidial assembly. Specifically, detailed analysis of eye imaginal discs stained with MAb22C10 indicated that pattern irregularities are present in the early stages of ommatidial assembly (i.e., the positioning and spacing of cells just posterior to the morphogenetic furrow).

In view of these observations, we proposed two possible models for the role of *rap* function in the developing eye imaginal disc (Karpilow et al., 1989). In the first model, *rap* gene function is important for the proper differentiation of photoreceptor cell R8. *rap* may be responsible for receiving or interpreting developmental cues that direct and determine the position of R8. Mutations in a gene product with these attributes would lead to the improper spacing/positioning of R8 cells behind the morphogenetic furrow and subsequent recruitment abnormalities. In the second model, the *rap* gene product serves as a signal for cellular interactions that lead to the recruitment and proper patterning of the cells R1–7. Such signals might also include inhibitory signals that prevent cells within a particular radius from differentiating into an R8. As is the case in the first model, mutations in such functions would lead to both abnormal distributions of R8 cells behind the morphogenetic furrow and to aberrant cluster sizes in the adult eye. In order to differentiate between the two possible alternatives, we have recently carried out a genetic, molecular, and electrophysiological analysis to determine the cellular types in *rap*$^{-}$ compound eyes. Our results indicate that *rap* gene function is critical for proper differentiation of R8 cells (Karpilow et al., in preparation).

SUMMARY

The idea that cell-cell interactions and positional cues are important in the determination of R cell fates first came from the genetic mosaic experiments by the Benzer group (see Ready et al., 1976) and later by the experiments of Lawrence and Green (1979). These experiments clearly established that in the developing ommatidium, cell fate is not decided by strict lineage mechanisms. Based on the sequential differentiation of the R cells and the relative positions of the various cells within developing ommatidium, Tomlinson and Ready (1987) proposed the sequential induction model for eye development. In this model specific cell position and cell-cell contacts are important in the determination of cell fate. Photoreceptor cell R8 is the first cell to differentiate immediately behind the morphogenetic furrow and it subsequently plays a key role in the sequential induction of other cells.

Thus, the initial step is the differentiation and the proper positioning of the cell R8, and gene functions such as *rap*, which are important in this first step, play a key role in pattern formation. Recent genetic and molecular studies on genes such as *sevenless (sev), bride of sevenless (boss),* and *rough (ro)* have further demonstrated the role of inductive interactions in the specification of cell fate and provided clues to the molecular nature of the developmental signals in the eye. In addition, these studies have also brought into focus the importance of the precise spatiotemporal regulation of these developmental cues in the determinative events. Molecular studies curently underway on *rap* should provide us information on the nature of the *rap* protein and thus might provide clues to the nature of the cellular interaction that *rap* mediates.

ACKNOWLEDGMENTS

We dedicate this paper to Prof. Seymour Benzer, whose contributions to this subject have been immense. The work presented here has been supported by NSF grants DMB 8608797 and BNS 8917632 (T.V.) and NIH training grant in genetics 5 T32 GM0743-13 (J.K.).

REFERENCES

Banerjee U, Renfranz PJ, Pollock JA, Benzer S (1987a): Molecular characterization and expression of *sevenless*, a gene involved in neuronal pattern formation in the *Drosophila* eye. Cell 49:281–291.

Banerjee U, Renfranz PJ, Hinton DR, Rabin BA, Benzer S (1987b). The *sevenless* protein is expressed apically in cell membranes of the developing *Drosophila* retina; it is not restricted to cell R7. Cell 51:151–158.

Banerjee U, Zipursky SL (1990): The role of cell-cell interaction in the development of the *Drosophila* visual system. Neuron 4:177–187.

Basler K, Hafen E (1988): Control of photoreceptor cell fate by the *sevenless* protein requires a function tyrosine kinase domain. Cell 54:299–311.

Braitenberg V (1967): Patterns of projection in the visual system of the fly. I Retina-lamina projections. Exp Brain Res 3:271–298.

Franceschini N, Kirschfeld K (1971): Etude optique in vivo des elements photoreceptors dan l'oeil compose de *Drosophila*. Kybernetik 8:1–13.

Hafen E, Basler K, Edstrom JE, Rubin GM (1987): *sevenless,* a cell-specific homeotic gene of *Drosophila*, encodes a putative transmembrane receptor with a tyrosine kinase domain. Science 236:55–63.

Harris WA, Stark WS, Walker JA (1976): Genetic dissection of the photoreceptor system in the compound eye of *Drosophila melanogaster*. J Physiol 256:415–439.

Karpilow J, Kolodkin A, Bork T, Venkatesh T (1989): Neuronal development in the *Drosophila* compound eye: *rap* gene function is required in photoreceptor cell R8 for ommatidial pattern formation. Genes Dev 3:1834–1844.

Kirschfeld K (1967): Die Projektion der optischen Umwelt auf das Raster der Rhabdomere im Komplexauge von Musca. Exp Brain Res 3:248–270.

Lawrence PA, Green SM (1979): Cell lineage in the developing retina of *Drosophila*. Dev Biol 71:142–152.

Ready DF (1989): A multifaceted approach to neural development. Trends Neurosci 12:102–109.

Ready DF, Hanson TE, Benzer S (1976): Development of the *Drosophila* retina, a neurocrystalline lattice. Dev Biol 530:217–240.

Reinke R, Zipursky SL (1988): Cell-cell interaction in the *Drosophila* retina: The bride of *sevenless* gene is required in photoreceptor cell R8 for R7 cell development. Cell 55:321–330.

Rubin GM (1989): Development of the *Drosophila* retina: Inductive events studied at single cell resolution. Cell 57:519–520.

Tomlinson A, Ready DF (1987a): Neuronal differentiation in the *Drosophila* ommatidium. Dev Biol 120:366–376.

Tomlinson A, Ready DF (1987b): Cell fate in the *Drosophila* ommatidium. Dev Biol 123:264–275.

Tomlinson A, Bowtell DDL, Hafen E, Rubin GM (1987c): Localization of the *sevenless* protein, a putative receptor for positional information in the eye imaginal disc of *Drosophila*. Cell 51:143–150.

Tomlinson A, Kimmel BE, Rubin GM (1988): *Rough*, a *Drosophila* homeobox gene required in photoreceptors R2 and R2 for inductive interactions in the developing eye. Cell 55:771–784.

Trujillo-Cenoz O, Melamed J (1966): Compound eye of dipterans: Anatomical basis for integration—an electron microscope study. J Ultrastruct Res 16:395–398.

Venkatesh TR, Zipursky SL, Benzer S (1985): Molecular analysis of the development of the component eye. Trends Neurosci 8:251–257.

Zipursky SL, Venkatesh TR, Teplow DB, Benzer S (1984): Neuronal development in the *Drosophila* retina: Monoclonal antibodies as molecular probes. Cell 36:15–21.

Cell-Cell Interactions in Early Development, pages 241–247

14. Epidermal Cell Redifferentiation: A Demonstration of Cell–Cell Interactions in Plants

Judith A. Verbeke
Department of Plant Sciences, University of Arizona, Tucson, Arizona 85721

INTRODUCTION

The formation of regular and predictable patterns during the development of multicellular plants requires that positional information be available to cells, which then differentiate accordingly. The mechanisms by which these patterns of discrete cell types are formed in response to position is a long-standing problem in developmental biology. A cell that reacts in a special way in consequence of its association with another can only do so if it acquires information from that other, information which must be conveyed through chemical or physical signals. The interacting cells then behave in their characteristic ways because they are programmed to transmit and receive their particular signals. The nature of the signals, their origin, and how they are transmitted are all unknown. However, the essence of the regulation of differentiation certainly must be the selective expression of one type of cellular specialization while leaving the other types unexpressed. Unraveling these complex interactions remains one of the major problems of modern biology.

There are several processes unique to the development of multicellular plants: 1) The presence of a rigid cell wall usually prohibits the morphogenetic cell migrations that are common occurrences in early animal development. 2) While embryonic development in plants resembles the progressive type of development seen in animals, postembryonic plant development is ongoing and repetitive in nature, as new structural units (e.g., leaves and branches) are continually added to the growing organism. 3) Throughout much of plant growth and development, there is no new contact between cells that have not arisen together as clones following cell division. Neighboring cells are truly "sister" cells. 4) During plant development, organs are sometimes initiated independently and then fused together to form the final structure. Fusing floral organs are good examples of this process. We have shown that diffusible factors initiate epidermal cell redifferentiation during carpel fusion in *Catharanthus roseus*. The overall goal of our work is to achieve a better understanding of the controls that govern differentiation and to characterize the communication factors that convey positional information to plant epidermal cells.

CATHARANTHUS AS AN EXPERIMENTAL SYSTEM

Because determination is a reflection of the state of differentiation, Lang (1973) suggested that it may be most useful to study those cells that display a high degree of determination, as indicated by a resistance to redifferentiate. Plant epidermal cells are such entities. In the absence of wounding (and usually in spite of wounding, plant epidermal cells are developmentally incompetent to redifferentiate. These cells show no redifferentiation (or grafting) response even after months of intimate contact with either wounded or nonwounded surfaces (Walker and Bruck, 1985; Moore, 1984). This stable differentiated state, which is exhibited by plant epidermal cells, arises early in ontogeny (Bruck and Walker, 1985). However, fusing floral organs are naturally occurring exceptions to this rule. Postgenital fusions are tissue unions that occur by adhesion of the outer epidermal cell walls of the fusing organs in the absence of protoplasmic union (Cusick, 1966). The fusion of floral organs is an important developmental event in the morphogenesis of many flowers and occurs in several families of higher plants (Boeke, 1971; Nishino, 1982). The course of the fusion does not vary much; the parts are forced together by growth, the epidermal cells interlock, and the enclosed cuticle, if present, disappears. In the fused epidermal layers, cell divisions—which are mostly periclinal—usually occur, which disturb the original cell pattern and obscure the exact place of fusion. In the region of coalescence, the identity of the originally separate epidermal layers is lost. These fusions involve a change in the developmental fate of the contacting epidermal cells.

Boke (1947, 1948, 1949) reported a clear example of postgenital fusion in the developing gynoecium of *Catharanthus roseus* (Madagascar periwinkle), and Walker (1975a,b,c) described the qualitative cytological changes that occur in the epidermal cells during this process. During normal floral ontogeny in *C. roseus*, the adaxial surfaces of the two carpels, which are originally separate, touch. Within 9 hr, approximately 400 contacting epidermal cells then convert, by redifferentiation, into parenchyma cells (Verbeke and Walker, 1985). The changes in these fusing cells are dramatic (Fig. 1). Prefusion epidermal cells have a thin cuticle, are rectangular in section and densely cytoplasmic, and divide strictly in the anticlinal plane (Walker, 1975b). Following contact, these cells still have the remnants of the cuticle, but become isodiametric and highly vacuolated and divide predominantly in the periclinal plane (Walker, 1975c).

CELL–CELL INTERACTIONS

Cell-to-cell communication has been implicated in the distinctive morphogenetic response seen in the fusing carpels of *Catharanthus* (Walker, 1978a). Preventing cell contact either by removal of one of the carpels (Walker, 1978a) or by insertion of a solid, impermeable barrier between the two carpels (Walker,

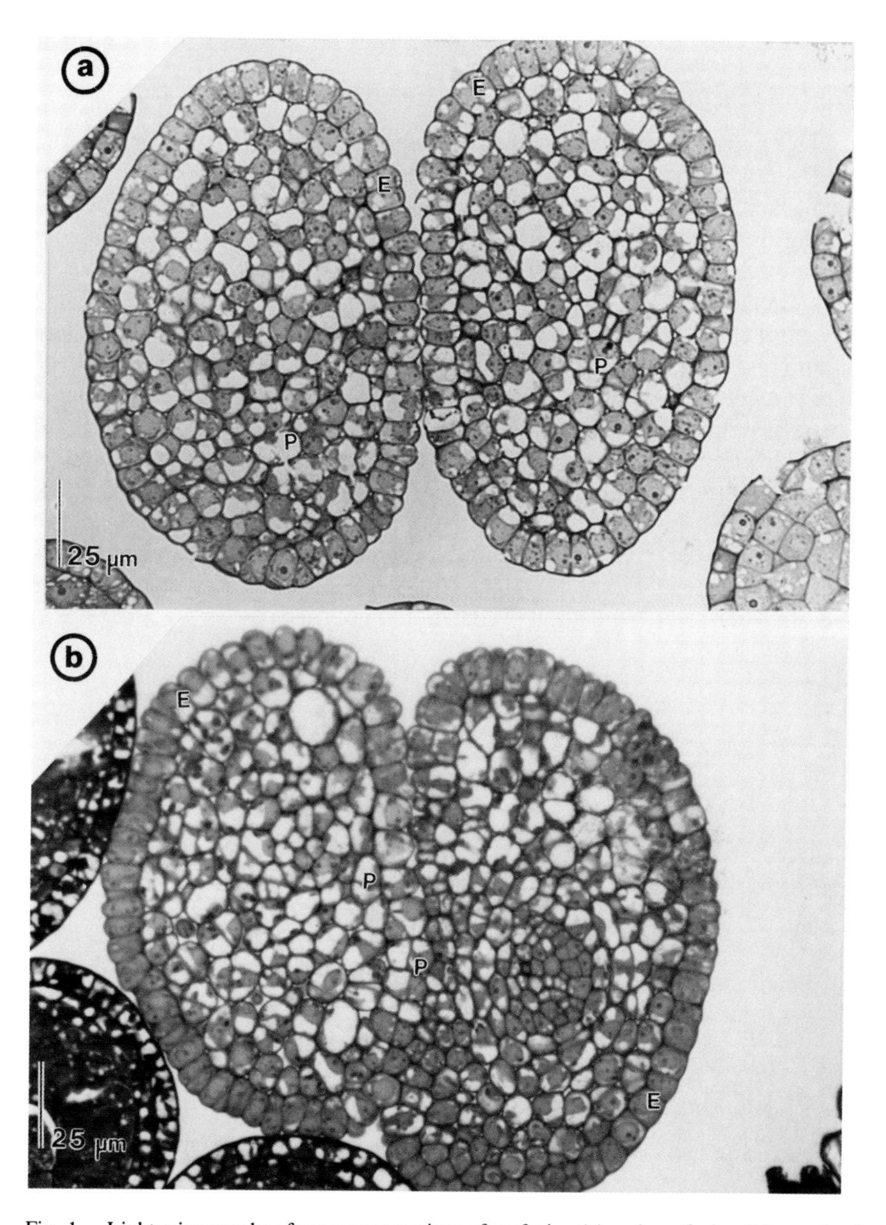

Fig. 1. Light micrographs of transverse sections of prefusion (**a**) and postfusion (**b**) carpels of *Catharanthus roseus*. E, epidermal cells; P, parenchyma cells.

1978b; Verbeke and Walker, 1986) prevents the change in cellular fate, and the cells remain epidermal.

Diffusible Factors?

When porous barriers are placed between the two carpels, epidermal cell redifferentiation occurs regardless of barrier composition or pore size (Verbeke and Walker, 1986), suggesting that water-soluble diffusible agents induce the specific morphogenetic response of epidermal cell redifferentiation. To further test the hypothesis that diffusible factors are involved in these responses, polycarbonate barriers impregnated with agar were placed between two carpels, prior to fusion. These agar-impregnated barriers do not block redifferentiation (Fig. 2); the factors diffuse through the agar from one carpel to the other (Siegel and Verbeke, 1989). Additionally, the factors can be trapped in agar; agar-impregnated barriers are placed between the carpels (so that the factors diffuse into the agar), and then transferred to the outer/abaxial epidermal cells of the carpel—cells that normally remain epidermal throughout development. In these experiments, epidermal cell redifferentiation is always induced (Siegel and Verbeke, 1989). Also, polycarbonate barriers without the agar can be loaded between the carpels and then transferred to the abaxial cells to induce epidermal cell redifferentiation (Verbeke, unpublished).

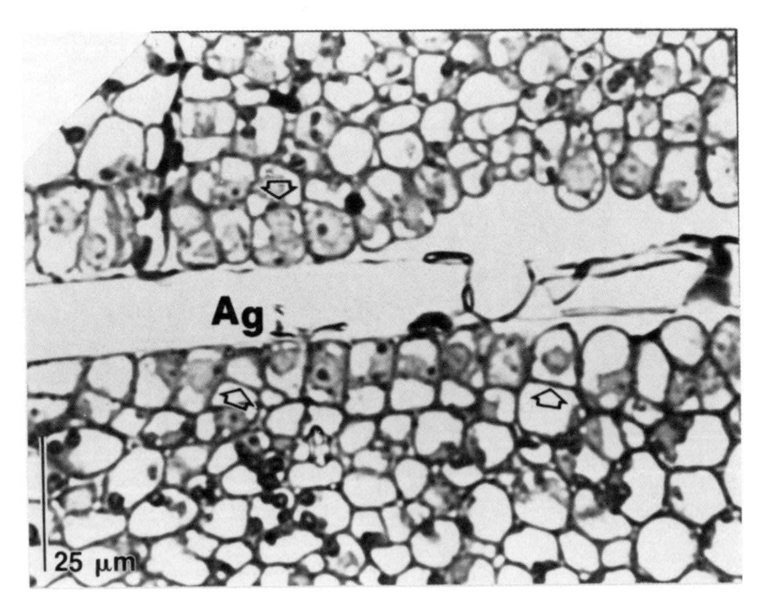

Fig. 2. Light micrograph of a transverse section showing redifferentiating (open arrow) epidermal cells in the carpels of *Catharanthus roseus*. Ag, agar-impregnated polycarbonate barrier.

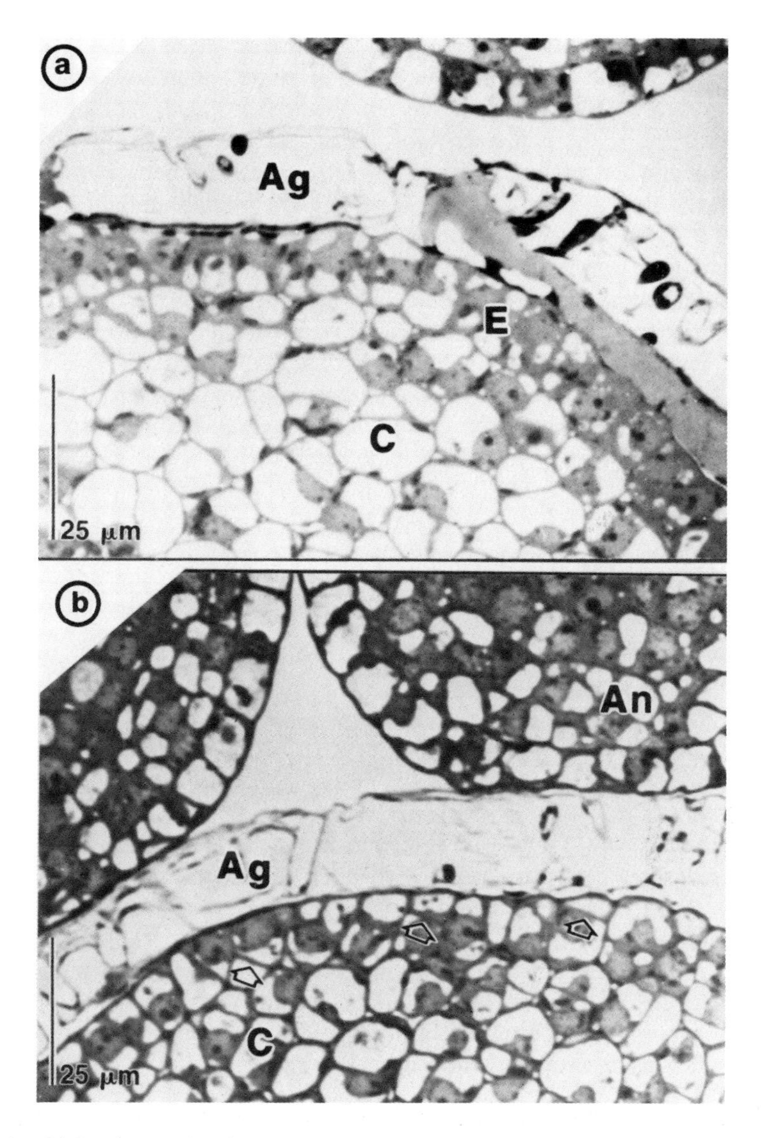

Fig. 3. Light micrographs of transverse sections of normally nonredifferentiating (i.e., abaxial) carpel surfaces of *Catharanthus roseus*. (**a**) Agar-impregnated polycarbonate barrier placed against a single carpel. (**b**) Agar-impregnated polycarbonate barrier, loaded at one carpel and transferred to this one. Ag, agar-impregnated polycarbonate barrier; An, anther; E, epidermal cells; C, carpel. Open arrows indicate redifferentiating epidermal cells.

Are the Carpels Identical?

Epidermal cell redifferentiation factors can be collected at the abaxial, or normally nonredifferentiating, surface as well. But all of our experiments have shown that factors from both carpels are essential to induce redifferentiation. If an unloaded barrier is placed at the abaxial surface of one carpel, the cells contacting that barrier remain epidermal. However, if that barrier is then transferred to the abaxial surface of the opposing carpel, redifferentiation occurs in 10–18 hr (Fig. 3). When the barrier is loaded at the abaxial surface of one carpel and then transferred to the abaxial carpel surface in another bud, redifferentiation occurs 50% of the time. We have also performed microsurgical manipulations in which a prefusion carpel is excised at its base and discarded; a carpel from another prefusion bud is then grafted in its place. Again, redifferentiation occurs 50% of the time in these surgical experiments. Our current working hypothesis is that the carpels in this system are not identical. Experiments are in progress to test this hypothesis.

SUMMARY

The postgenital fusion event in *Catharanthus roseus* provides a particularly useful system for studies of cell differentiation and cell-cell interaction:

1. Because the cells that ultimately redifferentiate are initially epidermal in nature, they are external and can be subjected to manipulation with a minimization of wounding. Observed changes, therefore, can be attributed unequivocally to the experimental conditions and not to a wound response.
2. Epidermal cell redifferentiation in this system is virtually complete within 9 hr of contact (Verbeke and Walker, 1985). The responses we monitor obviously are elicited very soon after cell contact and communication.
3. Because epidermal cells have a distinctive cytoplasm as well as a distinctive shape, it is possible to identify cytological changes as the cells redifferentiate.
4. The qualitative changes that occur in the redifferentiating cells have been well described and the events are well documented.
5. Perhaps most importantly, however, epidermal cell redifferentiation in any plant system is extremely rare.

We believe that it is only through the investigation of experimental systems like this one that we will begin to unravel the mechanisms underlying the basic aspects of plant growth and development.

ACKNOWLEDGMENTS

The author thanks S.N.-Yarkhan for excellent technical assistance. This work was supported by NSF grant DCB 86-15939 and DBC 90-04983.

REFERENCES

Boeke JH (1971): Location of the postgenital fusion in the gynoecium of *Capsella bursa-pastoris* (L.). Med Acta Bot Neerl 20:570–576.

Boke NH (1947): Development of the adult shoot apex and floral initiation in *Vinca rosea* L. Am J Bot 34:433–439.

Boke NH (1948): Development of the perianth in *Vinca rosea* L. Am J Bot 35:413–423.

Boke NH (1949): Development of the stamens and carpels in *Vinca rosea* L. Am J Bot 36:535–547.

Bruck DK, Walker DB (1985): Cell determination during embryogenesis in *Citrus jambhiri*. I. Ontogeny of the epidermis. Bot Gaz 46:188–195.

Cusick F (1966): On phylogenetic and ontogenetic fusions. In Cutter EG (ed): ''Trends in Plant Morphogenesis.'' London: Longmans, Green & Co., pp 170–183.

Lang A (1973): Inductive phenomena in plant development. In ''Basic Mechanisms in Plant Morphogenesis.'' Upton, NY: Brookhaven Natl Lab, pp 129–144.

Moore R (1984): Cellular interactions during the formation of approach grafts in *Sedum telephoides* (Crassulaceae). Can J Bot 62:2476–2484.

Nishino E (1982): Corolla tube formation in six species of Apocynaceae. Bot Mag Tokyo 95:1–17.

Siegel BA, Verbeke JA (1989): Diffusible factors essential for epidermal cell redifferentiation in *Catharanthus roseus*. Science 244:580–582.

Verbeke JA, Walker DB (1985): Rate of induced cellular dedifferentiation in *Catharanthus roseus*. Am J Bot 72:1314–1317.

Verbeke JA, Walker DB (1986): Morphogenetic factors controlling differentiation and dedifferentiation of epidermal cells in the gynoecium of *Catharanthus roseus*. II. Diffusible morphogens. Planta 168:43–49.

Walker DB (1975a): Postgenital carpel fusion in *Catharanthus roseus* (Apocynaceae). I. Light and scanning microscopic study of gynoecial ontogeny. Am J Bot 62:457–467.

Walker DB (1975b): Postgenital carpel fusion in *Catharanthus roseus*. II. Fine structure of the epidermis before fusion. Protoplasma 86:29–41.

Walker DB (1975c): Postgenital carpel fusion in *Catharanthus roseus*. III. Fine structure of the epidermis during and after fusion. Protoplasma 86:43–63.

Walker DB (1978a): Postgenital carpel fusion in *Catharanthus roseus*. IV. Significance of the fusion. Am J Bot 65:119–121.

Walker DB (1978b): Morphogenetic factors controlling differentiation and dedifferentiation of epidermal cells in the gynoecium of *Catharanthus roseus*. I. The role of pressure and cell confinement. Planta 142:181–186.

Walker DB, Bruck DK (1985): Incompetence of stem epidermal cells to dedifferentiate and graft. Can J Bot 63:2129–2132.

Cell-Cell Interactions in Early Development, pages 249–259

15. Cell Sorting and Pattern Formation in *Dictyostelium discoideum*

Ikuo Takeuchi

National Institute for Basic Biology, Myodaiji, Okazaki 444, Japan

INTRODUCTION

Within a multicellular organism in general, different types of cells are distributed in a definite pattern, in that each cell type occupies particular locations relative to the others. Such a pattern of differentiation is characteristic of the species and is achieved during development of the organism. The final positions of individual cell types are determined in two ways: 1) where the cells differentiate and 2) where the cells move after differentiation. In this chapter, I mainly focus on the second part, I show how differentiated slime mold cells move relative to each other to occupy certain positions in the tissue, and I discuss how the movement of cells contributes to formation of the pattern of differentiation.

Development of a cellular slime mold, *Dictyostelium discoideum*, leads to formation of a fruiting body, which is composed of three parts, a spore head, a supporting cellular stalk, and a basal disc. The spore cells and the stalk cells differ in chemical constituents as well as in morphology, whereas the basal disc cells are morphologically indistinguishable from the stalk cells. Development of the organism is initiated when growing cells deplete the food source. In due course, the cells aggregate to a collecting point and form a tissue of up to 100,000 cells. The aggregation starts when a few cells begin to emit a cAMP signal. The neighboring cells respond to the signal by chemotactically moving toward the signaling cells and then emitting the signal themselves to attract outlying cells. As the autonomous signaling cells in the center continue to emit the signal in pulses, the peripheral cells are radially collected, forming streams of cells. The tissue formed at the aggregation center first assumes the shape of a mound, but as more cells come in, a tip forms on the top and elongates to transform the tissue into a slug shape. The slug, either on the spot or after migrating over the substratum for a period, culminates to construct a fruiting body. During formation of the fruiting body, the anterior (about 20% of total) cells of the slug become stalk cells, while the posterior (about 80%) cells become spores.

MORPHOGENESIS AND PATTERN FORMATION DURING SLUG RECONSTRUCTION

Differentiation Pattern in Slugs

It has been shown that the prepattern of stalk and spore differentiation can be observed in the migrating slug. When a section of a slug is stained with fluorescein-labeled antispore immunoglobulin, the posterior presumptive spore cells are strongly stained, while the anterior presumptive stalk cells are almost devoid of staining (Fig. 1b) (Takeuchi, 1963). Such a pattern of differentiation is characteristic in two ways: 1) Both prestalk and prespore cells are located at specific sites within the tissue and 2) they are present in a certain proportion that is largely independent of the slug size.

The differentiation pattern of the slug is in fact a little more complicated than described above. When examined more carefully, there are cells in the posterior prespore region that are unstained by antispore immunoglobulin (Fig. 1b). Like anterior prestalk cells, these cells are strongly stained by vital staining such as neutral red and are therefore called anterior-like cells (Sternfeld and David, 1981a). They are very similar to prestalk cells not only in the staining property but also in biochemical characteristics (Devine and Loomis, 1985). When the posterior half of a slug containing vitally stained anterior-like cells is grafted onto the unstained anterior half of another slug, some stained cells appear in the rear part of the prestalk region within serveral hous. Actual measurements of individual anterior-like cells indicate that anterior-like cells are constantly exchanging with prestalk cells, at least in part, during slug migration (Kakutani and Takeuchi, 1986). This suggests that the anterior-like cells are in fact a kind of prestalk cell that is scattered in the prespore region.

Another deviation from the normal prestalk–prespore pattern is that, during slug migration, the prestalk region tends to extend toward the rear of the slug along the side adjacent to the agar substratum. When migration is prolonged, such dragging cells increase considerably in number (Fig. 1c), resulting in a decrease in the ratio of prespore to total cells, from the normal value of about

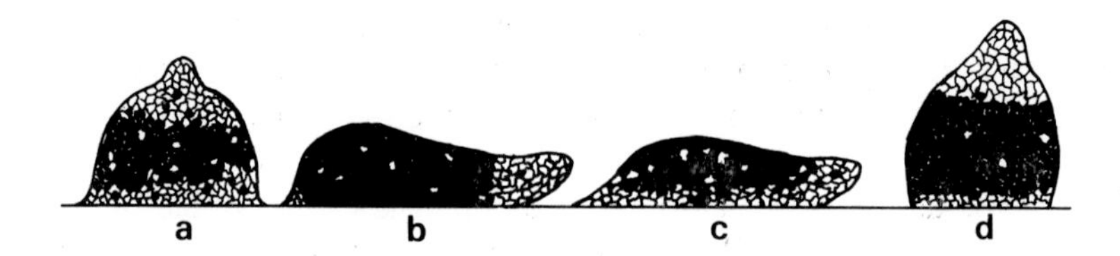

Fig. 1. Semidiagrammatic illustrations of distribution patterns of prespore cells in developing tissues, as examined by immunofluorescent antispore staining. **a**, late aggregation stage; **b**, a migrating slug soon after formation; **c**, a slug after long (30 hr) migration; **d**, early culmination stage. The baseline indicates the agar surface. Closed cells, stained prespore cells; open cells, unstained cells.

80% to about 60%. When such a slug is induced to culminate, however, these cells rapidly resynthesize the spore antigen and the ratio of prespore cells rapidly increases to the normal value within 1–2 h (Tasaka and Takeuchi, 1983). As Williams et al. (1989) recently showed that prestalk cells are divided into three subclasses that occupy different parts of the prestalk region, it would be interesting to know to which subclass of prestalk cells these cells in question belong.

Cell Sorting During Slug Reconstruction

It is known that when slugs are dissociated and dissociated cells are incubated on agar, they reaggregate and reconstruct slugs within a matter of a few hours. It has been shown by Bonner (1952), Sternfeld and David (1981a), and Takeuchi (1969) that during the process, prestalk cells are sorted out to the front of reconstructed slugs, while prespore cells are sorted out to the rear.

We have observed similar sorting out between prestalk and prespore cells in a cell mass that undergoes no morphogenesis. When cells dissociated from slugs are suspended in a salt solution and cultured in roller tubes, the cells stick together and form round agglomerates. Although the agglomerates themselves undergo no morphogenesis, the constituent cells follow a certain process of pattern formation. To examine this, the agglomerates were fixed at intervals and the sections stained with fluorescein-labeled antispore immunoglobulin (Fig. 2). Initially, both prestalk and prespore cells are dispersed in the agglomerates, but after a while prestalk cells are collected at the center of the agglomerates. Then, the prestalk cells move to the periphery and reconstitute the prestalk–prespore pattern, similar to the one observed in a migrating slug. In the meantime, cells in one or two outermost layers of the agglomerates differentiate into prestalk cells and then to mature stalk cells (Tasaka and Takeuchi, 1979).

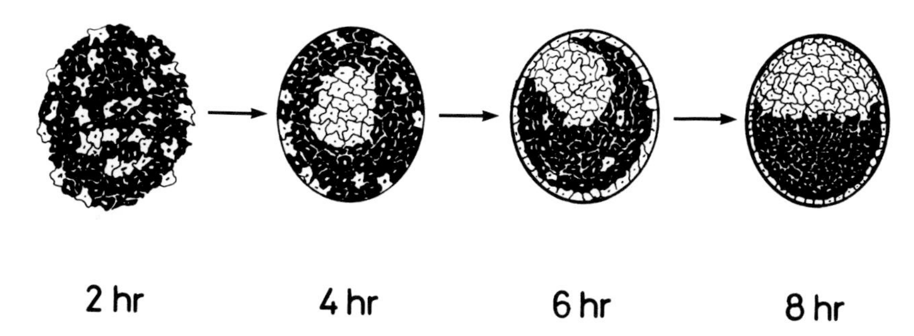

Fig. 2. Changes in distribution pattern of prespore and prestalk cells within submerged agglomerates. Migrating slugs were disaggregated and disaggregated cells were cultured in roller tubes in a salt solution. After indicated times, distribution of prespore (closed) and prestalk (open) cells was examined by immunofluorescent antispore staining. (From Takeuchi et al., 1988)

To confirm that the changes in distribution of prestalk and prespore cells are due to rearrangement of the cells while retaining their original cell type, prestalk cells labeled with ^{3}H-thymidine were mixed with unlabeled prespore cells and their locations in coagglomerates were followed by antoradiography. Labeled cells followed the same pattern of distribution as detected by immunocytochemistry, indicating that the majority of cells sorted out without changing their original cell type.

However, there was an exception to this—the case of cells in the outermost layers of agglomerates. Although these cells later differentiate into mature stalk cells, there was no preferential accumulation of labeled prestalk cells in these layers (Tasaka and Takeuchi, 1979). This indicates that any prespore cells that happen to be in the outermost layers lose the prespore antigen and become prestalk cells. In fact, during the course of the submerged culture, the ratio of prespore cells decreased from 80% to 60%. This decrease is probably due to the conversion of prespore cells to prestalk cells in the outermost layers.

This represents an example of cells differentiating according to their positions in the tissue. As mentioned above, similar position-dependent changes in cell type were also observed to occur during slug migration with cells adjacent to the agar surface in migrating slugs. Probably, in both cases, the changes are caused by loss of some substances from the cells to surrounding water. Apparently analogous phenomena are also observed at other stages of development. Both in the tissues at the late aggregation stage (Fig. 1a) and early culmination stage (Fig. 1b), the region next to the agar surface is devoid of prespore cells. However, whether this is caused by position-dependent cell-type changes or by sorting out of differentiated cells remains to be solved.

Mechanism of Sorting Out

Here, I would like to come back to the problem of sorting out between prestalk and prespore cells and consider the mechanism involved. For cell sorting in general, two possible mechanisms have been proposed: One is random movement with differential adhesiveness, and the other is directed movement or differential chemotaxis. In the former case, cells move randomly, but due to specific cell adhesion, cells are collected to certain spots within a tissue. In the latter case, however, the movement of a cell itself is directed through chemotaxis. We thought that we could discriminate between the two possibilities by tracing individual prestalk cells during the process of sorting out. For this purpose, slugs containing vitally stained prestalk and anterior-like cells were dissociated and the dissociated cells cultured in roller tubes. After a while, an agglomerate was picked up, sandwiched between two layers of agar and observed under an inverted microscope. Movement of individual stained cells was recored by a time-lapse videotape recorder and the positions of each cell were measured by a video writer.

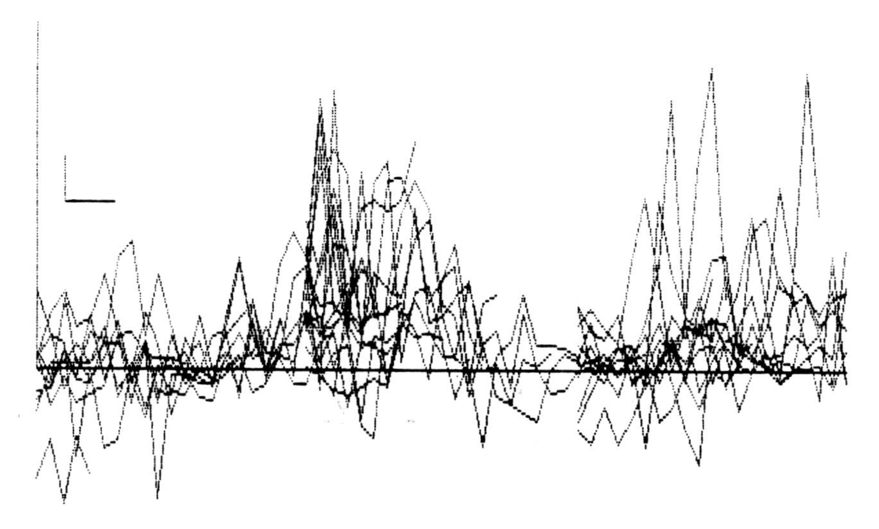

Fig. 3. Changes in time of the radial component of velocity of individual prestalk cells during sorting to the center of the final prestalk region in a submerged agglomerate. Longitudinal and horizontal scale bars represent 5 μm/min and 2 min, respectively. (From Takeuchi et al., 1988)

The results showed that prestalk cells did not move randomly but rather directionally toward the center, where prestalk cells were finally collected. Some cells moved in a row, forming a stream, which is characteristic of cell movement at the aggregation stage (Takeuchi et al., 1988). From these results, we calculated for each cell the radial components of the rates of movement toward the center of the final prestalk region; their changes in time are shown in Figure 3. Two things are worthy of note. First, most values are positive, indicating that the cells keep moving toward the center most of the time. Second, the rates of movement fluctuate in time, suggesting that their movement may be periodic.

As a considerable variation was noted among the cells, we questioned whether the fluctuations really represent cyclic changes and analyzed the data by applying the autocovariance function (Nisbet and Guney, 1982). From the observed fluctuation in velocity of prestalk cells, we calculated the autocovariance function. As shown in Figure 4, the function revealed that the observed fluctuation was surely cyclic in nature and that the periodicity of the cycle was about 15 min (Shigesada et al., unpublished observations). This and the fact that the prestalk cells move in a row with occasional formation of cell streams indicate that their movement resembles that of aggregating cells, except for a difference in the periodicity of movement. Aggregating cells move with a periodicity of about 5 min, prestalk cells with that of about 15 min.

At any rate, the fact that prestalk cells move directly to the center favors the idea that prestalk cells sort out from prespore cells due to differential chemo-

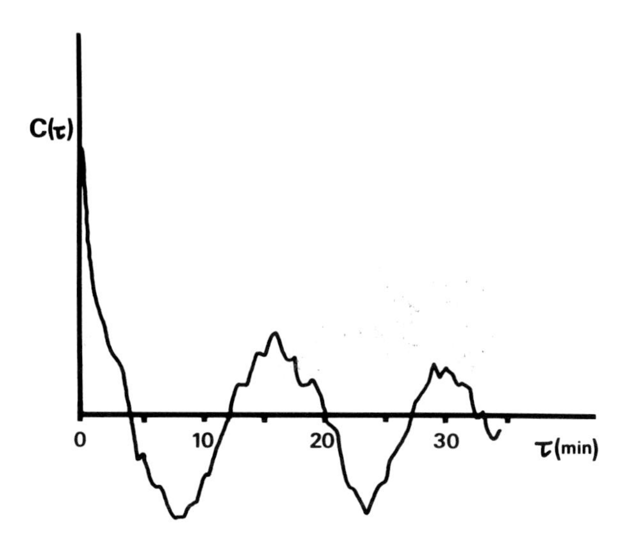

Fig. 4. Autocovariance function [c(τ)] determined from the observed fluctuation shown in Figure 3.

taxis rather than differential adhesiveness. In fact, it has been shown by Matsukuma and Durston (1979) and Sternfeld and David (1981a) that prestalk cells exhibit greater chemotactic response than prespore cells when cAMP is exogenously applied to the tissue. It was also shown by Mee et al. (1986) that separated prestalk cells respond chemotactically to cAMP, while prespore cells do not.

The reason why prestalk cells are sorted out to the center of the agglomerates is not clear, but this could be interpreted in the following way. Cells composing an agglomerate will produce cAMP, probably in pulses, under the direction of autonomous signaling cells located somewhere in the agglomerate. As cAMP diffuses away from the agglomerate to the surrounding water, a cAMP gradient will be established from the center to the periphery and will guide movement of chemotactic prestalk cells but not prespore cells.

Cell Movement After Cell Sorting

After prestalk cells are sorted out to the center of agglomerates, they begin to move in a mass, probably under the command of autonomous signaling cells contained in the prestalk region. They first move to the periphery of the agglomerates, probably guided by a gradient of oxygen concentration in agglomerates, since, as Sternfeld and David (1981b) showed, prestalk cells preferentially move toward a high oxygen concentration. After the prestalk cell mass comes to the periphery, it then begins to migrate in a mass along the circumference of the agglomerate. Although we were not able to observe movement of prespore cells that are unstained, we presume that prespore cells followed prestalk cells. This type of movement thus appears comparable to the migrating movement of slugs on the agar surface. In fact, the rates of rotation in agglomerates were comparable to those of migration of slugs on agar.

When we placed the agglomerates on the agar surface, they turned into slugs within a period of 3–4 h. First, an agglomerate assumes the shape of a hemisphere on the agar. Then the prestalk cell mass comes up to the top of the mound and is transformed into the tip. The tip then leads formation of a slug and thereafter movement of the slug (Fig. 5).

The reason why prestalk cells are colleted to the top of the mound may be explained as follows. In contrast to submerged agglomerates, where cAMP diffuses away from the whole surface, cAMP in the cell mound sitting on agar diffuses away from cells in contact with the agar surface. This will bring about a concentration of cAMP high on the top and low at the base. Chemotactic prestalk cells respond to the gradient of cAMP and are collected to the top. A gradient of oxygen concentration, high on the top and low at the base, may also function to attract prestalk cells.

MORPHOGENESIS AND PATTERN FORMATION DURING DEVELOPMENT

Now, I would like to compare morphogenesis of the submerged agglomerates as was described above with that of developing tissues observed during the normal course of development. The processes of pattern formation and morphogenesis of submerged agglomerates bear a close similarity to those occurring at the two stages of development: first at the late aggregation stage and second at the early culmination stage.

By the use of antispore immunoglobulin, we have examined when and where prespore cells first appear during development (Takeuchi et al., 1978). The results showed that we could first detect the prestalk–prespore pattern at the late aggregation stage, such as shown in Figure 1a. This pattern is in fact strikingly similar to what we have observed in tissues developing from submerged agglomerates (Fig. 5). Where prespore and prestalk cells first appear in a developing tissue has long been argued (see Williams et al., 1989). It is, however, probable that once prespore and prestalk cells appear in the tissue (regardless of locations where they first appear), prestalk cells sort out from prespore cells and form a tip, in a way similar to what we have observed with submerged agglomerates (where both cell types exist).

Similar processes of reformation of pattern proceed during the transition from the migration stage to the culmination stage. At the beginning of culmination, a slug changes its axis from horizontal to vertical. When a migrating slug is artificially induced to culminate by being exposed to overhead light,

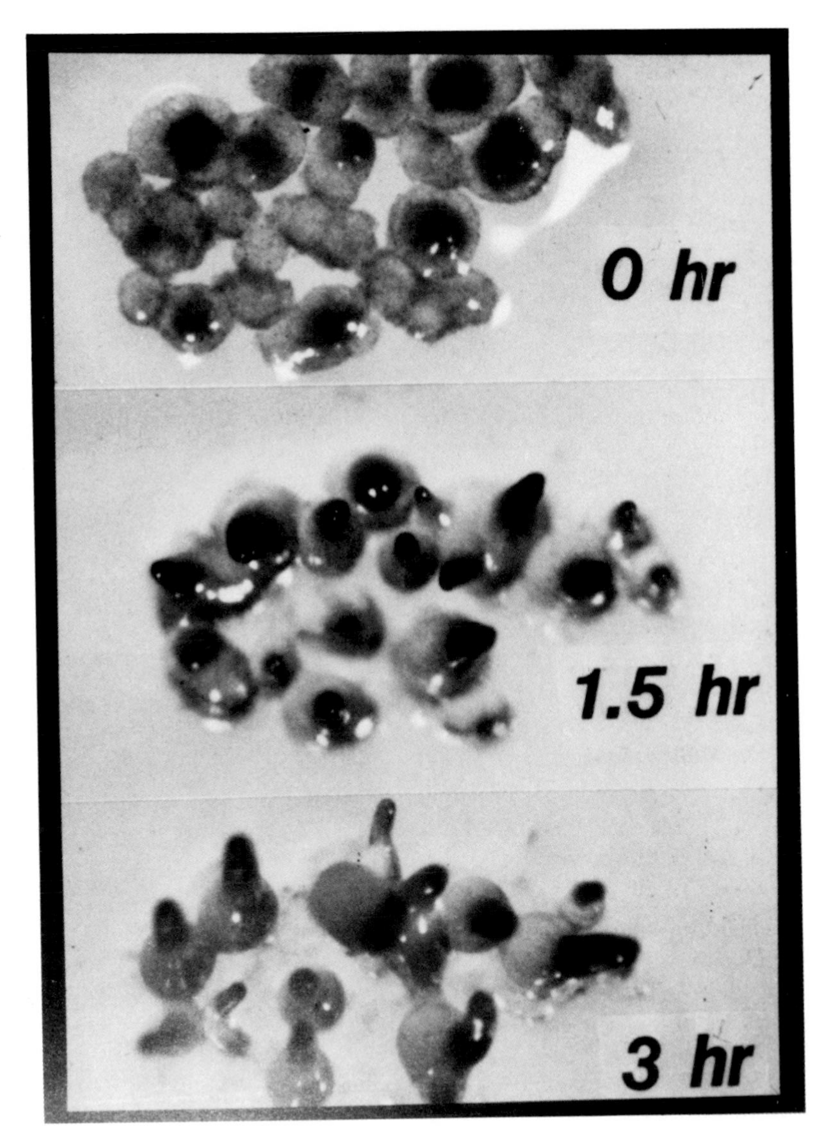

Fig. 5. Slug formation from submerged agglomerates on the agar surface. Migrating slugs containing neutral red-stained prestalk cells were disaggregated and disaggregated cells cultured in roller tubes. After sorting of prestalk cells was completed, the agglomerates were placed on the agar surface. (Kakutani, unpublished observations)

the slug stops migration and is transformed into a hemispherical mound, and then a tip appears on the top. During the first period, considerable mixing of prestalk and prespore cells occurred in the tissue (Tasaka and Takeuchi, 1983). The tip, containing only prestalk cells, elongates and leads a series of morphogenetic movements during formation of a fruiting body.

It is thus clear that prestalk cells are capable of leading morphogenetic movement throughout development. This is probably due to the fact that prestalk cells have a greater motive force than prespore cells. Measurement of a motive force of a slug or its segment was made by the use of a double-chamber method. A slug or its segment was allowed to migrate into an agar tunnel and the pressure needed to stop migration was measured. The measurements made with the anterior and posterior isolates of slugs as well as intact slugs revealed that the motive force of anterior isolates is three times as much as that of posterior isolates and the weighted mean of these values equaled that of whole slugs (Inouye and Takeuchi, 1980). As posterior isolates contain not only prespore cells but also 10% anterior-like cells, which are quite similar to prestalk cells, the real difference in motive force between prestalk and prespore cells would be larger than observed here.

Finally, I would like to briefly mention the relationship between regulation of cell differentiation and differentiation pattern. It is well known that in slime mold tissue, the proportion of prestalk to prespore cells is kept almost constant irrespective of the tissue size, and that when the normal proportion is perturbed by removing either cell type, the surplus type of cells is converted to the other cell type to restore the normal proportion (Raper, 1940).

It was, however, shown that the regulation of cell differentiation proceeds in the absence of the normal prestalk–prespore pattern. For example, when cells were shake-cultured in a solution containing cAMP, they formed small agglutinates of about 200 cells. In these agglutinates, prespore cells differentiated almost at random, i.e., there was no regional localization of prespore and prestalk cells. In spite of this, the ratio of prespore to total cells in the agglutinates was entirely normal (Oyama et al., 1983).

Another example is a temperature-sensitive mutant that affects the pattern. When the mutant was kept at the normal temperature, such as 21°C, the prestalk–prespore pattern was as normal as in the wild type. However, when the tempeature was raised to 27°C after cell aggregation, the pattern became deranged because of considerable mixing of prestalk cells with prespore cells. Nevertheless, the proportion of prespore to total cells was the same at both temperatures (Amagai et al., 1983).

These results indicate that the normal prestalk–prespore pattern is not a prerequisite for the regulation of differentiation. In other words, the particular location of prestalk and prespore cells in a tissue has no effect on the state of differentiation, at least for the majority of cells.

SUMMARY

1. Migrating slugs of *Dictyostelium discoideum* are composed of anterior prestalk cells, posterior prespore cells, and anterior-like cells scattered in the prespore region. Anterio-like and prestalk cells are quite similar and constantly converting to each other during slug migration.

2. When prestalk and prespore cells were isolated from slugs and cultured together in roller tubes, they formed coagglomerates in which prestalk cells sorted out from prespore cells to the center of the agglomerates. Prestalk cells moved directionally toward the center with a periodicity of about 15 min, indicating that the cells sorted due to differential chemotaxis rather than differential adhesiveness.

3. After the cell sorting, the prestalk cell mass moved to the periphery and then migrated in a mass along the circumference of an agglomerate, with a speed similar to that of slug migration.

4. Meanwhile, cells in the outermost layers differentiated in situ into prestalk cells and later to stalk cells. Similar site-dependent differentiation of prestalk cells was observed in the region adjacent to the agar surface of migrating slugs. Apparently analogous phenomena were also observed in aggregating and culminating tissues.

5. When submerged agglomerates were placed on agar, they assumed the shape of a hemisphere. The prestalk cells mass then came up to the top and was transformed into the tip, which led formation of a slug. Similar processes of pattern formation and morphogenesis were observed with tissues at the late aggregation and early culmination stages.

6. The capacity of prestalk cells for leading morphogenetic movement is attributable to the fact that they have a greater motive force than prespore cells.

7. Regulation of prestalk–prespore differentiation proceeded in the absence of the normal prestalk–prespore pattern within tissues.

REFERENCES

Amagai A, Ishida S, Takeuchi I (1983): Cell differentiation in a temperature-sensitive stalkless mutant of *Dictyostelium discoideum*. J Embryol Exp Morphol 74:235–243.

Bonner JT (1952): The pattern of differentiation in amoeboid slime molds. Am Nat 86:79–89.

Devine KM, Loomis WF (1985): Molecular characterization of anterior-like cells in *Dictyostelium discoideum*. Dev Biol 107:364–372.

Inouye K, Takeuchi I (1980): Motive force of the migrating pseudoplasmodium of the cellular slime mould *Dictyostelium discoideum*. J Cell Sci 41:53–64.

Kakutani T, Takeuchi I (1986): Characterization of anterior-like cells in *Dictyostelium* as analyzed by their movement. Dev Biol 115:439–445.

Matsukuma S, Durston AJ (1979): Chemotactic cell sorting in *Dictyostelium discoideum*. J Embryol Exp Morphol 50:243–251.

Mee JD, Tortolo DM, Coukell MB (1986): Chemotaxis-associated properties of separated prestalk and prespore cells of *Dictyostelium discoideum*. Biochem Cell Biol 64:722–732.

Nisbet RN, Guney WSC (1982): ''Modelling Fluctuating Populations,'' Ch. 1.4. New York: John Wiley, pp 6–10.

Oyama M, Okamoto K, Takeuchi I (1983): Proportion regulation without pattern formation in *Dictyostelium discoideum*. J Embryol Exp Morphol 75:293–301.

Raper KB (1940): Pseudoplasmodium formation and organization in *Dictyostelium discoideum*. J Elisha Mitchell Sci Soc 56:241–282.

Sternfeld J, David CN (1981a): Cell sorting during pattern formation in *Dictyostelium*. Differentiation 20:10–21.

Sternfeld J, David CN (1981b): Oxygen gradients cause pattern orientation in *Dictyostelium* cell clumps. J Cell Sci 50:9–17.

Takeuchi I (1963): Immunochemical and immunohistochemical studies on the development of the cellular slime mold *Dictyostelium mucoroides*. Dev Biol 8:1–26.

Takeuchi I (1969): Establishment of polar organization during slime mold development. In Cowdry EV, Seno PS (eds): ''Nucleic Acid Metabolism Cell Differentiation and Cancer Growth.'' Oxford: Pergamon Press, pp 297–304.

Takeuchi I, Okamoto K, Tasaka M, Takemoto S (1978): Regulation of cell differentiation in slime mold development. Bot Mag Tokyo (special issue) 1:47–60.

Takeuchi I, Kakutani T, Tasaka M (1988): Cell behavior during formation of prestalk/prespore pattern in submerged agglomerates of *Dictyostelium discoideum*. Dev Genet 9:607–614.

Tasaka M, Takeuchi I (1979): Sorting out behavior of disaggregated cells in the absence of morphogenesis in *Dictyostelium discoideum*. J Embryol Exp Morhol 49:89–102.

Tasaka M, Takeuchi I (1983): Cell patterning during slug migration and early culmination in *Dictyostelium discoideum*. Differentiation 23:184–188.

Williams JG, Duffy KT, Lane DP, McRobbie SJ, Harwood AJ, Traynor D, Kay RR, Jermyn KA (1989): Origins of prestalk–prespore pattern in *Dictyostelium* development. Cell 59:1157–1163.

Cell-Cell Interactions in Early Development, pages 261–272

16. Cell Sorting and Positional Differentiation During *Dictyostelium* Morphogenesis

Jeffrey G. Williams and Keith A. Jermyn
The Imperial Cancer Research Fund, Clare Hall Laboratory, South Mimms, Herts EN6 3LD, United Kingdom

INTRODUCTION

The two central processes in pattern formation are cellular differentiation and directed cell migration. Cell type-specific gene expression is now well understood, in a wide variety of developmental systems, but the mechanisms directing morphogenetic movement remain relatively obscure. During *Dictyostelium* development, there is a remarkable transition from unicellularity to multicellularity. This provides unique opportunities for analyzing the cellular interactions and directed cell migration required to assemble a patterned structure from an initially uniform population of cells.

Amoebae aggregate together in response to pulsatile cAMP signals emitted from the center of an aggregation territory, which may contain up to 100,000 cells. Because of the powerful combination of biochemical and genetic approaches that can be applied to *Dictyostelium*, considerable progress has been made in understanding chemotactic signaling by cAMP during aggregation (Devreotes, 1989; Firtel et al., 1989). It is intrinsically more difficult to analyze multicellular development because the aggregate transforms into a three-dimensional structure. Also, cells are differentiating into two distinct types, prestalk and prespore cells. The problem can, however, be reduced, in the first instance at least, to an analysis of just one of the cell types. The event that breaks the initial symmetry of the hemispherical aggregate is the formation of an apical tip, composed of prestalk cells. Again, at culmination, it is the movement of prestalk cells, and their expansion as they vacuolate and die, that lifts the spore mass off the substratum. Thus it is of central importance to identify prestalk cells at the moment they first differentiate and to follow their subsequent movements.

One great boon to analyzing multicellular development is the availability of a developmental intermediate, the migratory slug, in which cells remain "frozen" in a partially determined state of differentiation for extended periods of time. At the end of aggregation, the tip of the aggregate elongates to form a structure known as the first finger. Under conditions that are inappropriate for culmination, this topples onto its side to form the slug, which then migrates away. The anterior 20% of the slug is composed of prestalk cells and the rear 80% is predominantly composed of prespore cells. Prestalk cells stain selectively with vital dyes such as neutral red (Bonner, 1952). There are also scattered, neutral red-staining cells in the rear of the slug (Sternfeld and David, 1981) and these so-called "anterior-like cells" (ALC) are very similar in their properties to the prestalk cells in the anterior of the slug (Devine and Loomis, 1985).

The slug is a regulative structure. If it is divided into two by a cut between the prestalk and prespore region, and if sufficient time is allowed before culmination is induced, then both the anterior and posterior segment generate normally proportioned fruit. If the anterior segment is induced to culminate immediately, it generates a very stalky fruit (Raper, 1940), presumably because prestalk to prespore transdifferentiation is a slow process. The posterior segment regulates much more quickly because the ALC very rapidly migrate to the front to regenerate a prestalk zone (Sternfeld and David, 1981). This property of cells within the slug, to in some way "sense" the proportion of prestalk and prespore cells in the whole and to respond to alterations in this ratio, implies the existence of long-range signaling molecules with the power to direct cellular differentiation. Two molecules that might play such a role, and that might also induce initial cell type divergence, have been identified. These are cAMP and DIF.

In addition to its role as a chemotractant, cAMP acts to induce and maintain prespore cell differentiation (Kay et al., 1978; Barklis and Lodish, 1983; Mehdy et al., 1983; Oyama et al., 1983; Schaap and Van Driel, 1985; Oyama and Blumberg, 1986). Prestalk cell differentiation is induced by DIF, a chlorinated alkyl phenone (Town et al., 1976; Morris et al., 1987). It accumulates at the time of cell type divergence and acts in nanomolar concentration to repress prespore and induce prestalk differentiation (Brookman et al.,1982; Kay and Jermyn, 1983).

With the long-term aim of determining its mechanism of action, we isolated two genes that are dependent for their expression upon DIF (Jermyn et al., 1987; Williams et al., 1987). The two genes, pDd63 and pDd56, both encode proteins predominantly composed of a highly conserved, 24 amino acid repeat that is very rich in cysteine residues (Ceccarelli et al., 1987; Williams et al., 1987). They are localized to the extracellular matrices of the slug and the mature culminant. The slug synthesizes around itself a tube of protein and cellulose known as the slime sheath. This is left behind on the substratum as a trail marking the slug's progress. Stalk cells in the mature culminant are individually encased in a rigid cell wall and are further bounded by a discrete, outer matrix known as the stalk tube. Both proteins are present in slime sheath, the stalk cell wall, and the stalk tube (McRobbie et al., 1988a,b). We assume that they play some structural role in all three matrices. Since they encode extracellular matrix proteins, we have renamed the product of the pDd63 gene the ecmA protein and that of the pDd56 gene the ecmB protein.

GENERATION OF CELL AUTONOMOUS MARKERS FROM THE ecmA AND B GENES AND THEIR USE IN ANALYZING SLUG STRUCTURE

The ecmA and B mRNAs are highly enriched in prestalk over prespore cells and are unusual among prestalk-enriched gene products in being retained in mature stalk cells. The proteins themselves are, however, extracellular. In order to realize their potential as cell autonomous markers, we constructed fusion genes, containing the ecmA or ecmB promoters and 5′ noncoding regions fused to reporter genes. When fused to the chloramphenicol acetyl transferase (CAT) gene and stably transformed into *Dictyostelium*, the two promoters direct expression in a temporally correct, cell type-specific and DIF-inducible manner. A fragment of the SV40 T-antigen gene, containing the epitope for four monoclonal antibodies with a high affinity for T-antigen and the nuclear localization signal, was then inserted into the CAT gene (Jermyn et al., 1989). By staining transformants containing this construct with a mixture of the four antibodies, we were able to detect expression of the fusion genes, the fusion protein becoming localized to the nucleus. Cells expressing the ecmA fusion gene were localized in the front 10% of the length of the slug and and we termed these pstA cells. We initially believed that cells in the remainder of the prestalk zone, the 10% of the slug immediately behind the pstA cells, were not expressing the ecmA gene and so we termed them pst0 cells. However, we have now fused the ecmA promoter to the β-galactosidase gene and, using this much more sensitive reporter, we detect a low level of expression in pst0 cells (Jermyn and Williams, 1991). Interestingly, this anterioposterior gradient of ecmA gene expression disappears completely if slugs are allowed to migrate in the light, the gene being expressed at a high level throughout the prestalk zone. Culmination is triggered by overhead light, and it might be that more of the protein is needed as a preparatory step for fruit formation.

Cells expressing the ecmB fusion gene are localized in a central cone in the anterior part of the prestalk zone (Fig. 1). Hence, expression of the two genes defines different cell types. We term cells expressing the ecmB gene pstB cells. ALC also express the ecmA and ecmB genes, but at a lower level than cells in the anterior, prestalk region. Approximately one-third of ALC express the ecmA gene and one-third express the ecmB gene (Jermyn and Williams, 1991). Some of these cells may be expressing both markers, and there are cells in the rear of the slug that express the ecmA and/or the ecmB gene but do not stain detectably with vital dyes.

We have recently shown that pstB cells in the central core of the slug tip also express the ecmA gene (Jermyn and Williams, unpublished results). This

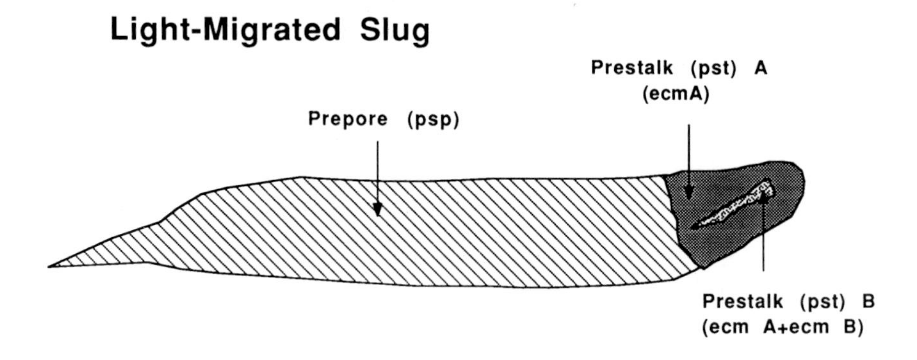

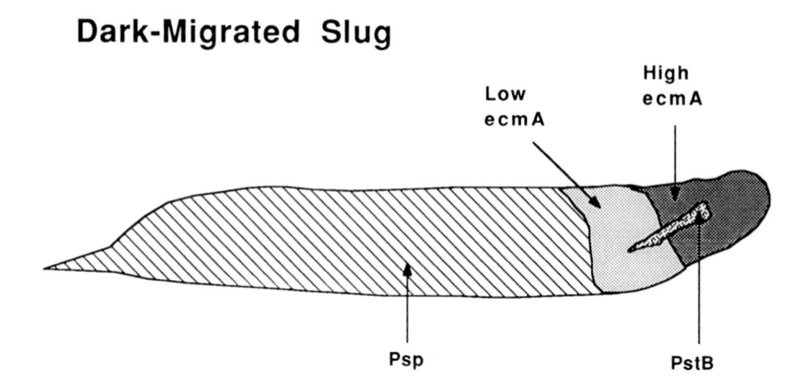

Fig. 1. A comparison of the structure of slugs migrating under different conditions of illumination. Light-migrated slugs were harvested from agar plates bearing aggregates and allowed to form in the presence of direct overhead illumination. Under these conditions, culmination occurs at a high frequency, so great care was taken to fix only those slugs that were clearly still in the migratory phase. Dark-migrated slugs were harvested from agar plates that were exposed to very low level, oblique illumination.

suggests a progression in cellular differentiation, with cells first expressing the ecmA gene and then activating the ecmB gene. Our analysis of slug formation and culmination strongly supports this view.

ANALYSIS OF ecmA AND ecmB GENE EXPRESSION DURING SLUG FORMATION

Aggregation generates a hemispherical mound of apparently similar cells. Formation of the apical tip is the initial pattern-forming event and there are, in principle, two ways it could be generated. Prestalk cells could differentiate within the aggregate and then migrate to the apex, or they could differentiate in situ within the tip. Prestalk cells and ALC are clearly able to sort from prespore cells under artificial conditions, such as removal of the anterior prestalk region or in submerged agglomerates (see Takeuchi, Chapter 15, this volume) but there has until very recently been no direct evidence to show this to be the actual mode of tip formation during normal development. Now, however, apical sorting of prestalk cells has been observed, using a fusion gene containing the promoter of a cysteine proteinase gene coupled to an immunologically detectable reporter (Gomer and Firtel, 1986), and we have used the ecmA fusion gene to show that pstA cells sort to the tip.

PstA cells are first detectable in very low number at the loose aggregate stage. In whole mounts of tight aggregates they appear to be randomly scattered, with no obvious clustering at the tip (Williams et al., 1989). As the tip appears and elongates, the pstA cells become predominantly localized to it, suggesting that they migrate to the tip through the prespore mass. Analysis of sections indicates that they are present throughout the depth of the aggregate, mixed in a true "salt and pepper" fashion with the prespore cells (D. Traynor, personal communication). By the time the tip is well extended, almost all of the pstA cells are localized to the anterior region, the area they occupy in the migrating slug. Thus pstA cells are *not* induced to differentiate at the apex; they first differentiate elsewhere within the aggregate and then migrate there, presumably in response to continued chemotactic cAMP signals emanating from the tip.

PstB cells are predominantly localized to the base of the aggregate at early stages in tip extension (Williams et al., 1989). When it topples onto its side, the base of the first finger becomes the rear of the migratory slug. There are no pstB cells at the rear of the slug after it has migrated away from its site of formation, hence the ecmB-expressing cells located at the base of the first finger must be lost into the slime trail. The extreme rear of older slugs is, however, often composed of prestalk cells. At culmination these "rearguard" cells form part of the basal disc, the expanded region at the base of the stalk that helps to support the fruit. If the slug forms a fruit in situ, without migrating away, then the basal pstB cells will be utilized as basal disc cells. It is only relatively late in slug formation that pstB cells can be detected in the center of the tip, the position they occupy in the migratory slug. We believe that the pstB cells in the tip derive from pstA cells that migrated into the tip and then further differentiated into pstB cells.

ANALYSIS OF ecmA AND ecmB GENE EXPRESSION AND OF pstA AND pstB CELL MOVEMENT DURING CULMINATION

When the migrating slug undergoes culmination, it sits on end and the rearguard cells become the base of the structure (Raper and Fennell, 1952). Prestalk cells within the tip synthesize and enter the stalk tube and this struc-

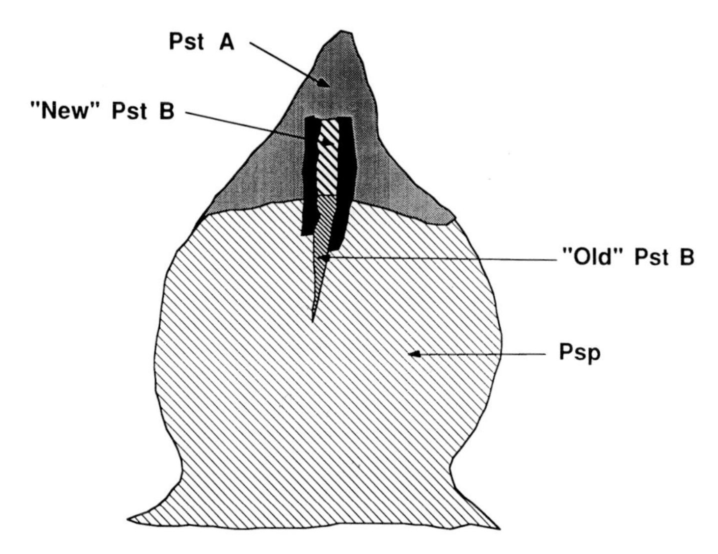

Fig. 2. Stalk cell differentiation in an early culminant. "Old" pstB cells are those cells that form a central core in the tip of the migratory slug. "New" pstB cells derive from pstA cells that have entered the stalk tube and differentiated into pstB cells by activating the ecmB gene.

ture extends down through the prespore mass by the continued accretion of prestalk cells. This extension, coupled with the swelling that accompanies vacuolization of the maturing stalk cells, lifts the prespore mass off the substratum (Fig. 2). The central funnel of pstB cells in the migrating slug lies at precisely the position where stalk tube formation is initiated at culmination and they act as the founders of this structure (Jermyn and Williams, 1991). They constitute the advancing tip of the stalk tube, eventually fusing with the rearguard cells at the base of the aggregate to form the central part of the basal disc.

During culmination, pstA cells in the papilla surrounding the stalk tube transform into pstB cells by activating the ecmB gene (Fig. 2; Jermyn and Williams, 1991). This occurs in a remarkably synchronous manner in a tightly defined region at and just above the entrance to the stalk tube. This is very similar to the situation in those Dictyostelids, such as *Dictyostelium mucoroides*, that form a stalk continuously during slug migration. Here prespore cells rapidly convert into stalk cells at the entrance to the stalk tube (Gregg and Davis, 1982). *Dictyostelium discoideum* differs from these species in two ways. First, there is a recognizable prestalk zone comprised of pstA cells, which will, under normal conditions, differentiate into stalk cells. Second, there is no ongoing differentiation of stalk cells at the stalk tube entrance. The behavior of pstA cells at culmination makes it tempting to believe that the central funel of pstB cells in the migratory slugs in some way reflects an abortive attempt at

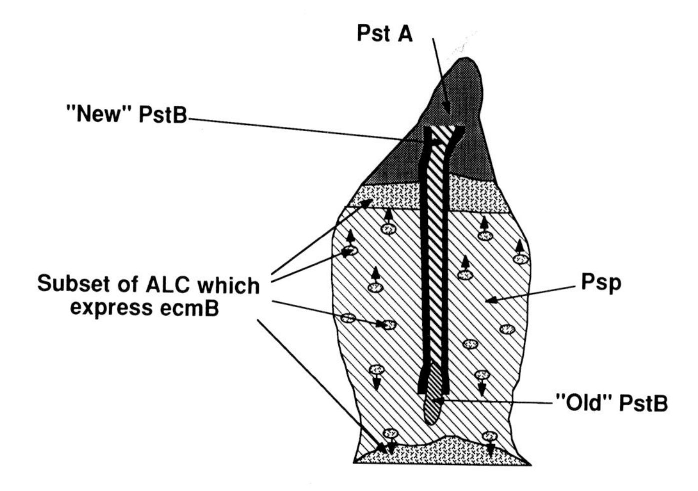

Fig. 3. Stalk cell differentiation midway through culmination.

stalk cell differentiation. There is support for this idea. Stalk cells label more strongly with fucose than prestalk cells and, in the slug, there is a region of high fucose incorporation at precisely the position occupied by the pstB cells (Gregg and Karp, 1977), suggesting that pstB cells are intermediates in stalk cell differentiation and that progression down the pathway is in some way blocked in the *D. discoideum* slug.

Induction of the ecmB gene at the entrance to the stalk tube almost certainly indicates the existence of a positionally localized signal. There is, however, a parallel induction of the ecmB gene in ALC at culmination and here a dispersed inductive signal must presumably be operative (Fig. 3). Our evidence for induction in ALC is in part direct and in part inferential. A fraction of the ALC in the slug express the ecmB gene at a low level and it seems logical to believe that the strongly expressing cells we see in culminants derive at least in part from them. Also, we have shown by grafting experiments in which the rear of ecmB transformant slugs were grafted onto the front of nontransformant slugs that the scattered cells expressing the gene in midculminants derive from the back of the slug (Jermyn and Williams, 1991). Further, indirect evidence derives from the fact that these cells have the same eventual fate as ALC. By fusing the back of a neutral red-stained slug onto the front of a Nile blue-stained slug, Sternfeld and David showed that ALC sort to surround the spore mass (Sternfeld and David, 1982). This is precisely the behavior we see for the ecmB cells outside the stalk tube. They come to constitute a cup above the spore head and a cup below it (Fig. 4).

Why is the gene induced in these cells? One clue may derive from the fact that the ecmB gene is also induced as cells enter the stalk tube. This is a rigid

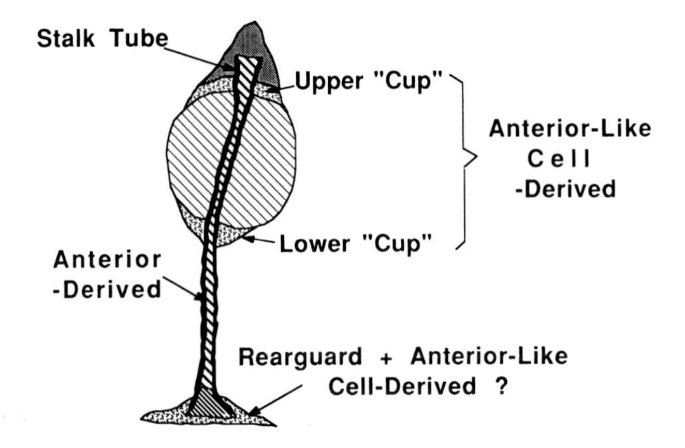

Fig. 4. Stalk cell differentiation late in culmination.

structure and we suspect that the ecmB protein plays a structural, supporting role in the stalk tube. Possibly, therefore, the ecmB protein in the cups above and below the spore head also plays a structural role, protecting the spore mass during its ascent of the stalk tube. Forming these structures at culmination might then be a primary role of ALC, although this does not exclude their playing an important role in cell type regulation during slug migration.

The remarkable behavior of the ALC, with some cells migrating to the base and some to the prestalk–prespore boundary, suggests the existence of two different signaling centers at opposite poles of the aggregate. If two such centers are generated during culmination, and given the similarity in the structure of the first finger and the preculminant, it raises the possibility that the pstB cells present in the base during slug formation also derive from cells that migrated there from elsewhere in the aggregate. *If* this is the case, then there are the following considerable similarities between the two processes.

Slug Formation

1) Activation of the ecmA and ecmB genes in scattered cells within the aggregate.

2) PstA cells sort to the tip and pstB cells sort to the base?

3) A subset of pstA cells, located at the position where the stalk tube will form, differentiate into pstB cells.

Culmination

1) Activation of the ecmA and ecmB genes in widely scattered ALC.

2) Some of the ALC expressing the ecmB gene sort to the prestalk–prespore boundary, while the remainder sort to the base.

3) Activation of the ecmB gene in pstA cells as they enter the stalk tube.

The essential unifying features of the two processes are:

1) Induction of the pstB gene both in widely dispersed cells in the aggregate *and* in positionally localized cells at the entrance to the stalk tube.

2) Some or all of the prestalk cells migrate downward to the base, while others migrate upward toward the tip.

These analogies are perhaps not too surprising given that slug migration is an optional phase in *D. discoideum* and, in other species, such as *P. pallidum*, stalk tube formation occurs simultaneously with primary tip formation. They do, however, raise two major questions.

1) Is the same inductive signal used to activate the ecmB gene in ALC as is used to activate the gene at the entrance to the stalk tube? Since the gene is dependent upon DIF for its expression in submerged cells in vitro, we assume that DIF is responsible for its induction in the intact slug. Here we face the major problem of studying low molecular weight morphogenetic signaling molecules such as DIF. They are diffusible and it is difficult to determine their effective concentrations in localized regions of the aggregate. The concentration of DIF in the back and front of the slug has been determined and, surprisingly, a somewhat higher concentration was found in the rear portion (Brookman et al., 1987). There is, however, debate over the significance of this result (Neave et al., 1986) and some method of analysis at the single cell level will ultimately be necessary to determine how the gene is activated in ALC.

This problem can perhaps be approached using the DIF-responsive genes. We have very recently shown that the ecmB promoter can be subdivided into a region that is necessary and sufficient to direct expression in ALC and another region that directs expression in the stalk tube (Ceccarelli et al., in press). This implies the existence of different inductive conditions in the two sites, although it does not prove that different morphogenetic signaling molecules are utilized. Given our belief that DIF is the primary activator of the gene, what might the proximal trigger(s) for induction be? The interpretation we favor is that there is a decrease in the effective concentration of an antagonist of DIF. There are two strong candi date molecules, cAMP and ammonia.

Transcription of the ecmA gene is very rapidly induced in submerged cells incubated in the presence of DIF, suggesting it to be directly responsive (Williams et al., 1987). Expression of the ecmB gene is induced relatively slowly by DIF under the conditions originally employed, where cAMP was present during the induction (Jermyn et al., 1987). If cAMP is removed at the time DIF is added, then the ecmB gene is much more rapidly induced, while expression of the ecmA gene is stimulated in the presence of cAMP (Berks and Kay, 1988; Berks and Kay, 1991). Thus cAMP and DIF are mutually antagonistic for expression of the ecmB gene. Possibly, therefore, the signal inducing its transcription is a drop in the effective concentration of cAMP.

Alternatively, induction of the ecmB gene may result from a drop in the effective concentration of ammonia. Ammonia is antagonistic to DIF in the induction of terminal stalk cell differentiation and expression of the ecmA and ecmB genes (Gross et al., 1983; Wang et al., 1990; Wang and Schaap, 1989). The slug may be maintained in its state of incomplete differentiation by ammonia. Enzymatic treatments that deplete ammonia stimulate culmination, and mutants that are hypersensitive to ammonia eschew culmination in favor of prolonged slug migration (Inouye, 1988; Newell and Ross, 1982; Schindler and Sussman, 1977).

2) How are the cells caused to undertake directed migration within the aggregate? The evidence for continued chemotaxis of cells to cAMP signals emanating from the tip of the aggregate is persuasive (reviewed by Schaap, 1986). It is, perhaps, the signal attracting pstA cells to the tip during slug formation and culmination. Also, it may be the signal that attracts the subset of ALC, which sort to the prestalk–prespore boundary. There are several potential explanations for the basal migration of pstB cells. The base may act as a competing source of cAMP signals, ALC in the lower part of the culminant may be negatively chemotactic to cAMP signals emanating from the tip, or perhaps the base is a source of a completely different signaling molecule.

All of these possibilities can more readily be addressed now that cells that migrate to the base can be stably marked by virtue of their expression of the ecmB gene.

REFERENCES

Barklis E, Lodish HF (1983): Regulation of *Dictyostelium discoideum* mRNAs specific for prespore or prestalk cells. Cell 32:1139–1148.

Berks M, Kay RR (1988): Cyclic AMP is an inhibitor of stalk cell differentiation in *Dictyostelium discoideum*. Dev Biol 125:108–114.

Berks M, Kay RR (1991): Combinatorial control of cell differentiation by cAMP and DIF-1 during development of *Dictyostelium discoideum*. Development 110:977–984.

Bonner J (1952): The pattern of differentiation in amoeboid slime molds. Am Nat 86:79–89.

Brookman J, Jermyn K, Kay R (1987): Nature and distribution of the morphogen DIF in the *Dictyostelium* slug. Development 100:119.

Brookman JJ, Town CD, Jermyn K, Kay RR (1982): Developmental regulation of a stalk-cell differentiation factor-inducing factor in *Dictyostelium discoideum*. Dev Biol 91:191–196.

Ceccarrelli A, McRobbie SJ, Jermyn KA, Duffy K, Early A, Williams JG (1987): Structural and functional characterization of a *Dictyostelium* gene encoding a DIF inducible, prestalk-enriched mRNA sequence. Nucl Acids Res 15:7463–7476.

Devine K, Loomis W (1985): Molecular characterization of anterior-like cells in *Dictyostelium discoideum*. Dev Biol 107:364.

Devreotes P (1989): *Dictyostelium discoideum*—a model system for cell-cell interactions in development. Science 245:1054–1058.

Firtel RA, van Haastert PJM, Kimmel RA, Devreotes PN (1989): G protein linked signal transduction pathways in development: *Dictyostelium* as an experimental system. Cell 58:253–259.

Gomer R, Firtel R (1986): Tissue morphogenesis in *Dictyostelium discoideum*. J Cell Biol 103:A436.

Gregg J, Davis R (1982): Dynamics of cell redifferentiation in *Dictyostelium mucoroides*. Differentiation 21:200.
Gregg JH, Karp GC (1977): An early phase in *Dictyostelium* cell differentiation revealed by 3H-1-fucose incorporation. In Cappuccinelli P, Ashworth JM (eds): ''Development and Differentiation in the Cellular Slime Moulds.'' New York: Elsevier/North-Holland, pp 297–310.
Gross J, Bradbury J, Kay R, Peacey M (1983): Intracellular pH and the control of cell-differentiation in *Dictyostelium discoideum*. Nature 303:244.
Inouye K (1988): Induction by acid load of the maturation of prestalk cells in *Dictyostelium discoideum*. Development 104:669.
Jermyn KA, Duffy K, Williams JG (1989): A new anatomy of the prestalk zone of *Dictyostelium*. Nature (London) 340:144–146.
Jermyn KA, Berks M, Kay RR, Williams JG (1987): Two distinct classes of prestalk-enriched mRNA sequences in *Dictyostelium discoideum*. Development 100:745–755.
Jermyn KA, Williams JG (1991): An analysis of culmination in *Dictyostelium* using prestalk and stalk-specific cell autonomous markers. Development 111:779–787.
Kay R, Garrod D, Tilly R (1978): Requirement for cell differentiation in *Dictyostelium discoideum*. Nature 271:58–60.
Kay R, Jermyn K (1983): A possible morphogen controlling differentiation in *Dictyostelium*. Nature 303:242.
McRobbie S, Jermyn K, Duffy K, Blight K, Williams J (1988a): Two DIF inducible, prestalk specific messenger RNAs of *Dictyostelium* encode extracellular matrix proteins of the slug. Development 104:275–284.
McRobbie S, Tilly R, Blight K, Ceccarelli A, Williams J (1988b): Identification and localization of proteins encoded by two DIF inducible genes of *Dictyostelium*. Dev Biol 125:59–63.
Mehdy MC, Ratner D, Firtel RA (1983): Induction and modulation of cell type specific gene expression in *Dictyostelium*. Cell 32:763–771.
Morris HR, Taylor GW, Masento MS, Jermyn KA, Kay RR (1987): Chemical structure of the morphogen differentiation inducing factor from *Dictyostelium discoideum*. Nature (London) 328:811–814.
Neave N, Kwong L, Macdonald J, Weeks G (1986): The distribution of the stlk cell-differentiation inducing factor and other lipids during the differentiation of *Dictyostelium discoideum*. Biochem Cell Biol 64:85.
Newell P, Ross F (1982): Genetic analysis of the slug stage of *Dictyostelium discoideum*. J Gen Microbiol 128:1639–1652.
Oyama M, Blumberg D (1986): Changes during differentiation in requirements for cAMP for expression of cell-type-specific specific mRNAs in the cellular slime mold, *Dictyostelium discoideum*. Dev Biol 117:550–556.
Oyama M, Okamoto K, Takeuchi I (1983): Proportion regulation without pattern formation in *Dictyostelium discoideum*. J Embryol Exp Morphol 75:293–301.
Raper K (1940): Pseudoplasmodium formation and organisation in *Dictyostelium discoideum*. J Elisha Mitchell Sci Soc 59:241–282.
Raper K, Fennell D (1952): Stalk formation in *Dictyostelium*. Bull Torrey Bot Club 79:25–51.
Schaap P (1986): Regulation of size and pattern in the cellular slime molds. Differentiation 33:1.
Schaap P, Van Driel R (1985): The induction of post-aggregative differentiation in *Dictyostelium discoideum* by cAMP. Evidence for the involvement of the cell surface cAMP receptor. Exp Cell Res 159:388–398.
Schindler J, Sussman M (1977): Ammonia determines the choice of morphogenetic pathways in *Dictyostelium discoideum*. J Mol Biol 116:161–169.
Sternfeld J, David C (1981): Cell sorting during pattern formation in *Dictyostelium*. Differentiation 20:10.

Sternfeld J, David C (1982): Fate and regulation of anterior-like cells in *Dictyostelium* slugs. Dev Biol 93:111.

Town C, Gross J, Kay R (1976): Cell differentiation without morphogenesis in *Dictyostelium discoideum*. Nature (London) 262:717–719.

Wang M, Roelfsema JH, Williams JG, Schaap P (1990): Cytoplasmic acidification facilitates but does not mediate DIF-induced prestalk gene expression in *Dictyostelium discoideum*. Dev Biol 140:182–188.

Wang M, Schaap P (1989): Ammonia depletion and DIF trigger stalk cell-differentiation in intact *Dictyostelium discoideum* slugs. Development 105:569–574.

Williams J, Ceccarelli A, McRobbie S, Mahbubani H, Kay R, Early A, Berks M, Jermyn K (1987): Direct induction of *Dictyostelium* prestalk gene expression by DIF provides evidence that DIF is a morphogen. Cell 49:185.

Williams J, Jermyn K, Duffy K (1989): Formation and anatomy of the prestalk zone of *Dictyostelium*. Development 7:91–97.

Cell-Cell Interactions in Early Development, pages 273–282

17. The p30 Movement Protein of TMV Alters Plasmodesmata Structure and Function

Patricia J. Moore, Carl M. Deom, and Roger N. Beachy
Department of Biology, Washington University, St. Louis, Missouri 63130

CELL-TO-CELL COMMUNICATION IN PLANTS

Plants face a unique problem in cell-to-cell communication. Most plant cells are enclosed by a rigid cell wall, consisting primarily of complex polysaccharides such as cellulose, which surrounds each plant cell in what essentially is a rigid box. Cell walls provide mechanical and structural support, and also are a source of regulatory molecules (Albersheim et al., 1980). This rigid enclosure, however, prevents direct interactions between the plasma membranes of neighboring cells. Thus, plant cells cannot form structures such as gap junctions, small pores that form between closely appressed plasma membranes of neighboring animal cells. Instead, plants have overcome this constraint by establishing thin strands of cytoplasm, which traverse the cell wall, interconnecting the cytoplasms of neighboring cells. These cellular connections are called plasmodesmata.

PLASMODESMATA

Plasmodesmata allow the passage of small molecules between cells and electrically couple connected cells. Thus, plasmodesmata can be considered analogues of gap junctions between animal cells. In addition to transporting nutrients and metabolites, plasmodesmata represent possible avenues of transport for regulatory molecules and developmental signals. Most cells of the plant are joined by the plasmodesmata into a continuum of living protoplasts termed the symplast (Gunning and Overall, 1983).

Plasmodesmata form channels through the cell walls of neighboring cells (Fig. 1). These channels are lined by the plasma membrane, so that the plasmalemma of adjoining cells is continuous. The channels have an internal diameter of about 25 to 30 nm (Gunning and Overall, 1983) and have an axial component, the desmotubule, which passes through the channel. The desmotubule is an extension of the endoplasmic reticulum (ER) (Gunning and Overall, 1983), although the lumen of the ER is closed between cells (Overall et al.,

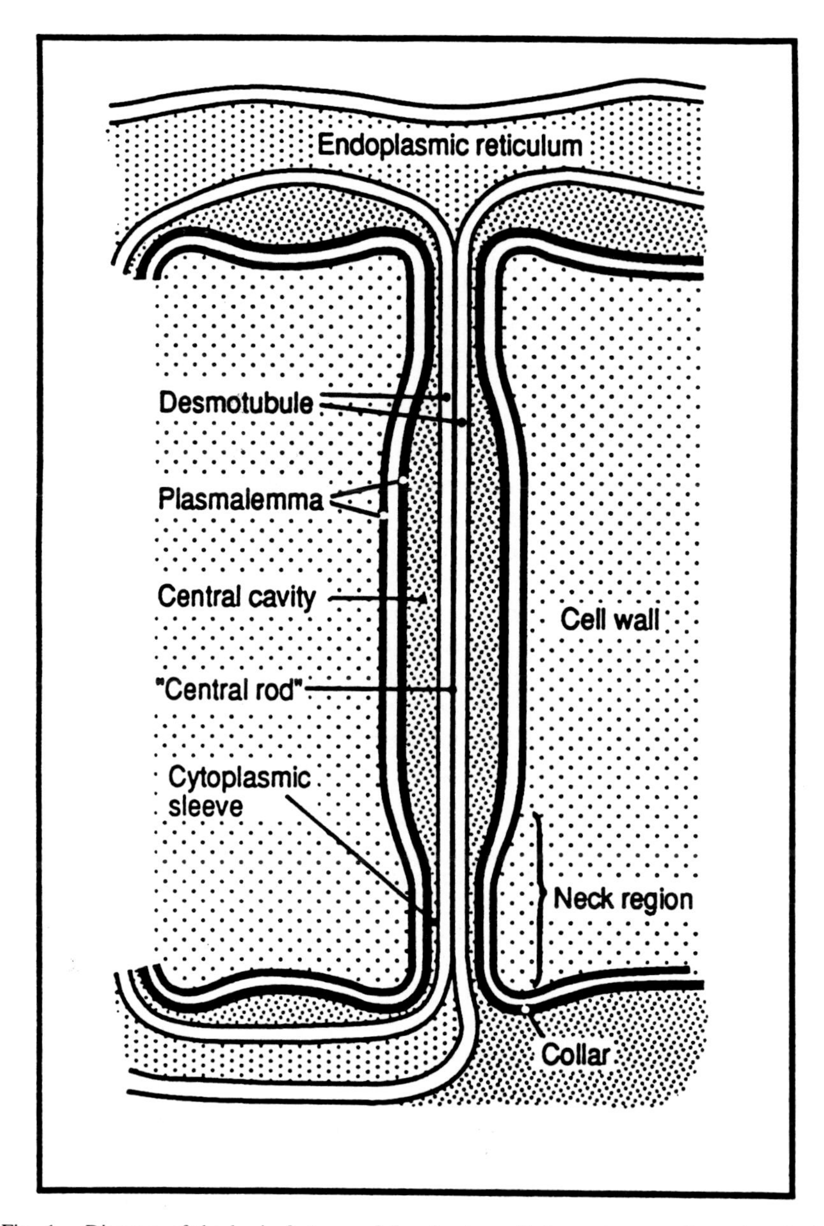

Fig. 1. Diagram of the basic features of the structure of plasmodesmata. Reproduced with permission from the Annual Review of Plant Physiology and Plant Molecular Biology, Vol. 41, © 1990 by Annual Reviews Inc.

1982). Thus, as in gap junctions, the only open pathway of transport between the cells is through the cytoplasm.

In plasmodesmata, the cytoplasmic annulus between the plasma membrane and the desmotubule (Fig. 1) is believed to be occluded by particles that may control transport through the plasmodesmata (Overall et al., 1982). The rate of diffusion of molecules through plasmodesmata is best explained by the division of the cytoplasmic annulus into several (9 to 20) discrete channels with diameters of about 3 nm (Terry and Robards, 1987). The size exclusion limit of plasmodesmata was determined to be between 700 and 900 daltons, similar to that of gap junctions (Robards and Lucas, 1990; Goodwin, 1983; Tucker, 1982).

Most plasmodesmata are formed during cell division by an incomplete fusion of the cell plate between sister cells. Strands of endoplasmic reticulum appear to become trapped as the new cell wall is formed. It is clear, however, that secondary plasmodesmata can occasionally form de novo between nonsister cells. For instance, plasmodesmata have been found between genetically unrelated cells of a chimera (Binding et al., 1987) and between undifferentiated cells, such as the dedifferentiated cells that form in grafts (Robards and Lucas, 1990).

The shape of plasmodesmata can change dramatically during differentiation of plant cells. For instance, desmotubules may anastomose, leading to plasmodesmata with multiple connections on one side or the other (Robards, 1976). An extreme example of change in plasmodesmata occurs during differentiation of guard cells of the epidermis. During guard cell maturation, the plasmodesmata between the forming guard cell and its sister cell are completely obliterated by cell wall material, isolating the mature guard cell from all others (Wille and Lucas, 1984). In most cases, the role of these shape changes in relation to the specific function of different cell types is not known.

PLASMODESMATA AND VIRAL MOVEMENT DURING INFECTION

In addition to their normal role in plant growth and development, plasmodesmata are exploited by some viruses to move from cell to cell. To develop a systemic infection, virus progeny must move from the initial site of infection and colonize new regions of the plant (Sulzinski and Zaitlin, 1982). In animal cells, viral progeny spread by budding from the infected cell and subsequently fuse with uninfected cells. However, plant viruses, because of the cell wall barrier, are unable to utilize this mechanism of movement.

Plant viruses can overcome the obstacle of the cell wall in one of two ways. They can create a new channel through the cell wall or they can utilize the existing channels. As viruses (and their genetic material) are much larger than the size exclusion limits of unmodified plasmadesmata, mechanisms to modify plasmodesmata to allow movement of viral progeny from cell to cell have evolved.

Movement Function Is Encoded by the Virus

Evidence from experiments carried out in the 1980s suggested that a specific virus-encoded function is important for cell-to-cell movement. Such a movement function could determine host range, since in a nonhost plant a virus would fail to move away from the site of infection even if it is capable of replicating in isolated protoplasts of that plant (Atabekov and Dorokhov, 1984; Ponz and Bruening, 1986). The movement function may also influence the virulence of a virus, influencing the rate of spread of the virus throughout the plant (Deom et al., 1987).

In the best-studied example of a virus-encoded movement function, there is conclusive evidence that a 30 kD protein in tobacco mosaic virus (TMV) is responsible for cell-to-cell spread of the infection. Evidence for the involvement of the 30 kD protein in viral movement came as a result of the isolation of a strain of TMV that was defective in cell-to-cell movement (Nishiguchi et al., 1978, 1980). At the nonpermissive temperature, this movement-deficient strain of TMV, the Ls1 strain, is unable to move from cell to cell although it replicates and assembles properly in inoculated leaves. Two-dimensional tryptic peptide mapping of the 30 kD proteins of Ls1 and its parental strain revealed a single amino acid change in the 30 kD protein of Ls1 (Leonard and Zaitlin, 1982). The 30 kD protein is one of the three known nonstructural proteins encoded by the TMV genome, and is produced by translation of a subgenomic messenger RNA (Watanabe et al., 1986).

To examine the specific biological function(s) of the TMV 30 kD protein, the 30 kD gene was introduced into tobacco plants by genetic transformation (Deom et al., 1987). Expression of the 30 kD protein in transgenic plants is able to complement the Ls1 mutation described above. At the nonpermissive temperature, none of the control plants develop systemic disease symptoms when inoculated with the Ls1 strain of TMV. However, in tobacco plants expressing the 30 kD protein, 90% of plants inoculated with Ls1 and held at the nonpermissive temperature showed symptoms within five days. These experiments provided direct evidence that the 30 kD protein functions in cell-to-cell movement of TMV. The protein has been referred to as the TMV movement protein (MP).

Two general mechanisms of action have been proposed for the virally encoded movement protein (Atabekov and Dorokhov, 1982). First, the movement protein might suppress a plant defense response that inhibits viral movement between cells. This mechanism was proposed based upon the localization of MP to the nuclei of protoplasts of cells infected with TMV. Localization to the nucleus may indicate that the movement protein acts at the transcriptional level by suppressing an induced cellular response to infection, or by altering the expression of genes that control the structure and function of plasmodesmata. Alternatively, the MP might itself modify plasmodesmata such that viruses can pass from cell to cell. The 30 kD MP of TMV is localized to plasmodesmata of intact plants during vital infection (Tomenius et al., 1987), supporting the idea that viruses alter plasmodesmata function.

In order to examine the effect that expression of the TMV 30 kD MP has on plasmodesmatal function, Wolf and coworkers (1989) measured and compared the size exclusion limits of plasmodesmata in control plants (MP−) and in plants expressing the MP gene (MP+). They found that in MP(+) plants that accumulated sufficient levels of MP, the size exclusion limits of the plasmodesmata were substantially altered when compared with MP(−) plants. In these studies, fluorescein isothiocyanate-conjugated dextran (F-dextran) of different sizes was injected into cells and the movement of the fluorescent dye away from the injected cell was monitored by fluorescent microscopy. F-dextran probes of less than 750 daltons moved with equal efficiency in MP(−) plants and MP(+) plants. F-dextrans of 3,900 daltons, however, were unable to move efficiently between cells in control plant lines, but did move efficiently from cell to cell in MP(+) plants. A 9,400 dalton dye probe was also able to move from cell to cell in nearly all injection experiments with MP(+) plants. However, a 17,200 dalton probe was unable to move from cell to cell even in MP(+) plants. Thus, expression of the TMV MP gene in transgenic tobacco plants alters the size exclusion limits of the plasmodesmata, increasing the size limit from 700 to 800 daltons to greater than 9,400 daltons, i.e., at least 10 times greater than control plants. These results strongly suggest that TMV MP acts by modifying plasmodesmata function to allow passage of viral progeny from cell to cell. The localization of MP in the nuclei of infected protoplasts may be due to the fact that removing the cell wall altered the targeting of the MP and led to an accumulation in the nucleus by default. It is, however, not yet known how MP interacts with the components of the plasmodesmata to alter the size exclusion limits.

Ability of the MP to Alter Plasmodesmata Is Developmentally Regulated

Deom et al. (1990) analyzed the accumulation of MP in several different tissues of plants that express the TMV MP gene and detected MP in young and old leaves, as well as in stems and roots. The MP isolated from these tissues had the same electrophoretic mobility in SDS-PAGE experiments. The highest level of MP was found in mature leaf tissues, and the subcellular distribution of MP changes as the leaf ages. Deom and coworkers (1990) demonstrated that in mature leaves most MP is found in the cell wall fraction, whereas in young leaves most MP is in a soluble fraction. These studies led to the hypothesis that the accumulation of MP in the leaves of transgenic tobacco is a function of leaf age.

Leaf age is also correlated to the ability of the MP to alter the size exclusion limits of the plasmodesmata. In young leaves, the 9,400 dalton F-dextran is restricted to the injected cell. In mature leaves, this dye probe moves readily from cell to cell. Thus the MP that is found in the young leaves is unable to modify the plasmodesmata to allow for the passage of this larger molecule. It is not clear why the MP does not modify plasmodesmata in young leaves. Perhaps plasmodesmata are more stringently controlled in young leaves. It is interesting to note that there are no gross morphological differences between MP(+) plants and MP(−) plants, although MP(+) plants tend to age more rapidly than MP(−) plants (Deom and Beachy, unpublished observations). Major perturbations in plasmodesmatal function might be expected to result in physiological abnormalities if plasmodesmata are involved, for example, in the transport of molecules that control metabolism. It is apparent, however, that the MP(+) plant lines produced to date control the influence of the MP in critical regions such as the meristem.

It is likely that a critical, threshold level of MP is required for modification of the plasmodesmata. TMV virus is able to move from cell to cell in very young leaves during viral infection, implying that during viral infection substantial local concentrations of MP are achieved. The concentration of MP in a whole, infected leaf and a leaf from a transgenicc plant is similar (Deom et al., 1990). However, because only a few cells of an infected leaf actually contain virus, the amount of MP per cell may be much higher than in a transgenic plant. It may be this localized concentration of MP that enables the TMV virus to move between cells.

MP Accumulates in Plasmodesmata of Mature Leaves

The accumulation of MP in a cell wall fraction from transgenic plants suggests that the MP has an affinity for and/or is stable in cell walls, as would be expected for a protein that modifies plasmodesmata. Indeed, MP has been localized to the cell wall during TMV infection (Tomenius et al., 1987). In order to further examine the relationship between MP and plasmodesmatal function, MP was localized in transgenic plants using immunogold localization (Atkins et al., 1990; Moore et al., 1990). Mature leaves were fixed and prepared for immunocytochemistry in the resin Lowicryl K4M. The MP antigen appears to be highly sensitive to heat, and processing had to be done at low temperature. Sections were immunolabeled with an antibody raised against an internal peptide of the MP followed by protein A-colloidal gold to localize the MP. Gold label was found exclusively over plasmodesmata (Fig. 2). No label was apparent over intercellular membranes or the cytoplasm or over the nucleus. The same result was obtained using a polyclonal antibody raised against MP produced in *E. coli* (Moore et al., 1990).

Most of the gold label, over 75%, was associated with the edges of the plasmodesmata, where the plasma membrane is expected to be located, suggesting that the MP interacts with the plasma membrane or a plasma membrane-

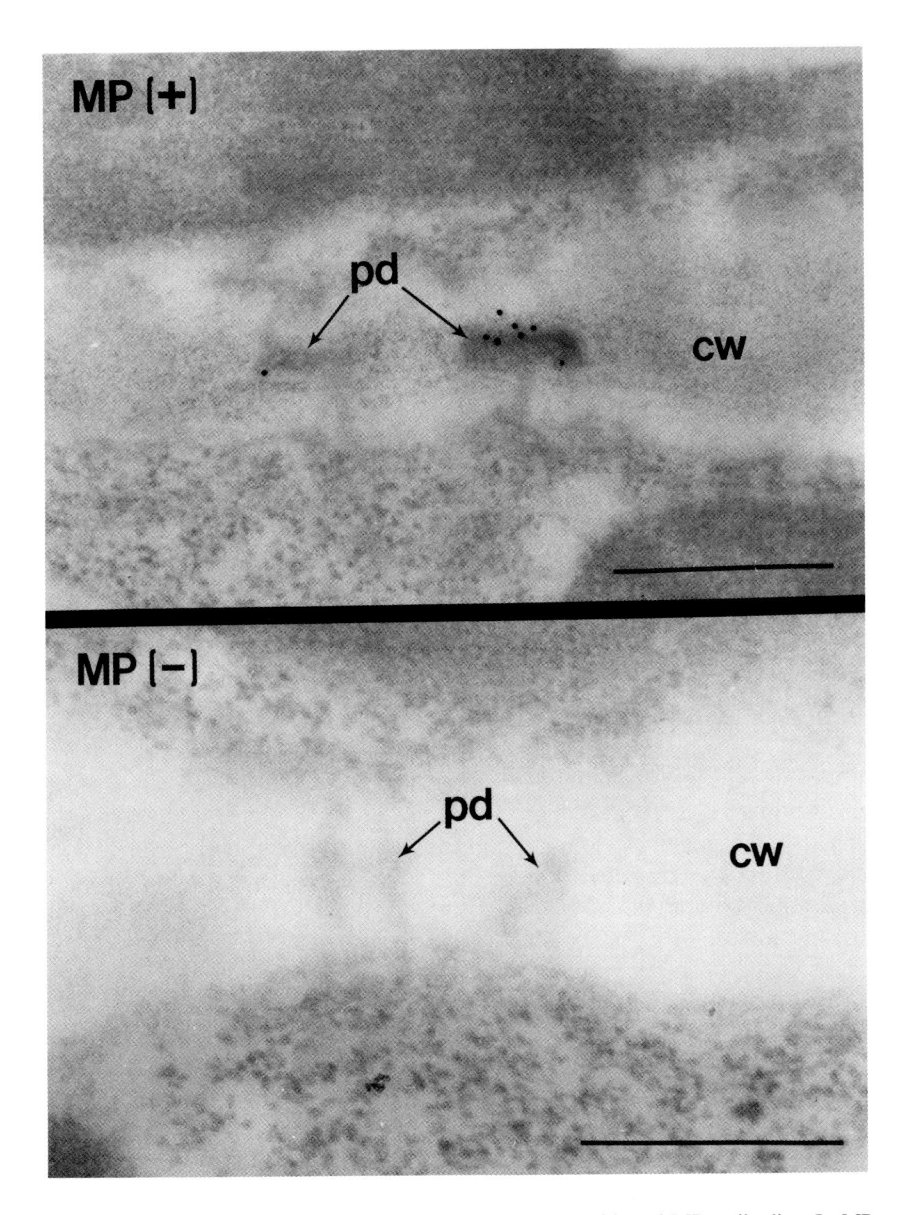

Fig. 2. Immunolabeling of fully expanded tobacco leaves with anti-MP antibodies. In MP-expressing plants (MP +), immunogold label is present over the plasmodesmata (pd) and not over other regions of the cell wall (cw) or the cytoplasm. In non-Mp-expressing plants (MP −), there is no immunogold label associated with the plasmodesmata or any other region of the cells.

associated protein. Little gold label was found over the plasma membrane outside of the plasmodesmata, suggesting that the compartment(s) with which MP is associated is also localized in the plasmodesmata.

Interestingly, no gold label was found over the neck region of the plasmodesmata (see Fig. 1), where control of transport through the plasmodesmata is proposed to be excised (Robards and Lucas, 1990). It is possible that only a very low level of MP, too little to be detected with immunoelectron microscopy, is in the neck region. This subset of MP may be responsible for the change in plasmodesmata function. The MP in the central cavity in that case would simply be excess. Alternatively, we may need to revise our ideas of how, and where, transport through the plasmodesmata is controlled.

MP AND PLASMODESMATA FUNCTION: MANY QUESTIONS REMAIN TO BE ANSWERED

It is of interest to know how MP accumulates in plasmodesmata and why it is stable there. The MP has no known signal or targeting sequences and is probably synthesized on soluble ribosomes (Deom and Beachy, unpublished data). Does soluble MP simply diffuse to plasmodesmata and bind, or is there a specific targeting mechanism?

It will be important to identify the specific component(s) of the plasmodesmata to which the MP binds. There is genetic evidence that there is a protein-protein interaction between the MP and a host factor (Meshi et al., 1989). The TM-2 gene in tomato confers resistance to TMV; however, resistance is only expressed in whole plants or leaf discs, not in protoplasts (Motoyoshi and Oshima, 1975, 1977; Stobbs and MacNeill, 1980). In plants homozygous for TM-2, the spread of TMV is restricted (Nishiguchi and Motoyoshi, 1987). A spontaneously occurring strain of TMV, TMV Ltb1, is able to systemically infect tomatoes with the TM-2 gene. The mutations in TMV Ltb1 that enable the virus to overcome the Tm-2 resistance are located in the 30 kD gene. These results suggest that the product of the Tm-2 gene is a factor required for cell-to-cell movement. Perturbing the molecular interactions between the Tm-2 gene product and the MP may prevent movement of the virus. Because the MP accumulates in the plasmodesmata, it is suggested that the Tm-2 gene product is a normal plasmodesmatal component. Future experiments should include 1) cross-linking studies to identify proteins in close proximity to the MP, and thus the associated plasmodesmata components; and 2) attempts to isolate the Tm-2 structural gene by genetic approaches.

Despite the strong conservation of plasmodesmatal structure, there may be differences in the biochemical composition of plasmodesmata. The host range restrictions of many viruses have been correlated to lack of cell-to-cell spread of virus and suggests that there are species-specific differences in the plasmodesmatal components with which the MP interacts. In addition, the change in subcellular distribution of MP as a function of leaf age indicates that there may be changes in plasmodesmatal composition during development or differentiation of plantcells. These changes in composition may relate to the specific function of plasmodesmata in different cell types.

In order to fully understand how the TMV MP acts to alter plasmodesmata function, we need to explore not only what plasmodesmatal components the MP is interacting with, but also the nature of those interactions. Once we begin to understand how the TMV MP alters plasmodesmata function, we will be on our way to understanding the control of transport of molecules through the plasmodesmata, and thus cell-to-cell communication in plants.

REFERENCES

Albersheim P, Darvill AG, McNeil M, Valent BS, Hahn MG, Lyon G, Sharp JK, Desjardins AE, Spellman MW, Ross LM, Robertsen BK, Aman P, Franzen LE (1980): Structure and function of complex carbohydrates active in regulating plant-microbe interactions. Pure Appl Chem 53:79–88.

Atabekov JG, Dorokhov YuL (1984): Plant virus-specific transport function and resistance of plants to viruses. Adv Virus Res 25:1–91.

Atkins D, Hull R, Wells B, Roberts K, Moore P, Beachy RN (1990): TMV 30K protein in transgenic tobacco plants is localized to plasmodesmata. J Gen Virol (in press).

Binding H, Witt D, Monzer J, Mordhorst G, Kollmann R (1987): Plant cell graft chimeras obtained by co-culture of isolated protoplasts. Protoplasma 141:64–73.

Deom CM, Oliver MJ, Beachy RN (1987): The 30-kilodalton gene product of tobacco mosaic virus potentiates virus movement. Science 337:389–394.

Deom CM, Shubert K, Wolf S, Holt C, Lucas WJ, Beachy RN (1990): Molecular characterization and biological function of the movement protein of tobacco mosaic virus in transgenic plants. Proc Natl Acad Sci USA 87:3284–3288.

Goodwin PB (1983): Molecular size limit for movement in the symplast of the Elodea leaf. Planta 157:124–130.

Gunning BES, Overall RL (1983): Plasmodesmata and cell-to-cell transport in plants. Bioscience 33:260–265.

Leonard DA, Zaitlin M (1982): A temperature-sensitive strain of tobacco mosaic virus in cell-to-cell movement generates an altered viral-encoded protein. Virology 117:416–424.

Meshi T, Motoyoshi F, Maeda T, Yoshiwoka S, Watanabe H, Okada Y (1989): Mutations in the tobacco mosaic virus 30-kD protein gene overcome Tm-2 resistance in tomato. The Plant Cell 1:515–522.

Moore PJ, Fenczik C, Beachy RN (1990): Changes in structure of plasmodesmata due to the expression of the TMV 30-kD movement protein in transgenic tobacco are developmentally regulated. (In preparation).

Motoyoshi F, Oshima N (1975): Infection with tobacco mosaic virus of leaf mesophyll protoplasts from susceptible and resistant lines of tomato. J Gen Virol 29:81–91.

Motoyoshi F, Oshima N (1977): Expression of genetically controlled resistance to tobacco mosaic virus infection in isolated tomato leaf mesophyll protoplasts. J Gen Virol 34:499–506.

Mishiguchi M, Motoyoshi F, Oshima N (1978): Behaviour of a temperature sensitive strain of tobacco mosaic virus in tomato leaves and protoplasts. J Gen Virol 39:53–61.

Nishiguchi M, Motoyoshi F, Oshima N (1980): Further investigations of a temperature-sensitive strain of tobacco mosaic virus: Its behaviour in tomato leaf epidermis. J Gen Virol 46:497–500.

Nishiguchi M, Motoyoshi F (1987): Resistance mechanism of tobacco mosaic virus strains in tomato and tobacco. In Evered D, Harnett S (eds): ''Plant Resistance to Viruses.'' New York: J. Wiley and Sons, pp 38–46.

Overall RL, Wolfe J, Gunning BES (1982): Intercellular communication in Azolla roots: I. Ultrastructure of plasmodesmata. Protoplasma 111:134–150.

Ponz F, Bruening G (1986): Mechanisms of resistance to plant viruses. Ann Rev Phytopathol 24:355–381.

Robards AW (1976): Plasmodesmata in higher plants. In Gunning BES, Robards AW (eds): ''Intercellular Communications in Plants: Studies on Plasmodesmata.'' Heidelberg: Springer-Verlag, pp 15–57.

Robards AW, Lucas WJ (1990): Plasmodesmata. Ann Rev Plant Physiol 41:369–419.

Stobbs LW, MacNeill BH (1980): Response to tobacco mosaic virus of a tomato cultivar homozygous for gene Tm-2. Can J Plant Path 2:5–11.

Sulzinski Ma, Zaitlin M (1982): Tobacco mosaicvirus replication in resistant and susceptible plants: In some resistant species virus is confined to a small number of initially infected cells. Virology 121:12–19.

Terry BR, Robards AW (1987): Hydrodynamic radius alone governs the mobility of molecules through plasmodesmata. Planta 171:145–157.

Tomenius K, Clapham D, Meshi T (1987): Localization by immunogold cytochemistry of the virus-encoded 30K protein in plasmodesmata of leaves infected with tobacco mosaic virus. Virology 160:363–371.

Tucker EB (1982): Translocation in the staminal hairs of *Setcreasea purpurea*. I. Study of cell ultrastructure and cell-to-cell passage of molecular probes. Protoplasma 113:193–201.

Watanabe T, Ooshika I, Meshi T, Okada Y (1986): Subcellular localization of the 30K protein in TMV-inoculated protoplasts. Virology 152:414–420.

Wille AC, Lucas WJ (1984): Ultrastructural and histochemical studies on guard cells. Planta 160:129–142.

Wolf S, Deom CM, Beachy RN, Lucas WJ (1989): Movement protein of tobacco mosaic virus modifies plasmodesmatal size exclusion limit. Science 246:377–379.

Cell-Cell Interactions in Early Development, pages 283–296

18. *lin-12* and *glp-1*: Homologous Genes With Overlapping Functions in *Caenorhabditis elegans*

Eric J. Lambie and Judith Kimble
Department of Biochemistry, College of Agriculture and Life Sciences, Graduate School, Laboratory of Molecular Biology, University of Wisconsin-Madison, Wisconsin 53706

INTRODUCTION

During the development of the nematode, *Caenorhabditis elegans*, cell fates are determined via a combination of cell-autonomous and cell-nonautonomous mechanisms. The latter, regulative phenomena, require the existence of one or more intercellular signaling pathways. In this review, we consider the functions of two genes, *lin-12* (*lin*eage abnormal) and *glp-1* (*g*erm *l*ine *p*roliferation defective), that are required for intercellular signaling during nematode development.

glp-1 FUNCTIONS IN INDUCTIVE SIGNALING

glp-1 is important for inductive signaling, i.e., communication between cells of dissimilar developmental potential. Two well-defined examples of *glp-1*-mediated induction have been identified (Austin and Kimble, 1987; Priess et al., 1987). The first involves the regulation of germline proliferation by the distal tip cell of the somatic gonad (Fig. 1A). The germline nuclei near the distal tip cell normally remain in the mitotic cell cycle, functioning as stem cells for the production of gametes in the adult (Nigon and Brun, 1955; Hirsh et al., 1976; Kimble and Hirsh, 1979). As germline nuclei move proximally, away from the distal tip region, they enter meiosis. The meiotic nuclei follow a progressive program of maturation as they move proximally, eventually differentiating into sperm or oocytes. All germline nuclei share a common cytoplasm (Hirsh et al., 1976). However, even in this syncytium, individual nuclei are compartmentalized into partially membrane-bound regions at the periphery of the gonad. The striking polarity of germline maturation has been shown to depend primarily on the distal tip cell. Removal of the distal tip cell (by laer microsurgery) at any time during development causes the distal germline nuclei to stop dividing mitotically and enter meiosis (Fig. 1B; Kimble and

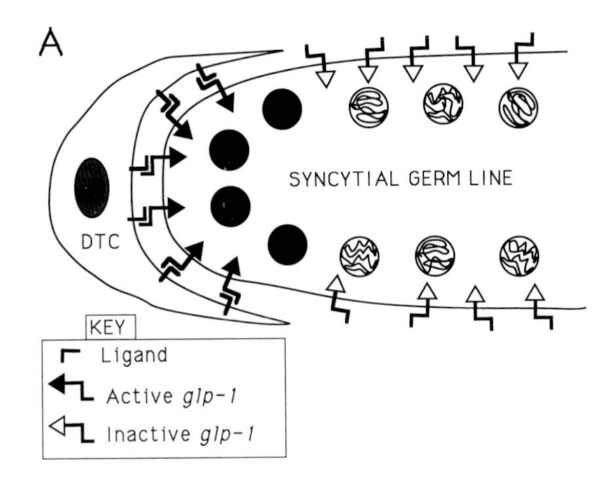
A
DTC
SYNCYTIAL GERM LINE
KEY
Ligand
Active glp-1
Inactive glp-1

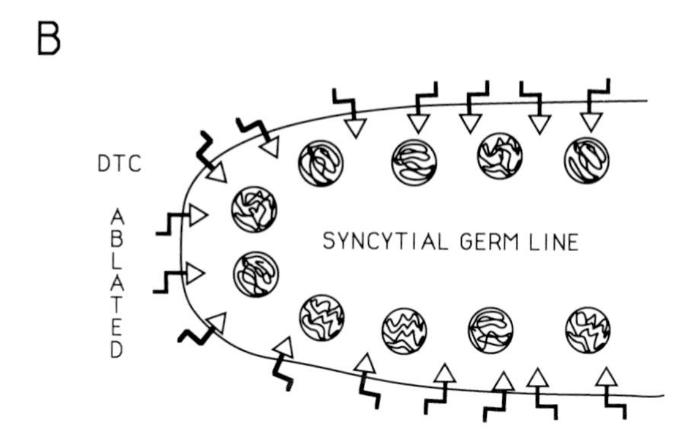
B
DTC
ABLATED
SYNCYTIAL GERM LINE

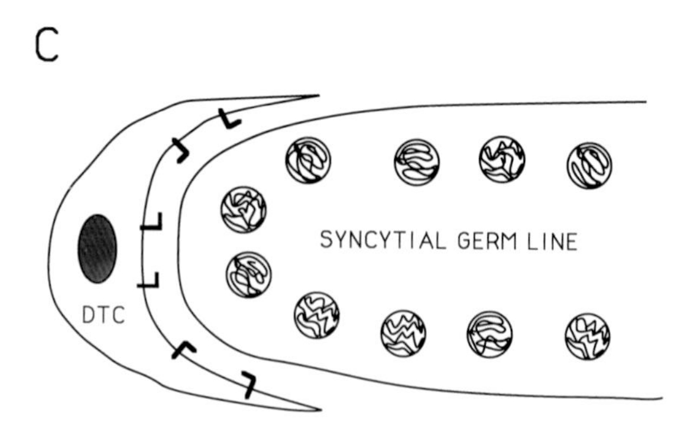
C
DTC
SYNCYTIAL GERM LINE

White, 1981). Thus, the distal tip cell maintains the mitotic stem cell population, either by inducing mitosis or inhibiting entry into meiosis. The zygotic phenotype of *glp-1* mutants strongly resembles that of animals in which the distal tip cell has been ablated (Fig. 1C). The germline nuclei in homozygous *glp-1* null (0) mutants undergo only one or two mitotic divisions before entering meiosis, and eventually differentiate into sperm (Austin and Kimble, 1987; Priess et al., 1987). Consequently, *glp-1* hermaphrodites are sterile; no germline nuclei are available for the production of oocytes. The results of genetic mosaic experiments indicate that *glp-1* is required in the germ line but not the distal tip cell for normal germline proliferation (Austin and Kimble, 1987). Thus, *glp-1* functions in the receiving mechanism in this case of inductive signaling.

The second inductive event that is dependent on *glp-1* occurs early in embryogenesis, between the 4- and 28-cell stages. Each blastomere of the early *C. elegans* embryo follows a unique and invariant lineage: Most of the hypodermis and nervous system are generated by the blastomere AB, while the blastomere EMS primarily produces muscles, gut, and the somatic gonad (Sulston et al., 1983). The pharynx is a neuromuscular feeding organ located at the anterior of the animal; the anterior section of the pharynx is derived from AB and the posterior from EMS. Micromanipulation experiments have revealed that the production of anterior pharyngeal mesoderm by the descendants of AB is dependent upon an inductive signal from EMS (or one of its descendants). When EMS is removed, AB produces only ectodermal tissues (Priess and Thomson, 1987). The role of *glp-1* in this inductive event has been studied by utilizing temperature-sensitive mutants (Priess et al., 1987; Austin and Kimble, 1987). At permissive temperature, homozygous *glp-1(ts)* hermaphrodites are self-fertile and produce normal offspring. However, the embryos produced by such animals do not develop normally if they are shifted to restrictive temperature during early development (prior to the 28-cell stage). These embryos fail to hatch and they exhibit various defects in morphogenesis. Most notably, they lack anterior pharyngeal mesoderm. These lethal developmental defects are strictly dependent upon maternal genotype: *glp-1(+)/glp-1(ts)* heterozygotes produced by crossing a *glp-1(ts/glp-1(ts)* hermaphrodite with a wild-type male do not survive, but *glp-1(0)/glp-1(0)* progeny of a heterozygous hermaphrodite proceed through embryogenesis normally.

Fig. 1. Induction of germline mitosis by the distal tip cell is regulated by *glp-1*. Mitotically proliferating nuclei are represented by black circles, and meiotic nuclei by circles containing wavy lines. The distal tip cell (DTC) nucleus is shown as a shaded oval. The production of a ligand by the distal tip cell, its localization at the cell membrane, and the germline-specific transmembrane location of *glp-1* are hypothetical. The association of *glp-1* with ligand is shown as activating *glp-1* to produce a local mitogenic signal. **A:** Wild type. **B:** Wild type, distal tip cell ablated. **C:** *glp-1(0)*.

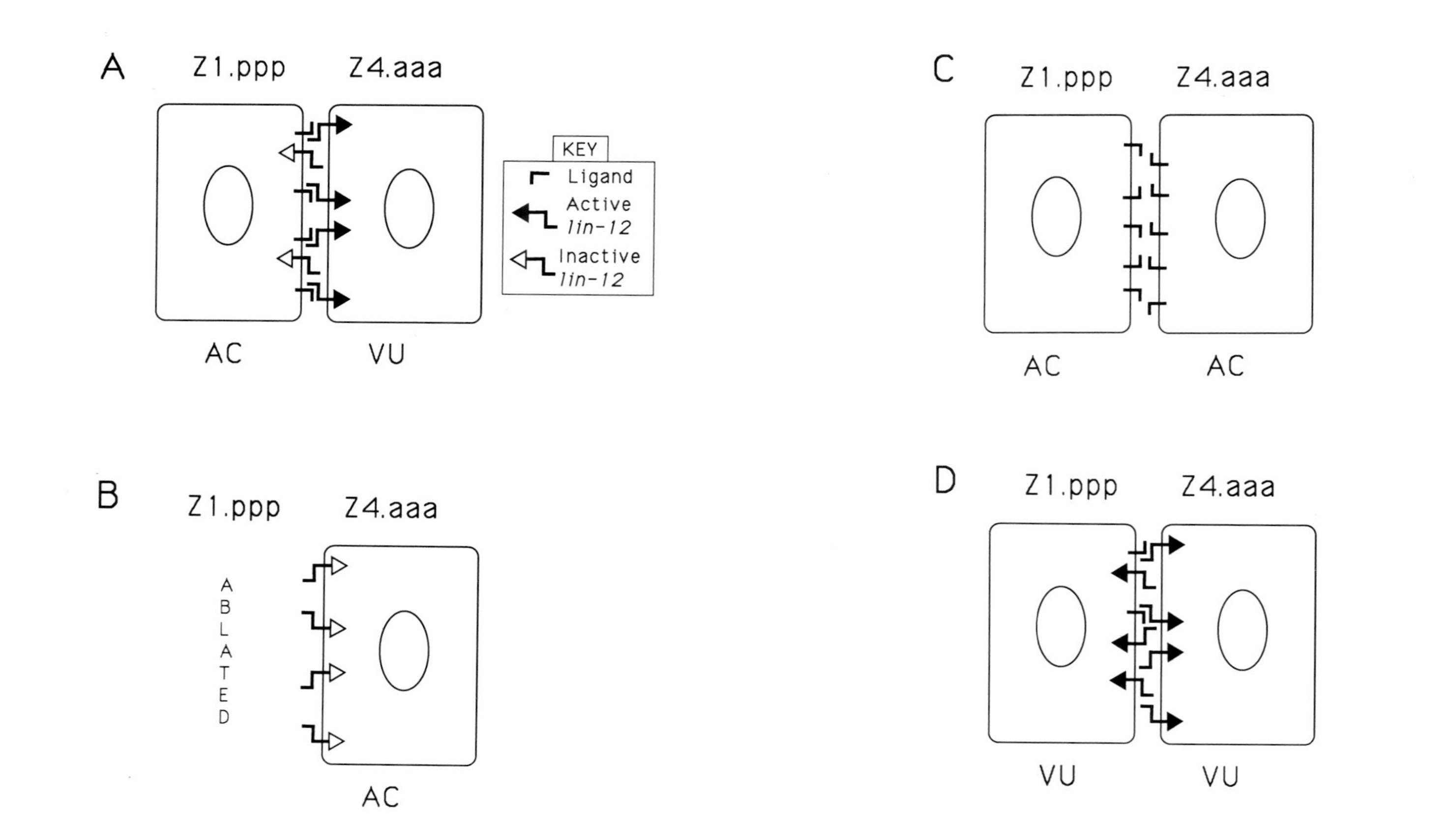
A
Z1.ppp
Z4.aaa
AC
VU
KEY
Ligand
Active
lin-12
Inactive
lin-12
B
Z1.ppp
Z4.aaa
ABLATED
AC
C
Z1.ppp
Z4.aaa
AC
AC
D
Z1.ppp
Z4.aaa
VU
VU

lin-12 FUNCTIONS IN LATERAL SIGNALING

Zygotic *lin-12* activity is important during the development of a number of different somatic tissues (Greenwald et al., 1983), but there is no apparent maternal requirement for *lin-12* (Lambie and Kimble, submitted). In most cases, *lin-12* has been implicated in lateral signaling, i.e., communication between cells of similar developmental potential. The best-characterized examples of *lin-12* function involve equivalence groups. An equivalence group, in its simplest form, consists of two cells with equivalent developmental potential, which cooperate such that one assumes a primary fate, while the other assumes a secondary fate (Sulston and White, 1980; Kimble, 1981). The primary fate is operationally defined as that adopted by a member of an equivalence group when its counterpart has been ablated by laser microsurgery. The restriction of fates within an equivalence group can be thought of as involving two separate decisions. First is the establishment of asymmetry, wherein one cell adopts the primary fate. Second is the propagation of asymmetry, in which a cell that has assumed the primary fate inhibits its neighbor from also adopting the primary fate. The role played by *lin-12* in each of these decisions appears to depend upon context. We consider three different equivalence groups here and how mutations in *lin-12* affect each.

The first two examples both involve the development of the vulva, the ventral opening in the hypodermis through which eggs are laid. The anchor cell (AC), which induces the underlying hypodermal cells to divide and produce the vulva, is derived from an equivalence group consisting of two cells of the somatic gonad, Z1.ppp and Z4.aaa (Fig. 2A; Kimble and Hirsh, 1979; Kimble, 1981). The primary fate is to become an AC, while the secondary fate is to become a ventral uterine precursor cell (VU) (Fig. 2B). The initial asymmetry in this equivalence group is apparently established stochastically, with either Z1.ppp or Z4.aaa adopting the primary fate. In recessive *lin-12* loss-of-function (lf) mutants, both Z1.ppp and Z4.aaa adopt the primary fate (Fig. 2C; Greenwald et al., 1983). Therefore, *lin-12* is required for the primary cell to signal its counterpart to assume the secondary fate. In dominant *lin-12* gain-of-function (gf) mutants, Z1.ppp and Z4.aaa both adopt the secondary fate (Fig. 2D; Greenwald et al., 1983). Moreover, if either Z1.aaa or Z4.ppp is ablated in a *lin12(gf)* mutant, the remaining cell adopts the secondary fate (Seydoux and Greenwald, 1989, 1990). This indicates that *lin-12* activity is sufficient to induce a cell to adopt the secondary fate. Genetic mosaic experiments have shown that *lin-12* activity is required only within the cell that adopts the

Fig. 2. Lateral signaling in the AC/VU equivalence group is regulated by *lin-12*. The production of a ligand by the AC cell, its localization at the cell membrane, and the transmembrane location of *lin-12* are hypothetical. **A:** Wild type. **B:** Wild type, Z1.ppp ablated. **C:** *lin-12(0)*. **D:** *lin-12(gf)*.

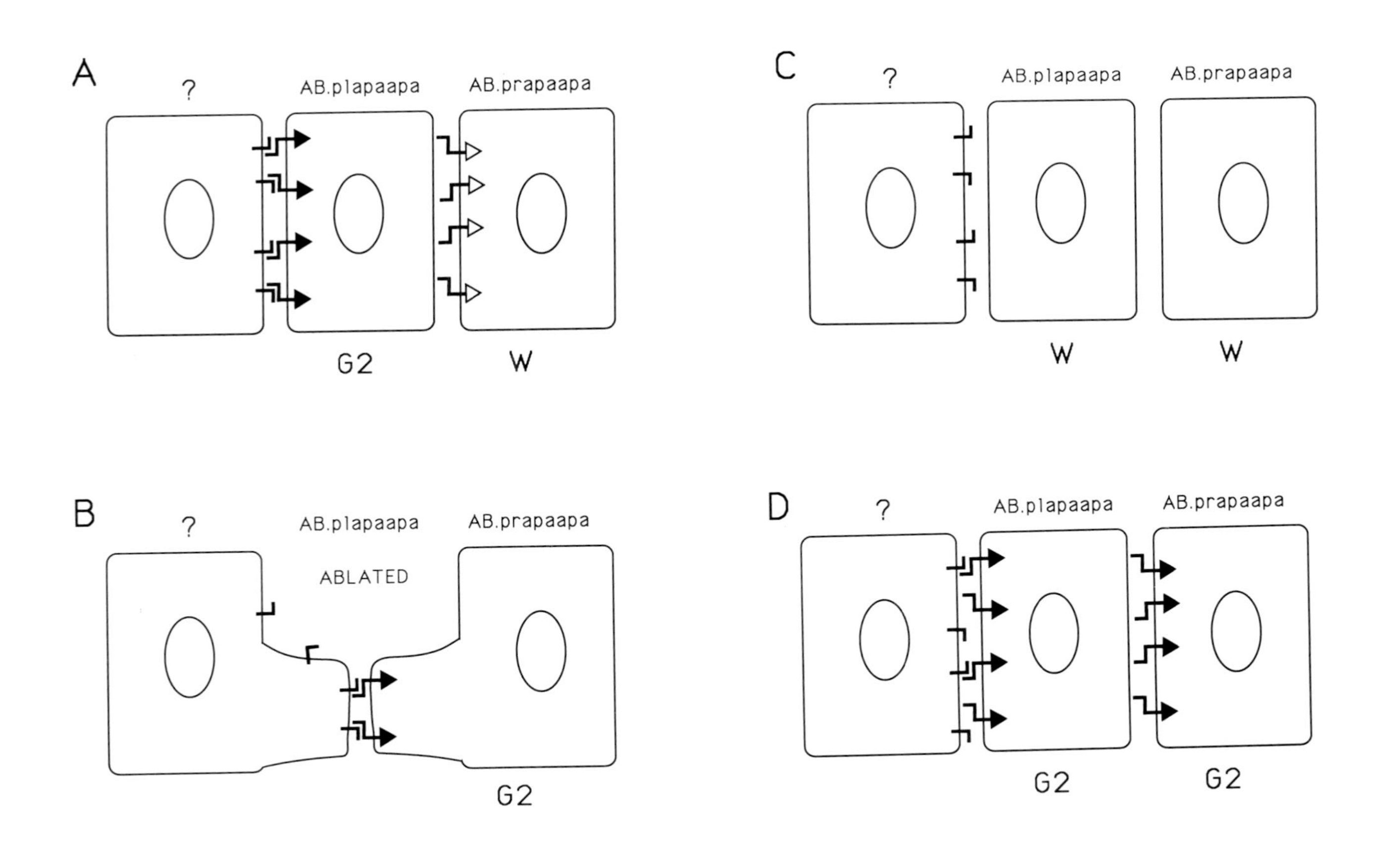
A
?
AB.plapaapa
AB.prapaapa
G2
W
B
?
AB.plapaapa
ABLATED
AB.prapaapa
G2
C
?
AB.plapaapa
AB.prapaapa
W
W
D
?
AB.plapaapa
AB.prapaapa
G2
G2

secondary fate, thus indicating that *lin-12* functions as part of the receiving mechanism in this case of lateral signaling (Seydoux and Greenwald, 1989).

The vulval precursor cells (VPCs) of the hypodermis are also members of an equivalence group. In this case, the primary fate is to execute a particular sublineage, while the secondary fate is to undergo a different sublineage (Sulston and White, 1980; Sternberg and Horvitz, 1986). The initial asymmetry is established by the AC, which induces adjacent VPCs to divide (Kimble, 1981). This induction is not dependent upon *lin-12* (Greenwald et al., 1983). The most proximal VPC assumes the primary fate and laterally signals its neighbors to assume the secondary fate (Sternberg, 1988). In *lin-12(lf)* mutants, this lateral signaling is defective and none of the VPCs adopt the secondary fate (Greenwald et al., 1983). In *lin-12(gf)* mutants, all members of the equivalence group adopt the secondary fate (Greenwald et al., 1983). Thus, in both of the cases considered so far, *lin-12* is essential for the assumption of the secondary fate; *lin-12(lf)* mutations apparently sever the communication between a primary cell and its neighbors, whereas *lin-12(gf)* mutations cause all cells to behave as if they are receiving a signal from an adjacent primary cell.

The situation is different in the third case considered here (Fig. 3). The embryonic blast cells, AB.plapaapa and AB.prapaapa, define an equivalence group (Fig. 3A,B; Sulston et al., 1983). The mechanism by which asymmetry is established is not known, but in wild-type animals, AB.plapaapa invariably assumes the primary fate (designated G2) and AB.prapaapa, the secondary fate (designated W). In *lin-12(lf)* mutants, both cells adopt the secondary fate, and in *lin-12(gf)* mutants both assume the primary fate (Fig. 3C,D; Greenwald et al., 1983). Thus, in this case *lin-12* does not appear to mediate the reception of a signal through which a primary cell induces its neighbor to assume the secondary fate. A simple explanation for the data would be that AB.plapaapa and AB.prapaapa compete for an extrinsically generated signal whose reception is mediated by *lin-12* and induces the primary fate. AB.plapaapa could be predisposed to adopting the primary fate if it intervened between AB.prapaapa and the source of the signal. This situation would be analogous to that observed in the development of the lateral hypodermis, where one member of an equivalence group can apparently intercept an extrinsic signal and thus prevent it from influencing other members of the same equivalence group (Waring and Kenyon, 1990).

In addition to its role in equivalence groups, *lin-12* regulates the fates of several pairs of nonequivalent, but homologous, cells (Greenwald et al., 1983).

Fig. 3. Lateral signaling in the G2/W equivalence group may be only indirectly regulated by *lin-12*. **A:** Wild type. **B:** Wild type, AB.plapaapa ablated. **C:** *lin-12(0)*. **D:** *lin-12(gf)*. Symbols are as in Figure 2. The production of a membrane bound ligand by an unidentified cell (labeled "?") that is ordinarily juxtaposed to AB.plapaapa, and the transmembrane location of *lin-12* are hypothetical.

For example, in male *C. elegans*, AB.plpppaaaa normally differentiates into a neuron (designated DA9), and its homologue, AB.prpppaaaa, executes a sublineage (designated Y). The fate of AB.plpppaaaa is unaffected by the ablation of AB.prpppaaaa, and vice versa; however, in a *lin-12(0)* mutant, both cells assume the DA9 fate, and, in a *lin-12(gf)* mutant, both execute the Y sublineage. *lin-12* may be required for intercellular signaling in such cases, but this signaling is not known to occur, either directly or indirectly, between the two cells whose fates are affected. Alternatively, *lin-12* may function in these contexts as part of an autocrine regulatory mechanism.

lin-12 AND *glp-1* ENCODE PUTATIVE TRANSMEMBRANE RECEPTORS

The *lin-12* and *glp-1* genes are located adjacently on chromosome III, separated by about 20 Kbp (Austin and Kimble, 1987; Yochem and Greenwald, 1989). They are transcribed convergently, and each gene produces a single major transcript (Austin and Kimble, 1989). *lin-12* and *glp-1* are remarkably similar in amino acid sequence (50–60% identical) and in overall molecular organization (Greenwald, 1985; Yochem et al., 1988; Yochem and Greenwald, 1989). It appears likely that they arose through the duplication of an ancestral gene. Moreover, both genes are homologous to the *Notch* gene of *Drosophila*, which is thought to function in a number of intercellular signaling events during fly development (Wharton et al., 1985; Kidd et al., 1986). *lin-12, glp-1*, and *Notch* are members of a gene family whose members mediate regulatory cell interactions during development. Additional members of this gene family have recently been isolated from several vertebrates, including *Xenopus* and *Homo sapiens* (Coffman et al., 1990; L. Ellisen and J. Sklar, personal communication).

A comparison of the predicted *lin-12* and *glp-1* proteins is diagramed in Figure 4. Each gene has an amino terminal hydrophobic region that could act as a signal sequence, and an internal hydrophobic region that could be a transmembrane domain. *Notch* has been shown to exist as a membrane-spanning protein, supporting the view that the region between the two hydrophobic segments exists as an extracellular domain, and the carboxy terminal portion as an intracellular domain (Kidd et al., 1989; Johansen et al., 1989; Fehon et al., 1990). The putative extracellular domains contain two main types of cysteine-rich repeated motifs: There are multiple tandem copies of a sequence homologous to epidermal growth factor (EGF), and three tandem copies of an "LNG" sequence, unique to the members of the *lin-12/Notch/glp-1* family (Greenwald, 1985; Yochem and Greenwald, 1989). The putative intracellular domain contains six tandem copies of a sequence first described in two yeast genes, the *cdc10* gene of *S. pombe*, and the *SWI6* gene of *S. cerevisiae* (Breeden and Nasmyth, 1987). This motif has also been found in other genes, e.g., the *fem-1* gene of *C. elegans* and the vertebrate erythrocyte ankyrin gene (Spence et al., 1990; Lux et al., 1990).

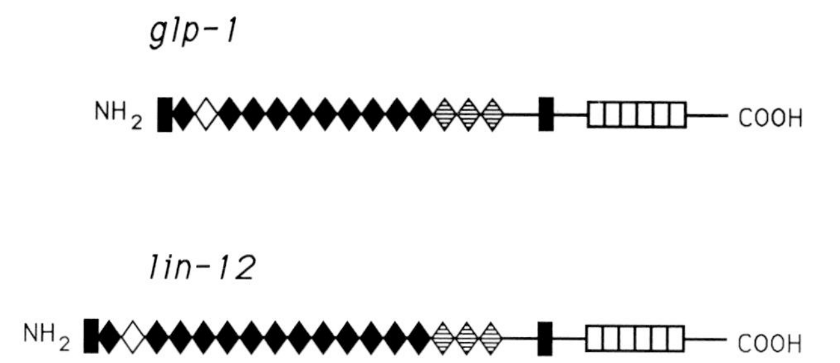

Fig. 4. The predicted protein products of *lin-12* and *glp-1* have extensive similarity. Sequence motifs shared by *lin-12* and *glp-1* are represented by the following symbols: filled rectangle, hydrophobic region; filled diamond, EGF-like repeat; open diamond, nonrepeated cysteine-rich element; striped diamond, LNG repeat; open rectangle, *cdc10/SWI6*-like repeat. Horizontal line indicates coding regions that are not strongly conserved.

The functions of these different domains are not known. However, the extracellular EGF-like motif has been found in a number of proteins that are thought to interact with transmembrane proteins and with components of the extracellular matrix (for review, see Davis, 1990). The *cdc10/SWI6* domain may mediate an interaction with the cytoskeleton, since there is evidence that erythrocyte ankyrin can anchor a membrane protein to the cytoskeleton (Bennett and Stenbuck, 1979).

The combined molecular and genetic data for *lin-12* and *glp-1* argue strongly that they function as transmembrane receptor proteins. Seydoux and Greenwald (1990) have proposed that *lin-12* and *glp-1* transduce intercellular signals by changes in their association state: Binding of an extracellular ligand would induce dimerization, which would activate the intracellular domain for signal transmission. As support for this model, they note that all eight of the *lin-12(gf)* mutations that have been sequenced are present within a restricted region of the putative extracellular domain, and could conceivably result in an increased, or ligand-independent, tendency toward dimerization. At present, there is no direct evidence that the *lin-12* protein forms dimers. In theory, *lin-12(gf)* mutations could affect any of a number of properties of the *lin-12* protein, including local conformation, posttranslational modification, turnover rate, and subcellular targeting.

The nature of the proposed extracellular ligand, whether diffusible, membrane bound, or affixed to an extracellular matrix, is unknown. It is possible that several ligands exist, and that they have different properties, or even opposing effects. Moreover, the effect of ligand binding could vary depending upon the identity of the cell expressing *lin-12* or *glp-1*. Hedgecock et al. (1990) have observed that lateral signaling between Z1.ppp and Z4.aaa is disrupted in mutants where the two cells are not closely apposed. This suggests that the AC to VU signal might not be readily diffusible.

Indeed, recent studies with transfected *Drosophila* tissue culture cells have revealed that the extracellular region of *Notch* protein binds in vivo to the extracellular region of another transmembrane protein, *Delta* (Fehon et al., 1990). *Delta* is one of five other zygotic neurogenic loci that have been identified in *Drosophila* (Lehman et al., 1983). Thus, *Delta* could act as an extracellular signal received by *Notch*. Other zygotic neurogenic loci have been found to encode a transmembrane channel protein (*big brain*: Rao et al., 1990), a G-protein (*groucho*: Preiss et al., 1988), and nuclear proteins (*Enhancer of split*: Klambt et al., 1989; *mastermind*: Smoller et al., 1990). These proteins are thought to act in conjunction with the products of the maternal neurogenic loci, *almondex* and *pecanex* (Perrimon et al., 1986; LaBonne et al., 1989), mediating one or more signal transduction pathways required for the specification of the nonneural ectodermal fate.

lin-12 AND *glp-1* ARE FUNCTIONALLY INTERCHANGEABLE

It is clear from the results of genetic studies that *lin-12* and *glp-1* play different roles during development. The presumed evolution of *lin-12* and *glp-1* from a common ancestor raises the question of how these homologous genes have assumed such disparate functions. In the early stages of their divergence, they must have interacted with the same proteins (and other factors) both extracellularly and intracellularly. Have these other proteins also evolved so that they interact specifically with either *lin-12* or *glp-1*, or is some other form of regulation responsible for dividing the functions of these two genes? One clue to this riddle lies in the observation that the bulk of *glp-1* RNA is expressed in the germline, whereas *lin-12* RNA is expressed primarily in the soma (Austin and Kimble, 1989). Thus, the separate roles of *lin-12* and *glp-1* could be instituted by limiting their expression to specific tissues. If so, *lin-12* and *glp-1* might actually be capable of functioning interchangeably, given the opportunity. Several lines of genetic and molecular evidence suggest that this is correct.

First, an unusual allele of *glp-1*, designated *q35*, has a semidominant multivulva phenotype resembling that seen in *lin-12(gf)* mutants (Austin and Kimble, 1987; S. Mango and J. Kimble, unpublished observations). This suggests that *glp-1(q35)* is expressed in a somatic tissue (the hypodermis), where it mimics the effect of excessive *lin-12* activity. The somatic effect of *glp-1(q35)* is surprising, but not irreconcilable with previous observations, since low levels of *glp-1* RNA have been detected in somatic tissues (Austin and Kimble, 1989). The *glp-1(q35)* mutation introduces a stop codon 122 amino acids from the normal carboxy terminus, and thus may affect some aspect of posttranscriptional regulation (E. Maine, personal communication). Indeed, since the homology between *lin-12* and *glp-1* is relatively low in the carboxy terminal region (about 28%), it may be that sequences contained within this region are involved in differentially regulating the activity or turnover rate of the two proteins.

A second indication that *glp-1* and *lin-12* may share functional interactions comes from analyses of the germline phenotype of *lin-12(lf)* mutants. In *lin-12(lf)* mutants, the germline nuclei in both the proximal and distal arms of the gonad undergo mitotic proliferation (Seydoux et al., 1990). The aberrant proximal germline proliferation is dependent upon the presence of the anchor cell and requires *glp-1* activity (presumably in the germline). This effect can be phenocopied by ablating the somatic gonadal cells that normally separate the AC from the germline. Seydoux et al. (1990) speculate that the lateral signal from the AC to the VU is capable of inducing mitosis in the germline by binding to and activating *glp-1*. In wild-type animals this signal would be intercepted by *lin-12* molecules expressed on the surface of the intervening somatic cells. This model is similar to that depicted in Figure 3, regarding the effect of *lin-12* mutations on the G2/W equivalence group.

Double mutant analysis has provided the most direct evidence that *lin-12* and *glp-1* are capable of functioning interchangeably. Unlike either single mutant, the *lin-12(lf) glp-1(lf)* double mutant invariably arrests and perishes in the first larval stage (Lambie and Kimble, submitted). These lethally arrested larvae exhibit a distinctive combination of abnormalities, collectively termed the Lag phenotype (*lin-12* and *glp-1*). Most or all of the defects observed appear to result from embryonic cell fate transformations; some structures are duplicated (e.g., the excretory pore) and others are absent (e.g., the rectum). The duplication of the excretory pore is thought to result from a failure in communication between the members of an equivalence group. Overall, these results are interpreted to mean that there are certain embryonic cell fate decisions where the zygotic expression of either *lin-12* or *glp-1* is sufficient for normal development. Given the molecular similarity between *lin-12* and *glp-1*, it seems reasonable to suppose that in these cases, *lin-12* and *glp-1* are capable of interacting with common intercellular and extracellular components of the signaling pathway.

IDENTIFICATION OF GENES THAT INTERACT WITH *lin-12* AND/OR *glp-1*

Several genetic screens have been performed in attempts to identify genes that interact with *lin-12* and/or *glp-1*. Suppressor screens have been successfully used to identify genes that modify the phenotypes produced by *lin-12(lf)* and *lin-12(gf)* mutations (M. Sundaram and I. Greenwald, personal communication; F. Tax, J. Thomas, and R. Horvitz, personal communication). Similar strategies have been used to isolate suppressors of *glp-1(ts)* mutants (Maine and Kimble, 1989; E. Maine, personal communication; J. Priess, personal communication). Two distinct types of unlinked recessive suppressors and a set of dominant suppressors that are linked to *glp-1* have been identified (for review, see Maine and Kimble, 1990). Several of the recessive suppressor loci have been found to encode collagens (Kramer et al., 1988; J. Kramer, personal communication). To explain these observations, Maine and Kimble (1989) have suggested that the extracellular matrix of the gonad regulates the activity state of *glp-1* via an interaction with the extracellular domain of *glp-1*, and that mutations in collagen genes could alter the structure of the gonadal basement membrane in such a way as to perturb this regulation.

A different approach for identifying genes that interact with *lin-12* and *glp-1* has been to search for genes that are required for the activities of both *lin-12* and *glp-1* (Lambie and Kimble, submitted). Two genes, *lag-1* and *lag-2*, have been identified in this way as being required for both *lin-12* and *glp-1* to function. Loss-of-function mutations in either gene are recessive and result in a Lag phenotype that is identical with that of *lin-12 glp-1* double mutants. Leaky or temperature-sensitive alleles of *lag-1* and *lag-2* produce postembryonic phenotypes resembling those seen in both *lin-12(lf)* (two anchor cells) and *glp-1(lf)* (failure of germline proliferation) single mutants. The functions of *lag-1* and *lag-2* are not known; however, in view of the similarity between *lin-12*, *glp-1*, and *Notch*, it seems possible that *lag-1* and/or *lag-2* are homologous to the zygotic neurogenic loci of *Drosophila*.

FUTURE DIRECTIONS

As our understanding of the molecular mechanisms that underlie the functioning of *lin-12* and *glp-1* has expanded, the apparent differences between inductive signaling and lateral signaling have correspondingly diminished. Detailed comparisons of the temporal and spatial distribution of the *lin-12* and *glp-1* gene products should soon be possible through the use of gene-specific antibodies. Transformation-mediated reverse genetics will permit the dissection of the various coding and noncoding sequences at the *lin-12* and *glp-1* loci. The molecular isolation and characterization of the various genes that are required for *lin-12* and/or *glp-1* to function will provide important insights regarding the functioning of *lin-12* and *glp-1*, and will also allow comparisons to be made between the molecular hardware of cell-cell communication as it exists in *C. elegans* and in other organisms.

ACKNOWLEDGMENTS

We wish to thank L. Ellisen, J. Sklar, M. Sundaram, I. Greenwald, R. Horvitz, F. Tax, J. Thomas, E. Maine, and S. Mango for communicating unpublished results, and Diane Church for comments on the manuscript. Research done in the authors' laboratory was funded by NIH grant GM31816 to J.K. E.J.L. was supported by postdoctoral fellowship DRG-989 from the Damon Runyon-Walter Winchell Cancer Fund.

REFERENCES

Austin J, Kimble J (1987): *glp-1* is required in the germ line for regulation of the decision between mitosis and meiosis in *C. elegans*. Cell 51:589–599.

Austin J, Kimble J (1989): Transcript analysis of *glp-1* and *lin-12*, homologous genes required for cell interactions during development of *C. elegans*. Cell 58:565–571.

Bennett V, Stenbuck PJ (1979): The membrane attachment protein for spectrin is associated with band 3 in human erythrocyte membranes. Nature 280:468–473.

Breeden L, Nasmyth K (1987): Similarity between cell-cycle genes of budding yeast and fission yeast and the *Notch* gene of *Drosophila*. Nature 329:651–654.

Coffman C, Harris W, Kintner C (1990): *Xotch*, the *Xenopus* homolog of *Drosophila Notch*. Science 249:1438–1441.

Davis CG (1990): The many faces of epidermal growth factor repeats. New Biol 2:410–419.

Fehon RG, Kooch PJ, Rebay I, Regan CL, Xu T, Muskavitch M, Artavanis-Tsakonas S (1990): Molecular interactions between the protein products of the neurogenic loci *Notch* and *Delta*, two EGF-homologous genes in *Drosophila*. Cell 61:523–534.

Greenwald I (1985): *lin-12*, a nematode homeotic gene, is homologous to a set of mammalian proteins that includes epidermal growth factor. Cell 43:583–590.

Greenwald IS, Sternberg PW, Horvitz HR (1983): The *lin-12* locus specifies cell fates in *Caenorhabditis elegans*. Cell 34:435–444.

Hedgecock E, Culotti JG, Hall DH (1990): The *unc-5, unc-6*, and *unc-40* genes guide circumferential migrations of pioneer axons and mesodermal cells on the epidermis in *C. elegans*. Neuron 2:61–85.

Hirsh D, Oppenheim D, Klass M (1976): Development of the reproductive system of *Caenorhabditis elegans*. Dev Biol 49:200–219.

Johansen KM, Fehon RG, Artavanis-Tsakonis S (1989): The *Notch* gene product is a glycoprotein expressed on the cell surface of both epidermal and neuronal precursor cells during *Drosophila* development. J Cell Biol 109:2427–2440.

Kidd S, Kelley MR, Young MW (1986): Sequence of the *Notch* locus of *Drosophila melanogaster*: Relationship of the encoded protein to mammalian clotting factor and growth factors. Mol Cell Biol 6:3094–3108.

Kidd S, Baylies MK, Gasic GP, Young MY (1989): Structure and distribution of the *Notch* protein in developing *Drosophila*. Genes Dev 3:1113–1129.

Kimble J (1981): Alterations in cell lineage following laser ablation of cells in the somatic gonad of *Caenorhabditis elegans*. Dev Biol 87:286–300.

Kimble J, Hirsh D (1979): The postembryonic cell lineages of the hermaphrodite and male gonads in *Caenorhabditis elegans*. Dev Biol 70:396–417.

Kimble JE, White JG (1981): On the control of germ cell development in *Caenorhabditis elegans*. Dev Biol 81:208–219.

Klambt C, Knust E, Tietze K, Campos-Ortega JA (1989): Closely related transcripts encoded by the neurogenic gene complex *Enhancer of split* of *Drosophila melanogaster*. EMBO J 8:203–210.

Kramer JM, Johnson JJ, Edgar RS, Basch C, Roberts S (1988): The *sqt-1* gene of *C. elegans* encodes a collagen critical for organismal morphogenesis. Cell 55:555–565.

LaBonne SG, Sunitha I, Mahowald AP (1989): Molecular genetics of *pecanex*, a maternal-effect neurogenic locus of *Drosophila melanogaster* that potentially encodes a large transmembrane protein. Dev Biol 136:1–16.

Lehman R, Jimenez F, Dietrich U, Campos-Ortega JA (1983): On the phenotype and development of mutants of early neurogenesis in *Drosophila melanogaster*. Roux Arch Dev Biol 192:62–74.

Lux SE, John KM, Bennet V (1990): Analysis of cDNA for human erythrocyte ankyrin indicates a repeated structure with homology to tissue-differentiation and cell-cycle control proteins. Nature 344:36–42.

Maine E, Kimble J (1989): Identification of genes that interact with *glp-1*, a gene required for inductive cell interactions in *Caenorhabditis elegans*. Development 103:133–143.

Maine E, Kimble J (1990): Genetic control of cell communication in *C. elegans* development. Bioessays 12:265–271.

Nigon V, Brun J (1955): L'Evolution des structures nucleaires dans l'ovogenese de *Caenorhabditis elegans* Maupas 1900. Chromosoma (Berlin) 7:129–169.

Perrimon N, Mohler D, Engstrom L, Mahowald AP (1986): X-linked female-sterile loci in *Drosophila melanogaster*. Genetics 113:695–712.

Preiss A, Hartley DA, Artavanis-Tsakonas S (1988): The molecular genetics of *enhancer of split*, a gene required for embryonic neural development in *Drosophila*. EMBO J 7:3917–3927.

Priess JR, Schnabel H, Schnabel R (1987): The *glp-1* locus and cellular interactions in early *C. elegans* embryos. Cell 51:601–611.

Priess JR, Thomson JN (1987): Cellular interactions in early *C. elegans* embryos. Cell 48:241–250.

Rao Y, Jan LY, Jan YN (1990): Similarity of the product of the *Drosophila* neurogenic locus *big brain* to transmembrane channel proteins. Nature 345:163–167.

Seydoux G, Greenwald I (1989): Cell autonomy of *lin-12* function in a cell fate decision. Cell 57:1237–1245.

Seydoux G, Greenwald I (1990): Analysis of gain-of-function mutations of the *lin-12* gene of *Caenorhabditis elegans*. Nature 346:197–199.

Seydoux G, Schedl T, Greenwald I (1990). Cell-cell interactions prevent a potential inductive interaction between soma and germline in *C. elegans*. Cell 61:939–951.

Smoller D, Friedel C, Schmid A, Bettler D, Lam L, Yedvobnick B (1990): The *Drosophila* neurogenic locus *mastermind* encodes a nuclear protein unusually rich in amino acid homopolymers. Genes Dev 4:1688–1700.

Spence AW, Coulson A, Hodgkin J (1990): The product of *fem-1*, a nematode sex-determining gene, contains a repeated motif found in cell cycle control proteins and receptors for cell-cell interactions. Cell 60:981–990.

Sternberg PW (1988): Lateral inhibition during vulval induction in *Caenorhabditis elegans*. Nature 335:551–554.

Sternberg PW, Horvitz HR (1986): Pattern formation during vulval development in *Caenorhabditis elegans*. Cell 44:761–772.

Sulston JE, Schierenberg E, White JG, Thomson JN (1983): The embryonic cell lineage of the nematode *Caenorhabditis elegans*. Dev Biol 100:64–119.

Sulston JE, White JG (1980): Regulation and cell autonomy during postembryonic development of *Caenorhabditis elegans*. Dev Biol 78:577–597.

Waring DA, Kenyon C (1990): Selective silencing of cell communication influences anteroposterior pattern formation in *C. elegans*. Cell 60:123–131.

Wharton KA, Johansen KM, Xu T, Artavanis-Tsakonas S (1985): Nucleotide sequence from the neurogenic locus *Notch* implies a gene product that shares homology with proteins containing EGF-like repeats. Cell 43:567–581.

Yochem J, Greenwald I (1989): *glp-1* and *lin-12*, genes implicated in distinct cell-cell interactions in *C. elegans*, encode similar transmembrane proteins. Cell 58:553–563.

Yochem J, Weston K, Greenwald IS (1988): *C. elegans lin-12* encodes a transmembrane protein similar to *Drosophila Notch* and yeast cell cycle gene products. Nature 335:547–550.

Cell-Cell Interactions in Early Development, pages 297–319

19. The Cytoskeletons of Gametes, Eggs, and Early Embryos

Brian K. Kay, Janice P. Evans, Elizabeth C. Raff, Edwin C. Stephenson, Mary Lou King, David L. Gard, Michael W. Klymkowsky, Richard P. Elinson, Jon M. Holy, and Susan Strome

University of North Carolina, Chapel Hill, North Carolina 27599-3280 (B.K.K., J.P.E.); Indiana University, Bloomington, Indiana 47405 (E.C.R., S.S.); University of Rochester, Rochester, New York 14627 (E.C.S.); University of Miami, Miami, Florida 33101 (M.L.K.); University of Utah, Salt Lake City, Utah 84112 (D.L.G.); University of Colorado, Boulder, Colorado 80309-0347 (M.W.K.); University of Toronto, Toronto, Ontario (R.P.E.); University of Wisconsin, Madison, Wisconsin 14627(J.M.H.)

OVERVIEW

In this mini-symposium, the major cytoskeletal proteins systems have been discussed in the context of early development. The systems include microfilaments and associated proteins, microtubules (MT), and intermediate filaments. While these proteins have been well characterized biochemically in certain adult tissues and their tissue culture cells, little is known about their role in oocytes, eggs, and early embryos. It has long been presumed that the cytoskeleton is involved in numerous important functions, such as changes in cell shape and motility during morphogenesis, the distribution and inheritance of morphogens and other cytoplasmic components, and determination of cell lineages. This mini-symposium presented the work of nine different laboratories, all on the various roles of cytoskeletal proteins during the development of animals such as *Drosophila melanogaster, Xenopus laevis, Cynops pyrrhogaster, Strongylocentrotus purpuratus*, and *Caenorhabditis elegans*.

GAMETOGENESIS

Elizabeth C. Raff: Microtubule Function in *Drosophila* Development

It is clear that in development, microtubules have important missions in regulating cell division, cell shape, and cell motility. Microtubules are assembled from heterodimers of alpha and beta tubulin; both subunits are encoded in multigene families. In *Drosophila*, there are four genes encoding different beta tubulin subunits: two of the isoforms (β1, β2) have a predicted sequence that is highly conserved among eukaryotes, while two of the isoforms (β3, β4) are very diverged (Rudolph et al., 1987; Michielis et al., 1987; Diaz and Raff, 1987). During development, the β1 subunit predominates in many tissues, the β2 subunits is expressed only in male germ cells, and the β3 and β4 subunits are expressed in different tissues, in a somewhat reciprocal, temporal fashion (Kemphues et al., 1979, 1982; Raff et al., 1982; Biajolin et al., 1984; Gasch et al., 1988, 1989; Kimble et al., 1989, 1990). The existence of multiple tubulin isoforms and the complexity of their differential expression raises the question of whether different isoforms possess different intrinsic functional properties, which may in part be responsible for the specificity of microtubule function in vivo. This question has been experimentally addressed by a comparison of the microtubule assembly capacities of the two *Drosophila* beta tubulins, the divergent developmentally regulated isoform β3, and the conserved sequence testis-specific isoform β2 (Hoyle and Raff, 1990).

The β3 gene is transiently expressed in the embryo and again in the pupa in the mesodermally derived musculature, and in the pupa is also expressed in several different tissues of ectodermal origin; adult expression is confined to specific somatic cells in the gonads (Raff et al., 1982; Biajolin et al.,1984; Gasch et al., 1988, 1989; Kimble et al., 1989, 1990). Overall, the temporal and spatial pattern of β3 expression suggests that this isoform may play a specialized functional role, being primarily utilized in assembly of transient cytoskeletal microtubules involved in determining cell shape or tissue organization (Kimble et al., 1989, 1990). For example, during embryonic development, the β3 subunit is only expressed for a short period in midembryogenesis in developing muscle, and appears to function solely in the transient microtubule scaffold, which has been shown to precede differentiation of the sarcomeres during myogenesis in both vertebrates and invertebrates (Crossley, 1972, 1978; Warren, 1974; Tassin et al., 1985). β3 is not expressed in other tissues in the embryo, nor do the transient myogenic microtubules contain the conserved sequence isoform β1 (Gasch et al., 1988), which is the beta tubulin subunit utilized for other embryonic microtubules, including both their transient spindle and cytoskeletal microtubules in the rapid cell cycles in early embryos, and the stable microtubule cytoskeleton of the nervous system formed later in embryogenesis.

Similarly, the conserved sequence testis-specific isoform β2 is also utilized in both transient and stable microtubule arrays, and for assembly of all morphological classes of microtubules, including both singlet and multiple-walled tubules. β2 expression is confined to the postmitotic cells of the male germline. β2 is the only beta tubulin isoform in these cells, and has multiple roles in microtubule function, including assembly of the motile axoneme of the sperm flagellum, the meiotic spindles, and the cytoskeletal microtubules required for mitochondrial elongation and for nuclear shaping (Kemphues et al., 1979, 1982; Fuller et al., 1987, 1988).

Wild-type β3 expression in the testis is confined to the nondividing somatic cyst cells, two of which enclose each group of syncytially developing germ cells (Kimble et al., 1989). During the course of spermatogenesis, as each single germ cell differentiates into 64 mature sperm, the two surrounding somatic cells undergo enormous changes in cell size and shape, suggesting rapid changes in the cytoskeleton consistent with the putative paradigm for β3 function. In wild-type males, β3 is not expressed in the germ cells. In order to experimentally test whether β3 can supply the multiple microtubule functions in the germline normally performed by β2, a hybrid gene was constructed in which the regulatory regions of the β2 gene drive expression of the β3 gene coding region (Hoyle and Raff, 1990). *Drosophila* transformed with the hybrid gene express β3 in the postmitotic male germ cells.

As shown in Figure 1, when β3 is expressed in the absence of β2 in the testes of β2 null males, β3 supports assembly of only one class of microtubules, the transient cytoskeletal tubules involved in elongation of the mitochondrial derivatives (compare panels A and B). These mitochondrial microtubules function normally, but axoneme assembly, meiosis, and other microtubule-mediated processes do not occur. Thus β3 alone can support only one of the functions normally performed by β2. When β3 is coexpressed with β2 in the germline of males carrying the hybrid gene but wild type for the β2 gene, β3 acts as a dominant poison of axoneme assembly, causing dominant male sterility (compare panels C and D). In all such males, spindles and all classes of cytoplasmic microtubules are assembled and function normally. In males in which β3 is less than 20% of the total testis beta tubulin pool, axonemes of normal morphology are assembled and motile functional sperm are produced. However, males in which β3 exceeds 20% of the total testis beta tubulin pool produce no motile sperm and are therefore sterile. In the axonemes assembled in the developing spermatids in such males, the doublet tubules acquire the morphological character of the singlet microtubules of the central pair and accessory tubules of wild-type axonemes, perhaps reflecting the normal function of β3 in assembly of singlet cytoskeletal microtubules.

The results of the hybrid gene experiment unambiguously demonstrate that β2 and β3 tubulin are not functionally interchangeable (Hoyle and Raff, 1990). Taken together with these data, analysis of the patterns of expression of the *Drosophila* beta tubulins suggests the possibility that the isoforms that are highly conserved in sequence may have similar, and general, microtubule assembly capacity, while the assembly properties of divergent isoforms may be specialized or restricted. Thus the specific developmental role played by β3-tubulin in the embryonic cytoskeleton may reflect intrinsic microtubule assembly properties that are unique to this isoform.

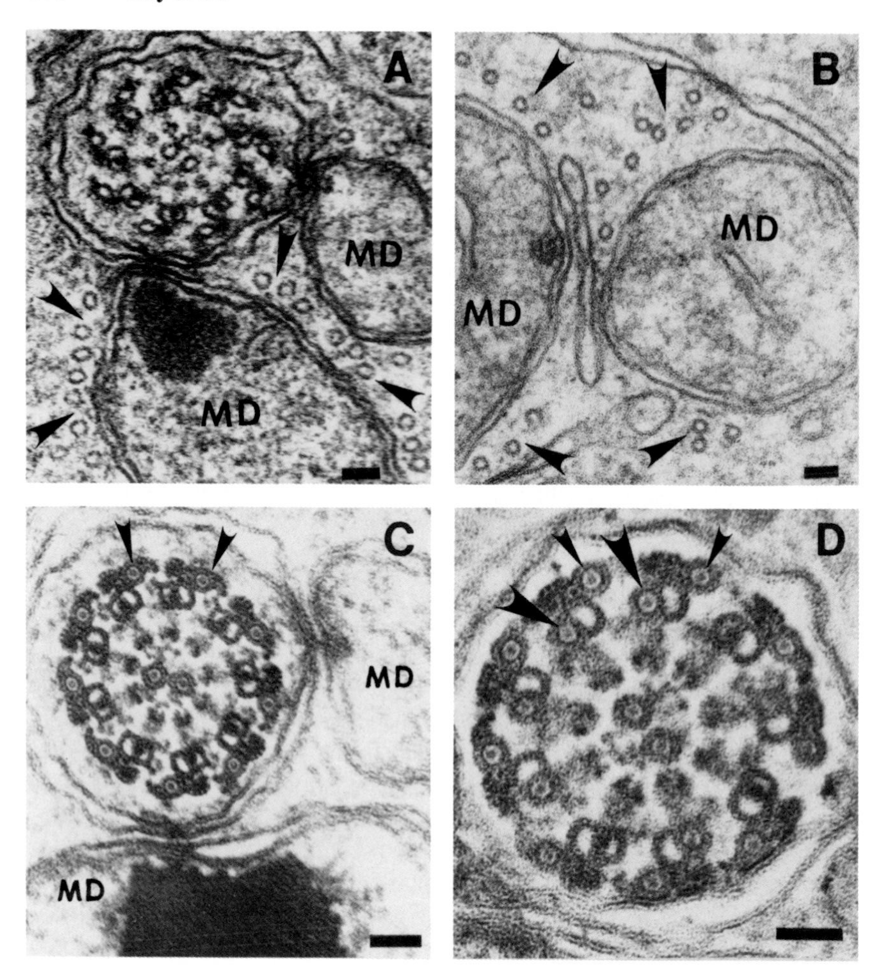

Fig. 1. Microtubules in spermatids in wild-type males and in males that express the divergent developmental isoform β3-tubulin in the germline driven by a hybrid β2β3 gene. Cross sections through the tails of developing spermatids, showing development of the axonemes and mitochondrial derivatives (MD): **A:** Intermediate stage spermatid in a fertile wild-type male that expresses the normal testis specific β2-tubulin isoform in the postmitotic germ cells. The axoneme exhibits the "9 + 2 + 9" morphology typical of insect sperm, consisting of nine outer doublet tubules surrounding a central pair of singlet microtubules, plus nine accessory tubules, which are singlet microtubules associated with the B tubules of each doublet. Mitochondrial derivatives are associated with the cytoplasmic microtubules (arrowheads), which are required for mitochondrial elongation. **B:** Intermediate stage spermatid in a sterile male carrying two copies of the hybrid β2β3 gene but homozygous for a null mutation in the β2 gene. In males of this genotype, β3 but not β2 is expressed in the postmitotic germ cells; construction of the hybrid gene and isolation of transformed stocks is described in Hoyle and Raff (1990). This spermatid is of a stage similar to that shown in panel A, but the only functional microtubules are the mitochondrial-associated microtubules (arrowheads). No axonemes are assembled, and other

Edwin C. Stephenson: Localization of the *bicoid* mRNA at the Anterior Pole of *Drosophila* Oocytes

During oogenesis of *Drosophila*, the primary oogonium undergoes four cell divisions, without the completion of cytokinesis, to yield 16 interconnected cells within the egg chamber. One of these cells is determined to form the oocyte and the remaining 15 cells to differentiate into nurse cells that will aid the growth of the oocyte. The nurse cell nuclei become polyploid and are very active metabolically, synthesizing proteins, mRNAs, and ribosomes. These macromolecules travel through the cytoplasmic bridges into the oocyte. During the last few hours of oogenesis, the nurse cells squeeze via actin contraction the remainder of their cytoplasmic contents through the cytoplasmic bridges into the oocyte. Eventually, the mature oocyte fills the egg chamber.

Do the nurse cell macromolecules mix homogeneously in the oocyte cytoplasm or do they take assigned positions to create an asymmetric arrangement? From experiments by Nüsslein-Volhard and others, it is clear that the anterior–posterior and dorsal–ventral axes of the developing embryo are predetermined in the oocyte. Genetic analysis has shown that the anterior position is determined by the *bicoid* gene product and that the posterior position is determined by the *nanos* gene product (Nüsslein-Volhard et al., 1988). In each case the *bicoid* and *nanos* mRNAs are highly localized in the oocyte. The *bicoid* mRNA is synthesized by the nurse cells and is deposited at the anterior pole of the oocyte, near the cytoplasmic bridges (Berleth et al., 1988). Experiments involving synthetic *bicoid* mRNAs have shown that a nucleotide sequence in the 3′ untranslated region is responsible for this localization in the oocyte (MacDonald and Struhl, 1988).

What proteins are responsible for anterior localization of *bicoid* mRNA? At least two maternal effect genes, *swallow* and *exuperantia* are necessary for localization of *bicoid* activity. In *swallow* homozygotes, the *bicoid* mRNA is

---

microtubule-mediated processes, including meiosis and nuclear shaping, also fail to occur. **C:** Late stage spermatid in a fertile wild-type male. Cytoplasmic mitochondrial-associated microtubules are no longer present, and the mitochondrial and axoneme morphology are similar to that in the mature motile sperm. In the mature axoneme, the lumen of each of the singlet tubules of the axoneme contains an electron dense core (small arrowheads). **D:** Late stage spermatid in a sterile male carrying two copies of the hybrid β2β3 gene and a single functional copy of the β2 gene. In males of this genotype, both β3 and β2 are expressed in postmitotic germ cells, and β3 comprises approximately 25% of the total testis β-tubulin pool. Processes mediated by singlet microtubules—meiosis, nuclear shaping, and mitochondrial elongation—occur normally, but in the axonemes the A tubules of many of the doublets contain an electron-dense luminal core (large arrowheads), such as is normally present only in the singlet central and accessory tubules (small arrowheads). In the axoneme shown, five of the doublets have filled A tubules. This morphological defect correlates with the failure to produce mature motile sperm. For all panels, bars = 50 nm. Micrographs A and C were taken by J.C. Caulton; B and D by F.R. Turner.

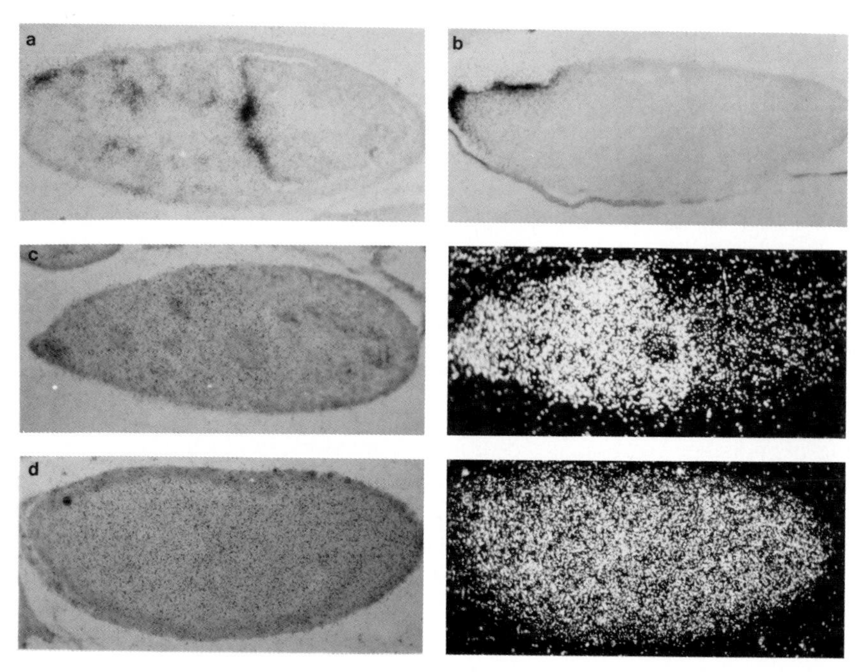

Fig. 2. *bicoid* message localization requires microtubules in *Drosophila*. Egg chambers were dissected from control or treated females and hybridized in situ with a probe that detects *bicoid* messenger RNA. Panels a and b are control egg chambers, dissected from untreated females. Panels c and d are egg chambers from females that had been injected with colchicine to give an approximate whole-body concentration of 20 μg/ml, and held overnight at 25°C before dissection. In all figures anterior is left; posterior, right. **a:** Stage 10 egg chamber, untreated. The anterior half of the egg chamber is comprised of nurse cells, which are transcriptionally active and the source of most or all RNA bound for the oocyte. The oocyte occupies the posterior half of the egg chamber. Note the presence of *bicoid* message in a perinuclear position in nurse cells, and the accumulation of *bicoid* message at the anterior end of the oocyte. **b:** Mature oocyte, untreated. *bicoid* message is restricted to the anterior tip of the oocyte. **c:** Stage 10 egg chamber, colchicine treated, bright-field and dark-field pairs. Note the loss of *bicoid* message localization around nurse cell nuclei and at the anterior oocyte margin. **d:** Mature oocyte, colchicine treated, bright-field and dark-field pairs. *bicoid* message is uniformly distributed throughout the oocyte.

evenly distributed in the oocyte cytoplasm, as detected by in situ hybridization of sections of egg chambers (Fig. 2). Recently, the *swallow* gene has been cloned, and a DNA fragment, when introduced by P-element transformation into flies, will rescue the *swallow* phenotype (Stephenson et al., 1988). Computer analysis of the *swallow* gene nucleotide sequence suggests that *swallow* is a novel protein, with a 90 amino acid repeat domain at the NH_2-terminus, which resembles the repeats present in RNA-binding proteins, and a heptad repeat, with an amphipathic helical character, at its COOH-terminus (Chao et al., in preparation). To localize the *swallow* protein in the oocyte, current efforts have been directed to the preparation of fusion proteins and antibodies.

Drug inhibitor studies have been useful in investigating the potential role of MTs and actin filaments (Pokrywka and Stephenson, in preparation). Agents (colchicine, nocodazole) or conditions (0°C) that disrupt MT arrays in vivo (Wilson and Bryan, 1974) lead to the loss of *bicoid* mRNA localization in the developing oocyte (Fig. 2). Interestingly, the disruption by nocodazole is not permanent, as it can be reversed by washing away the inhibitor, although it is not clear whether there is relocalization of previously synthesized *bicoid* mRNA or that the dispersed *bicoid* mRNA is degraded and only the de novo synthesized *bicoid* message is localized. When egg chambers are incubated with cytochalasin D, an agent that disrupts actin filaments (MacLean-Fletcher and Pollard, 1980), there is no effect on *bicoid* mRNA localization. Thus, it appears that microtubules, but not actin filaments, are also involved in the localization of *bicoid* mRNA. An attractive model to consider is that microtubules are involved in both the transit of *bicoid* mRNA from the nurse cells to the oocyte and the sequestration within the oocyte at the cytoplasmic bridge connections. However, we cannot exclude the possibility that microtubules play an indirect role in *bicoid* message localization, perhaps by maintaining cell polarity.

Mary Lou King: Maternal mRNAs and the Cytoskeleton in *Xenopus* Oocytes and Early Embryos

Embryonic determination is mediated at least in part by molecules regionally localized in the cytoplasm of the oocyte that become disproportionately segregated into specific blastomeres during cleavage. One such molecule is the Vg1 mRNA (Rebagliati et al., 1985), which is localized to the vegetal cortex of *Xenopus* stage VI oocytes and potentially encodes a TGF-β-like protein (Weeks and Melton, 1987). Ultrastructural analysis of the frog oocyte cortex reveals a region rich in microfilaments and and cytokeratin filaments (Gall et al., 1983). An attractive hypothesis is that Vg1 mRNA is anchored to the vegetal cortex by a cytoskeletal linkage, as has been suggested for mRNAs in other animal systems (Jeffery, 1989).

We have isolated a high salt (500 mM KCl) dependent insoluble cytoskeletal fraction from fully grown *Xenopus laevis* oocytes and have shown that it is highly enriched in cytokeratins and Vg1 mRNA (Pondel and King, 1988). This fraction, which contains 4–5 times more total poly A + RNA than the soluble fraction, is also enriched in the mRNA for the 56 kD cytokeratin protein but not histone H3 mRNA. During oogenesis, Vg1 RNA becomes progressively concentrated in the cytoskeletal fraction coincidently with its progressive localization to the vegetal cortex, suggesting that Vg1 RNA in the cytoskeletal fraction corresponds to the localized message anchored at the cortex (Fig. 3).

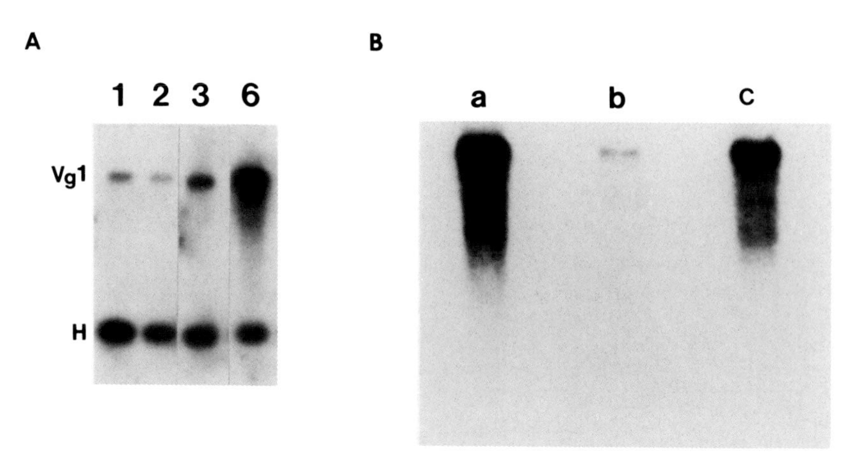

Fig. 3. **A:** Vg1 RNA is progressively concentrated in the cytoskeletal fraction during oogenesis. Northern blots of RNA extracted from the cytoskeletal fraction of stage I (lane 1), II (lane 2), III (lane 3), and VI (lane 6) oocytes and probed with [^{32}P]-labeled pVg1 and histone H3 DNA. Vg1 RNA is concentrated 20–30-fold by stage III and 60–100-fold by stage VI. **B:** Vg1 RNA is released form the cytoskeletal fraction within minutes of germinal vesicle breakdown (GVBD). Northern blot of RNA extracted form the cytoskeletal fraction of (lane a) stage VI oocytes; progesterone-treated stage VI oocytes (lane b) at the time of GVBD or (lane c) just before GVBD. Blots were probed with [^{32}P]-labeled pVg1 DNA. Note that Vg1 RNA is released only in oocytes that have undergone GVBD.

Upon oocyte maturation, Vg1 mRNA (but not histone mRNA) distribution changes and the message is now found in the soluble fraction. Vg1 RNA release can be triggered in vitro by progesterone and inhibited by the phosphodiesterase inhibitor, theophylline. Although the cytokeratin network also breaks down at maturation, a process that takes several hours (Klymkowsky, 1987) (coinciding with the solubilization of the 56 kD cytokeratin protein), we have found that the Vg1 mRNA is released within minutes of germinal vesicle breakdown (GVBD) (Fig. 3). The germinal vesicle itself is not involved in the cytoskeletal releases, as Vg1 mRNA will still appear in the soluble fraction in enucleated oocytes after progesterone exposure. Interestingly, the release of Vg1 mRNA can be induced in stage VI oocytes by the addition of cytoplasm from a matured oocyte. An interesting possibility is that Vg1 mRNA is anchored to the vegetal cortex through an RNA/cytokeratin binding protein that is phosphorylated during GVBD. In this regard, we have found a novel protein kinase activity in the cytoskeletal fraction that requires Mg^{2+} and is fully active at temperatures that inhibit protein kinase activities in the soluble fraction.

David L. Gard: Confocal Microscopy of Microtubules in *Xenopus* Oocytes and Eggs

Visualization of microtubules during the important stages of oogenesis in *Xenopus* has proven to be difficult due to the large size of oocytes and eggs. Early studies by electron microscopy found very few recognizable microtubules in oocytes (Heidemann et al., 1985), despite a biochemically measured pool of tubulin sufficient to assemble more than 1.5 km of microtubule (Gard and Kirschner, 1987). Furthermore, estimates suggest that as much as 20% of this tubulin is in polymeric form (Jessus et al., 1987; Gard, unpublished observation), corresponding to nearly 300 m of microtubules within a single oocyte!

One of the major obstacles to identifying microtubules in *Xenopus* oocytes has been achieving adequate fixation of these very large cells. Thus, previous studies have been unable to identify individual microtubules (Palecek et al., 1985; Wylie et al., 1985) without resorting to techniques that destroy the three-dimensional organization (Huchon et al., 1988). By using fixatives containing formaldehyde and glutaraldehyde, combined with recently developed techniques for clearing *Xenopus* oocytes (Andrew Murray, as cited by Dent and Klymkowsky, 1987), and confocal immunofluorescence microscopy, we have been able to visualize microtubule organization in whole-mounted oocytes, with resolution of individual microtubules (Gard, in preparation).

Stage I oocytes contain a poorly organized network of microtubules, with a concentration of microtubules loosely associated with the oocyte cortex, nucleus, and mitochondrial mass (Fig. 4). Surprisingly, optical sectioning of entire stage I oocytes failed to identify a distinct focus of microtubule organization indicative of a microtubule organizing center (MTOC). The lack of discrete MTOCs in stage I oocytes was also suggested by the results of microtubule regrowth after cold-induced depolymerization.

During oogenesis, there is a progressive increase in microtubule number and organization, with a concomitant increase in tubulin protein. By stage VI, oocytes contain dense arrays of microtubules, which appear to be organized by a region of amorphous tubulin- and microtubule-containing material localized to the vegetal surface of the nucleus (Fig. 4). Microtubules originate from the perinuclear region, and extend to oocyte cortex in both animal and vegetal hemispheres. Estimates of the number of microtubules reaching the oocyte cortex suggest that single oocytes contain as many as 10^6 microtubules, which is sufficient to account for the biochemically measured amount of polymer.

Many microtubules in stage I and VI oocytes stain with antibody specific for acetylated α-tubulin. Previous studies in cultured cells suggest that tubulin acetylation is indicative of a nondynamic, or stable, population of microtu-

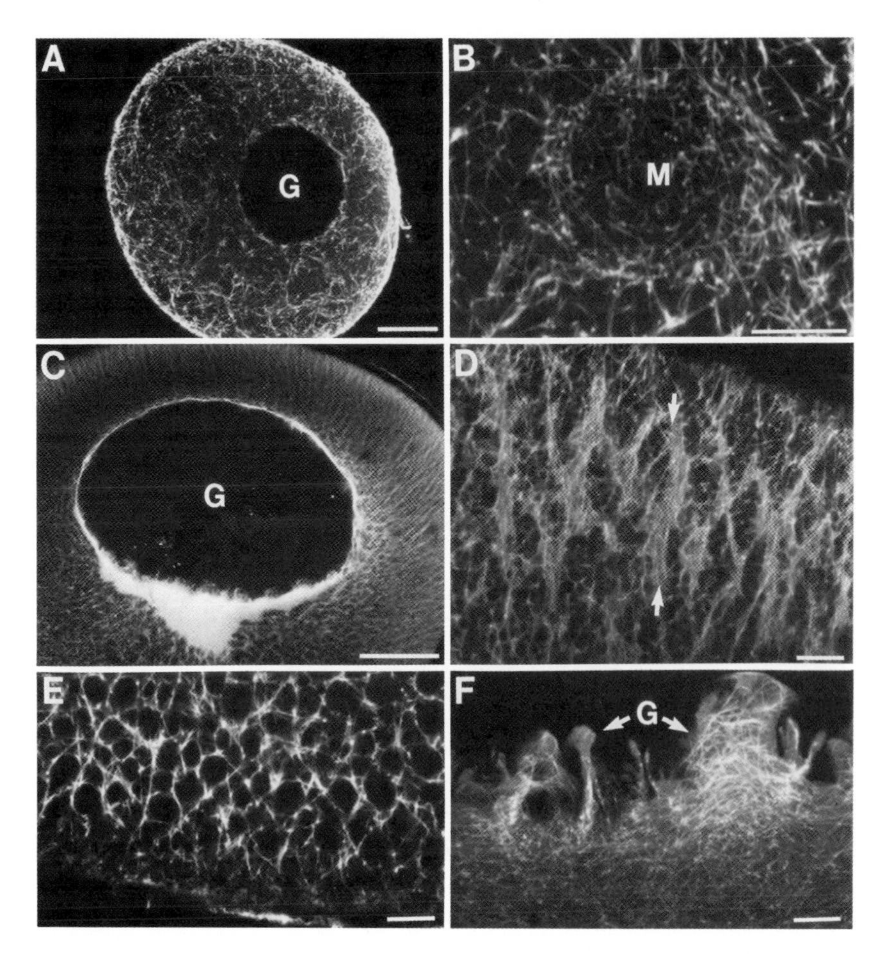

Fig. 4. Distribution of microtubules in frog oocytes. **A:** A medial optical section of a stage I oocyte fixed with formaldehyde and glutaraldehyde (without taxol) and processed for immunofluorescence with antitubulin. Numerous cytoplasmic microtubules (G, germinal vesicle). Bar = 25 μm. **B:** Higher magnification of a medial section of the mitochondrial mass in a stage I oocyte. Note the microtubules surrounding and penetrating the mitochondrial mass (M). Bar = 10 μm. **C:** A medial section of a stage VI oocyte, showing the germinal vesicle (G) and yolk-free cytoplasm associated with its vegetal surface. A poorly resolved fibrous array extends from the GV toward the oocyte cortex. Bar = 100 μm. **D:** Higher magnification of the cortical cytoplasm of the animal hemisphere, showing bundles of radially organized microtubules (arrows). Bar = 10 μm. **E:** The vegetal hemisphere contains a less-well-organized network of cytoplasmic microtubules. Bar = 10 μm. **F:** Higher magnification of the convoluted vegetal surface of the GV (G), showing the dense microtubule network and matrix of the yolk-free region of cytoplasm. Bar = 10 μm.

bules (Webster and Borisy, 1989). Oocyte microtubules might thus provide a stable framework for the intracellular transport of developmentally important signaling molecules (Yisraeli et al., 1990). We are extending these studies by examining the complex reorganization of the microtubule cytoskeleton during meiotic maturation.

THE TRANSITION FROM OOCYTE TO EGG

Michael W. Klymkowsky: Cytokeratins and Vimentin in *Xenopus* Oocytes and Early Embryos

In contrast to other vertebrates, there are two distinct vimentin-type intermediate filament proteins in *Xenopus laevis*. The 55 kD form is recognized specifically by the monoclonal antibody RV202; the monoclonal antibody 14h7 recognizes both the 55 kD and a 57kD form of vimentin (Dent et al., 1989). 14h7 has been used to isolate full-length vimentin cDNAs from a *Xenopus* expression library. These cDNA fall into two classes that are identified VIM2 and VIM4 vimentin cDNAs characterized by Hermann et al. (1989). The VIM1/VIM4 cDNAs, while highly homologous, differ throughout their coding and noncoding regions (Hermann et al., 1989; Dent and Klymkowsky, unpublished observations). In vitro transcription/translation of these cDNAs indicates that VIM1 clones encode the 55 kD form of vimentin, while VIM4 clones encode the 57 kD form. Both vimentins are found in all cell lines and in all embryos ($n > 30$) examined and are differentially regulated in adult tissue (Dent and Klymkowsky, in preparation). *Xenopus laevis* is known to have undergone a genomic duplication event some 30 million years ago (Kobel and Du Pasquier, 1986). It is likely that the two vimentin genes were produced by this event. Their differential expression indicates that they have evolved to fulfill different roles within the organism.

Vimentin-like immunoreactivity has been found associated with the mitochondrial mass (Balbiani body) of the stage I oocyte (Godsave et al., 1984a; Dent and Klymkowsky, 1989). Wylie and colleagues (1985) have also found vimentin-like immunoreactivity associated with the germ plasm of the late stage oocyte and early embryo. To study vimentin's function in the oocyte, mutant forms of the 55 kD vimentin protein have been constructed, following the strategy pioneered by Albers and Fuchs (1987). Expression of this mutant vimentin in either mammalian or *Xenopus* cells leads to the disruption of the endogenous vimentin filament system (Fig. 5). If the expression of mutant vimentin disrupts the organization of the mitochondrial mass or the integrity of germ plasm, it would be direct evidence that vimentin plays a role in maintaining the integrity of this organelle.

A distinct cytokeratin-type intermediate filament (IF) system also exists in the *Xenopus* oocyte (Franz et al., 1983; Godsave et al., 1984b). In the late stage oocyte, these cytokeratin filaments are localized in the cortex and form

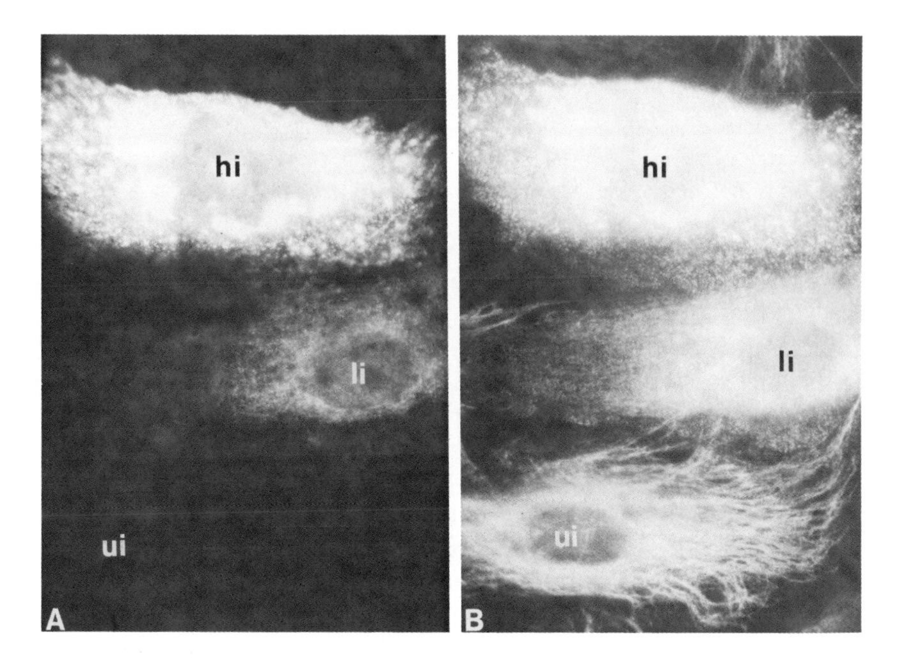

Fig. 5. Expression of altered vimentin molecules in cultured *Xenopus* cells. *Xenopus* XR1 cells were injected with pSRα-vimentin-1Δc20.myc DNA. This plasmid drives the expression of a truncated form of vimentin. After 20 h, the expression of the mutant vimentin protein was visualized by staining fixed cells with the monoclonal antibody 9E10 (which reacts with a C-terminal tagging sequence derived from the human c-myc protein; Munro and Pelham, 1984); the organization of the endogenous vimentin system was visualized using a rabbit antivimentin antibody. The level of vimentin-1Δc20.myc expression varies significantly between injected cells, but even in cells expressing relatively low levels of the mutant vimentin, the organization of the endogenous vimentin system is altered (ui, marks uninjected cell; hi, marks injected cell expressing high levels of the mutant vimentin protein; li, marks cell expressing low levels of the mutant vimentin protein). Panels A and B were stained with the anti-c-myc (9E10) and anti-vimentin (14h7) antibodies, respectively.

a polar asymmetric system (Klymkowsky et al., 1987). During oocyte maturation, the majority of the cytokeratin filaments disappear (Klymkowsky and Maynell, 1989); upon fertilization or activation of the egg, cytokeratin filaments reappear in the embryonic cortex (Klymkowsky et al., 1987). The injection of maturation promoting factor (MPF) into the oocyte induces disappearance of cytokeratin filaments (Klymkowsky and Maynell, 1989). Like the disassembly of the nuclear lamina during meiosis, the disappearance of cytokeratin filaments involves their disassembly into soluble oligomers. However, unlike its effect on the IF-like nuclear lamins, MPF's effect on cytokeratin filaments is indirect. Protein synthesis is required for MPF to induce the disassembly of cytokeratin filaments (Klymkowsky and Maynell, 1989). Interestingly, protein synthesis is also required to maintain cytokeratins in the disassembled state; if protein synthesis is inhibited, cytokeratin filaments reassemble (Klymkowsky et al., submitted). This suggests that the ''cytokeratin filament disassembly factor'' is biosynthetically labile. In the mature oocytes, cytokeratins have a more acidic pI; however, there is no definitive evidence that this acidification is involved in the disassembly of the cytokeratin filaments (Klymkowsky et al., submitted).

The dependence of the maturation-induced disassembly of the cytokeratin filaments on protein synthesis enabled the relationship between the disassembly of cytokeratin filaments and the solubilization of the vegetally localized maternal mRNA Vg1 during oocyte maturation (see above) to be examined. In oocytes injected with MPF, cytokeratin filaments disappear and Vg1 becomes soluble. Cycloheximide blocks the MPF-induced disruption of cytokeratin filaments, but has only a partial inhibitory effect on MPF's ability to solubilize Vg1 mRNA (Klymkowsky et al., submitted). This indicates that the disassembly of the cytokeratin filament system is not required for the solubilization of a substantial proportion of the Vg1 mRNA within the oocyte. It indicates that the linkage between Vg1 mRNA and the ''cytoskeleton'' is sensitive to MPF.

Janice P. Evans: Recruitment of Vinculin Into the Cortex of *Xenopus* Eggs

To expand the understanding of the cytoskeleton's role in *Xenopus laevis* development, we have focused on two particular cytoskeletal proteins, talin and vinculin. These proteins were first isolated from chicken gizzard, and have been subsequently identified in other animal cells. Often these two protein species are colocalized, particularly at adhesion plaques, at dense plaques of smooth muscle, at myotendinous junctions, and at postsynaptic neuromuscular junctions. While the functions of these proteins are not completely understood, talin has been found to bind both vinculin and the fibronectin receptor, integrin. There is evidence that vinculin can interact with α-actinin, providing the basis for a model for the connection between the extracellular matrix and the cytoskeleton. (For a review, see Burridge et al., 1988.)

Antibodies to the chicken forms of talin and vinculin were used in double-labeling immunofluorescence experiments. These experiments showed that talin and vinculin are colocalized in several different *Xenopus* cell types: adhesion plaques of cultured kidney (A6) cells, the cell peripheries of oviduct cells, and the postsynaptic neuromuscular junctions of tadpole tail muscle fibers. These antibodies also identify cognate proteins of the appropriate sizes on immunoblots of A6 cell and oviduct lysates (Evans et al., 1990). Using these antibodies on *Xenopus* oocytes, we find talin to be highly localized at the cortices of oocytes and vinculin to be in the oocyte cytoplasm and absent from the oocyte cortex (Fig. 6). Vinculin is first detectable as a cytoskeletal compo-

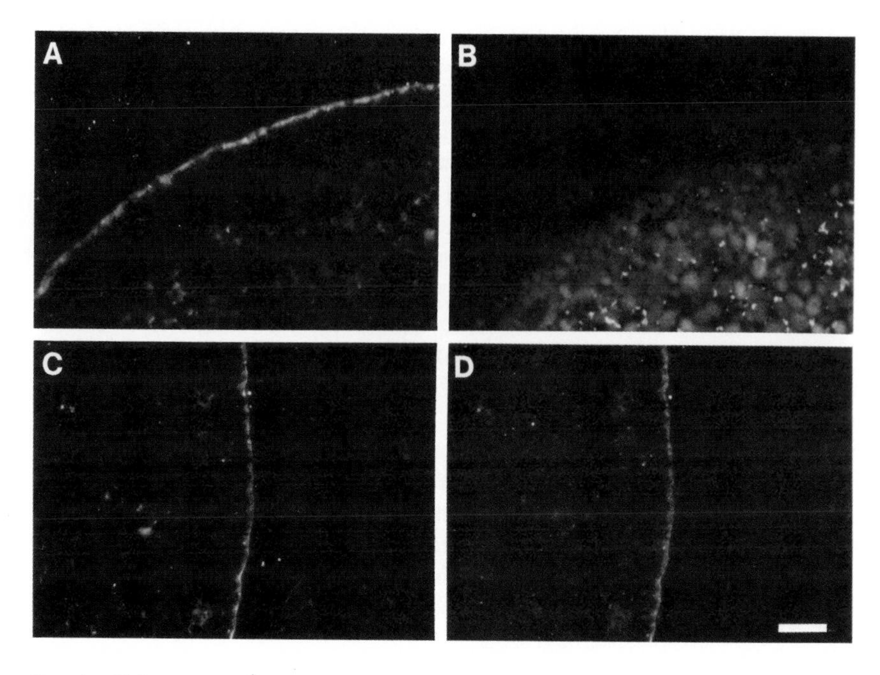

Fig. 6. Talin and vinculin become colocalized upon the transition from oocyte to egg. One micrometer-thick sections of *Xenopus* oocytes or eggs were double-labeled with a rabbit antitalin (Panels A and C) and mouse antivinculin (Panels B and D) antibodies, then reacted with rhodamine- or fluorescein-conjugated secondary antibodies. Panels **A** and **B** show a section of an oocyte, isolated from ovary via manual dissection and collagenase treatment. Note that antitalin antibodies stain the oocyte cortex (A), while the antivinculin antibodies do not (B). Panels **C** and **D** show a section of an egg, ovulated and oviposited by a hormonally stimulated female frog. Note that both antibodies label the egg cortex.

nent in eggs, appearing sometime during or between oocyte maturation and oviposition (Fig. 6). During early embryo development, talin and vinculin are colocalized in the cortices of cleavage furrows and blastomeres. Thus, the change from unlinked localization to colocalization appears to be developmentally regulated, occurring during the transition from oocyte to egg.

The focus of this project is now on pinpointing the timing of this change in localization. The oocyte-to-egg transition occurs in three basic steps: maturation, ovulation, and oviposition. From immunoblots on oocytes that have been matured in vitro, it appears that oocyte maturation is not sufficient to localize vinculin to the cortices of oocytes. Considering vinculin's possible role in the transmembrane link between the extracellular matrix (ECM) and the cytoskeleton, it is possible that changes in the oocyte's extracellular matrix that occur during ovulation (e.g., conversion of the coelomic envelope to vitelline envelope, or deposition of jelly coat) could trigger vinculin's recruitment to the cortex. Consequently, attention is being turned to experiments to determine if ovulation is effective at localizing vinculin to the cortex. In an attempt to identify molecules that could mediate signals from the oocyte's ECM to the cytoskeleton, immunoprecipitation on metabolically labeled oocytes and immunoblots on oocyte lysates are being done using antibodies to peptides corresponding to carboxy termini of several human integrin subunits. The *Xenopus* oocyte appears to contain antigens immunologically related to these human proteins.

Finally, cDNA sequences that code for *Xenopus* vinculin are being cloned using polymerase chain reaction (PCR) on cDNA synthesized directly from RNA or in cDNA–λ recombinant libraries. One hundred ninety-seven base pairs of coding region from oocyte cDNA has been isolated and sequenced; this region shows 86% homology to chicken vinculin at the nucleotide level. Of particular interest is a putative talin-binding domain, located at the 5′ end of the coding region of a clone isolated from chick embryo fibroblast (Jones et al., 1989). PCR is being used to determine if the *Xenopus* oocyte's vinculin mRNA lacks this region, based on the observation that talin and vinculin are not colocalized at the oocyte cortex. From immunoprecipitation experiments of metabolically labeled oocytes, it is clear that the oocyte is actively synthesizing vinculin. Thus, posttranslational modifications may play an important role in regulating the talin–vinculin interactions at the cortex.

FERTILIZATION AND EARLY CLEAVAGE

Richard P. Elinson: Cytoskeleton Events in the First Cell Cycle of Amphibian Eggs

In many amphibians, there is neither a fast block nor a slow block to polyspermy in fertilization. Whereas anurans are generally monospermic, in urodeles there is physiological polyspermy (Elinson, 1986). Typically, 1–20 sperm enter the eggs of the urodele, *Cynops pyrrhogaster*; yet, none of the accessory nuclei survive to hinder proper development. Two different mechanisms have been hypothesized. Bataillon and Tchou Su (1930) suggested that there is a small region of ''active cytoplasm'' in the egg, and any sperm nucleus that strays into that region is capable of fusing with the female pronucleus. The diploid nucleus then initiates mitosis first, ahead of the lagging male pronuclei. The latter degenerate because they are out of synchrony with the metabolism of the egg, which is tied to the metabolism of the successful nucleus. A contrasting hypothesis was set forth by Fankhauser (1948), who suggested that an ''activator substance'' emanating from the female pronucleus attracts one male pronucleus to it. Once pronuclear fusion has occurred, the zygotic pronucleus then in turn releases an ''inhibitor substance'' that directs the degeneration of the accessory nuclei.

Experiments in collaboration with Yasuhiro Iwao (Yamaguchi University, Japan) have been performed to test these hypotheses. In this work, the assays for survival of the accessory sperm nuclei are the replication of its centromeres, the formation of multiple spindles, and numerous cleavage furrows at the time of first cleavage. When hydrostatic pressure (7,000 psi) is imposed on fertilized eggs, many accessory sperm nuclei survive (8% multipolar furrows in controls versus 41–56% at 7,000 psi). High pressure inhibits astral growth and should equalize sperm nuclear migration, eliminating a single principal sperm nucleus. Another approach has been to inject a number of different substances into polyspermic eggs to test for their ability to rescue accessory sperm nuclei. A list is shown below:

Substance	Multipolar furrowing
Cynops GV	+
Xenopus GV	+
Cynops egg cytosol	+
Cynops egg cytosol + Ca^{2+}	–
Fertilized *Cynops* egg cytosol	–
Unfertilized *Xenopus* egg cytosol	+

From these results, MPF is a candidate for the active component, since MPF is sensitive to Ca^{2+}, absent in fertilized eggs, and is not species-specific. In agreement with these findings, accessory sperm nuclei survive when crude MPF fractions are injected into polyspermic urodele eggs (Fig. 7). It is presently unclear whether or not MPF is the only natural activator.

These experiments and the MPF hypothesis support both Bataillon's and Fankhauser's hypotheses. Any sperm that reaches the center of the egg will be exposed to the greatest concentration of MPF, which will lead to centrosomal duplication and a decrease in the concentration of active MPF in the surrounding cytoplasm. As MPF is inactivated in the egg, the accessory sperm nuclei will become quiescent and later degenerate.

Jon M. Holy: Unequal Cell Division at the Fourth Cleavage in Sea Urchin Embryos

The first three cleavage divisions of sea urchins are equal, and give rise to an eight-cell embryo consisting of animal and vegetal tiers of four blastomeres each. During the fourth division, cleavage of animal blastomeres is again equal; in many species, however, vegetal blastomeres divide unequally to produce macromeres and micromeres. The first step in this unequal division is migration of the vegetal nuclei toward the vegetal pole, and spindles subsequently assemble adjacent to the vegetal pole plasma membrane (Dan, 1979, 1984). Because centrosomes are involved in spindle assembly and organization, the

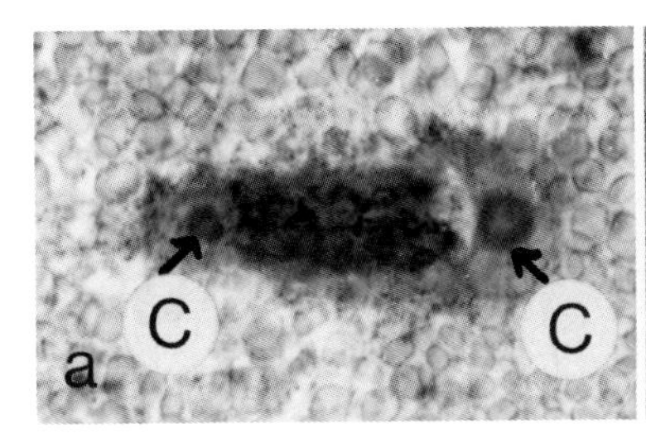

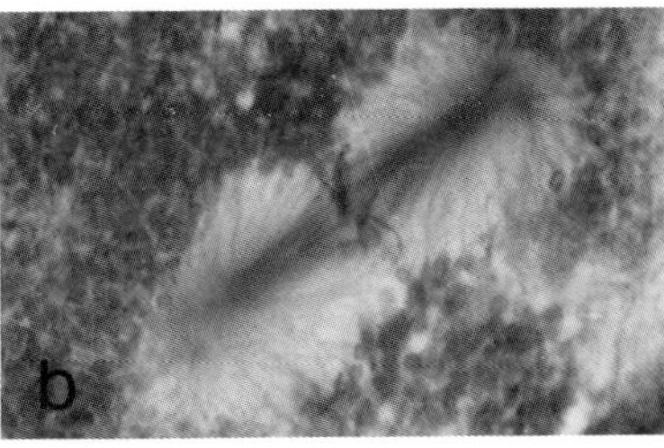

Fig. 7. Rescue of accessory sperm nuclei in polyspermic urodele fertilization. **a:** Centrosome (C) duplication in an accessory sperm nucleus following injection of an MPF fraction from *Xenopus* eggs. **b:** Extra spindle of accessory sperm origin following injection of unfertilized egg cytoplasm. See Iwao and Elinson (1990).

structural dynamics of immunofluorescently labeled centrosomes were examined by laser scanning confocal light microscopy (LSCM) and three-dimensional computer reconstruction methods in order to assess the roles of centrosomes in the unequal fourth cleavage of sea urchin embryos.

Embryos of *Lytechinus pictus* were cultured to fourth division, fixed with ice-cold methanol, and immunolabeled with a monoclonal anticentrosomal antibody raised against purified sea urchin centrosomes. This antibody (termed 4D2) appears to bind to a 45 kD polypeptide present in pericentriolar material (Thompson-Coffe et al., 1990). Stacks of optical sections through labeled blastomeres were obtained with a Bio-Rad MRC-600 LSCM, and data set volumetrically rendered with a Silicon Graphics 4D/70GT Iris graphics workstation using Voxel View software (Vital Images Inc., Fairfield, Iowa). Comparisons between the amounts of immunoreactive material within mesomere, macromere, and micromere centrosomes were obtained by projecting stacks of optical sections into single composite images and measuring pixel values (i.e., fluorescence intensities) and calculating the volumes of labeled spherical centrosomes.

Within vegetal blastomeres, micromere spindle poles were found to contain less immunoreactive centrosomal material (36% of the total amount) than macromere poles (64%) (Fig. 8), whereas centrosomal material was evenly apportioned between the daughter cells of the equally dividing animal blastomeres. This observation raises the possibility that the amount of centrosomal material a daughter cell inherits may be linked to the amount of cytoplasm it receives. Micromere centrosomes also exhibited a unique structural change during fourth cleavage, becoming elongate and filiform during late anaphase and telophase (Fig. 8). The dissimilar behavior of centrosomes within unequally cleaving vegetal blastomeres is mirrored by the asymmetric appearance of the mitotic spindle of these cells: The micromere spindle pole lies subjacent to the vegetal pole plasma membrane and displays a flattened aster. Dan and Ito (1984) proposed that spindle–membrane interactions of the microtubular cytoskele-

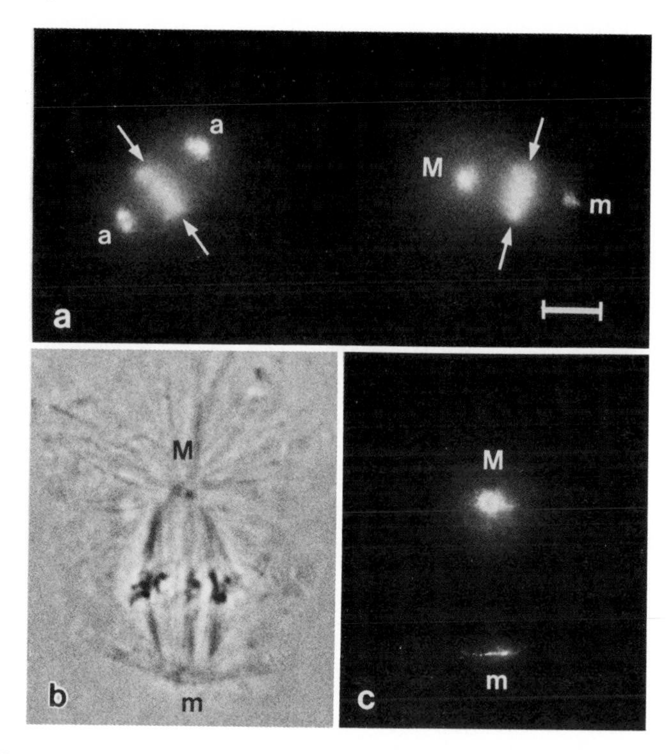

Fig. 8. Centrosomal antigen during fourth cleavage of sea urchin embryos. **a:** Double-labeled animal (**left**) and vegetal (**right**) blastomeres fixed at metaphase of fourth division. Centrosomes are labeled with monoclonal antibody 4D2; DNA is stained with Hoechst 33258. Chromosomes are aligned on the metaphase plate (arrows) in both blastomeres. Note that the micromere centrosome (m) appears smaller than macromere (M) centrosome or mesomere (a) centrosomes. **b:** Phase contrast micrograph of spindle isolated from vegetal blastomere at metaphase of fourth cleavage. The macromere aster (M) is radiate, whereas the micromere aster (m) is flattened. **c:** Immunolabeled vegetal blastomere centrosomes at telophase of fourth cleavage. The macromere centrosome (M) is essentially spherical, whereas the micromere centrosome (m) has undergone a filiform elongation. Bar = 10 μm for panel a, 5 μm for panels b and c.

ton, micromere centrosome, and vegetal cortex occurs during the unequal fourth cleavage of sea urchin embryos.

Susan Strome: Cytoplasmic Segregation in *C. elegans* Embryos

The topic of this presentation is the control of cleavage patterns, cytoplasmic partitioning, and cell fate determination in *C. elegans*. P granules provide useful markers for monitoring asymmetric partitioning and for analyzing the relationship between partitioning patterns and cleavage patterns. These cytoplasmic granules, visualized using monoclonal antibodies, are maternally supplied to the embryos and are progressively segregated to the germ lineage during the early embryonic cleavages (Strome and Wood, 1982, 1983). During the first cell cycle, the granules are segregated to the end of the embryo that will form the F1 cell, the first germline blastomere. They are subsequently passed to P2, P3, and finally P4, which gives rise to all the eggs and sperm in the adult worm. By injecting fluorescently labeled anti-P granule antibodies into the gonads of adult hermaphrodites, it has been possible to label and monitor P granules in living embryos. Such fluorescently labeled P granules are seen to move into the region of cytoplasm destined for the next P-cell daughter. Thus, P granule movement is a component of the segregation process; differential stabilization/destabilization of P granules in different regions of cytoplasm may also play a role.

From drug inhibitor studies, it is clear that the segregation of P granules depends on actin filaments, and is independent of microtubules (Strome and Wood, 1983). By "pulsing" embryos with microfilament inhibitor, the period of sensitivity has been determined to be the 5–10 min window during pronuclear migration. Embryos pulsed with cytochalasin D during this time fail to segregate P granules to the posterior pole and fail to undergo the first unequal, stem-cell-like division; instead they usually divide equally and distribute P granules to both daughters (Hill and Strome, 1988) (Fig. 9). Interestingly, some of these unusual two-cell embryos undergo subsequent divisions and segregate P granules as if their polarity has been reversed or else their anterior or posterior

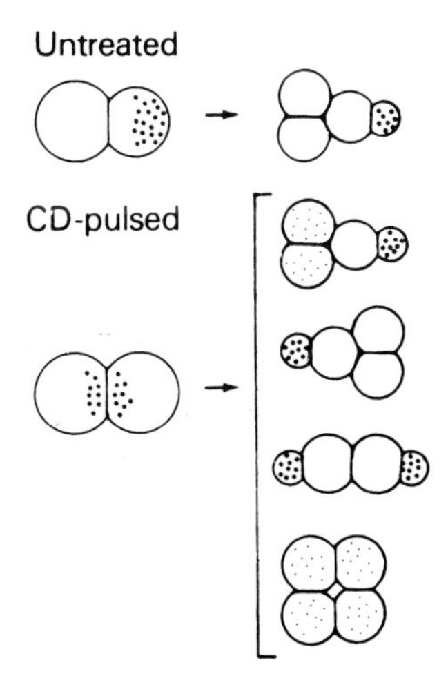

Fig. 9. The effects of cytochalasin D on embryonic cleavage and P granule distribution. Schematic representation of the four cleavage patterns observed after a cytochalasin D pulse during pronuclear migration in *C. elegans*. These are, from top to bottom: "normal," "reverse polarity," "posterior duplication," and "anterior duplication." Anterior is to the left, and posterior is to the right. The dots illustrate the distribution of P granules in these types of embryos.

end has been duplicated (Hill and Strome, 1990) (Fig. 9). In all embryos examined, unequal division and unequal P granule partitioning appeared to be coupled; all blastomeres that underwent an unequal division had segregated P granules to the smaller daughter. During equal divisions, P granules were distributed to both daughters. This result suggests that unequal divisions and P granule partitioning are mechanistically coupled. Elucidation of the control of partitioning patterns may come from analysis of mutants defective in partitioning (see below).

Although P granules are germline-specific structures, it is unknown whether they serve an essential role in germline development. To identify genes that encode maternal components required for germline development, screens for maternal-effect sterile (*mes*) or ''grandchildless'' mutants have been carried out. The phenotype of such mutants is the production of sterile but otherwise healthy progeny by homozygous *mes* mothers. The mutants analyzed thus far fall into two classes: 1) *mes-1* appears to be defective in partitioning P granules during embryogenesis. All alleles are leaky; *mes* mothers produce both sterile and fertile progeny. Defective P granule partitioning is observed in only some embryos, and there is preliminary evidence that those embryos that missegregate P granules grow up into sterile adults. Analysis of this and other partitioning-defective mutants (Kemphues et al., 1988) may contribute to our understanding of partitioning events and the role of P granules in germ-cell development. 2) *mes-2,-3*, and *-4* do not show defects in P granule partitioning. Almost all alleles are tight; all of the progeny produced by *mes* mothers are sterile. Almost all alleles result in defects in postembryonic proliferation of the germline, leading to the production of adults that either lack or have a greatly reduced number of germ cells. Molecular analysis of these *mes* mutants will reveal the nature and localization of some of the maternal products that control germline development.

ACKNOWLEDGMENTS

J.P.E. and B.K.K. would like to acknowledge the grant support of the American Cancer Society (CD-263). E.C.R. acknowledges grant support from the United States Public Health Service (HD16739) and from the American Cancer Society (CD-375). M.L.K. was supported from NIH GM 33932. M.W.K. acknowledges the contributions of Laurie Maynell, Joe Dent, Robert Cary, Jeff Bachant, Dave Shook, and Alberto Domingo. D.G. thanks the NIHGMS, the University of Utah Research Committee, and the College of Arts and Sciences for support. S.S. thanks NIH GM34059 for support. J.H. was supported by a NIH NRSA Postdoctoral Training Grant, and thanks Cathy Thompson-Coffe for the gift of antibody and Gerald Schatten for encouragement.

REFERENCES

Albers K, Fuchs E (1987): The expression of mutant epidermal keratin CDNAs transfected in simple and squamous cell carcinoma lines. J Cell Biol 105:791–806.

Bataillon E, Tchou Su (1930): L'analyse expérimentale de la fécondation et sa définition par les processus cinétiques. Ann Sci Natl Zool 17:9–36.

Berleth FS, Burri M, Thoma G, Bopp D, Richstein S, Frigerio G, Noll M, Nüsslein-Volhard C (1988): The role of localization of *bicoid* RNA in organizing the anterior pattern of the *Drosophila* embryo. EMBO J 7:1749–1756.

Biajolin S, Falkenburg D, Renkawitz-Pohl R (1984): Characterization and developmental expression of β tubulin genes in *Drosophila melanogaster*. EMBO J 3:2543–2548.

Burridge K, Fath K, Kelly T, Nuckolls G, Turner C (1988): Focal adhesions: Transmembrane junctions between the extracellular matrix and the cytoskeleton. Ann Rev Cell Biol 4:487–525.

Crossley AC (1972): Ultrastructural changes during transition of larval to adult intersegmental muscle at morphogenesis in the blowfly *Calliphora erythrocephela*. I. Dedifferentiation and myoblast fusion. J Embryol Exp Morphol 27:43–74; II. The formation of adult muscle. Ibid; 27:75–101.

Crossley AC (1978): The morphology and development of the *Drosophila* muscular system. Ashburner M, Wright TRF (eds): "The Genetics and Biology of *Drosophila*," Vol 2b. New York: Academic Press, pp 499–560.

Dan K (1979): Studies on unequal cleavage in sea urchins. I. Migration of the nuclei to the vegetal pole. Dev Growth Diff 21:527–535.

Dan K (1984): The cause and consequence of unequal cleavage in sea urchins. Zoo Science 1:151–160.

Dan K, Ito S (1984): Studies of unequal cleavage in molluscs: I. Nuclear behavior and anchorage of a spindle pole to cortex as revealed by isolation technique. Dev Growth Diff 26:249–262.

Dent J, Klymkowsky M (1989): Whole mount analysis of cytoskeletal reorganization and function during oogenesis and early embryogenesis in *Xenopus*. Shatten H, Shatten G (eds): "The Cell Biology of Development." New York: Academic Press, pp 64–107.

Dent JA, Polson AG, Klymkowsky MW (1989): A whole mount immunocytochemical analysis of the expression of the intermediate filament protein vimentin in *Xenopus*. Development 105:61–74.

Diaz HB, Raff EC (1987): Structure and evolution of the *Drosophila melanogaster* β4-tubulin gene. J Cell Biol 105:201a.

Elinson RP (1986): Fertilization in amphibians: The ancestry of the block to polyspermy. Int Rev Cytol 101:59–100.

Evans JP, Page BD, Kay BK (1990): Talin and vinculin in the oocytes, eggs, and early embryos of *Xenopus laevis*: A developmentally regulated change in distribution. Dev Biol 137:403–413.

Fankhauser G (1948): The organization of the amphibian egg during fertilization and cleavage. Ann NY Acad Sci 49:684–708.

Franz JK, Gall L, Williams MA, Picheral B, Franke WW (1983): Intermediate-size filaments in a germ cell: Expression of cytokeratins in oocytes and eggs of the frog *Xenopus*. Proc Natl Acad Sci USA 80:6254–6258.

Fuller MT, Caulton JH, Hutchens JA, Kaufman TC, Raff EC (1987): Genetic analysis of microtubule structure: The β-tubulin subunits have different effects on structurally different microtubule arrays. J Cell Biol 104:385–394.

Fuller MT, Caulton JH, Hutchens JA, Kaufman TC, Raff EC (1988): Mutations that encode partially functional β2 tubulin subunits have different effects on structurally different microtubule arrays. J Cell Biol 107:141–152.

Gall L, Picheral B, Gounon P (1983): Cytochemical evidence for the presence of intermediate filaments and microfilaments in the egg of *Xenopus laevis*. Biol Cell 47:331–342.

Gard D, Kirschner M (1987): Microtubule assembly in cytoplasmic extracts of *Xenopus* oocytes and eggs. J Cell Biol 105:2191–2201.
Gasch A, Hinz U, Leiss D, Renkawitz-Pohl R (1988): The expression of β1 and β3 genes of *Drosophila melanogaster* is spatially regulated during embryogenesis. Mol Gen Genet 211:8–16.
Gasch A, Hinz U, Renkawitz-Pohl R (1989): Intron and upstream sequences regulate the expression of the *Drosophila* β3 tubulin gene in the visceral and somatic musculature respectively. Proc Natl Acad Sci USA 86:3215–3218.
Godsave SF, Anderton BH, Heasman J, Wylie CC (1984a): Oocytes and early embryos of *Xenopus laevis* contain intermediate filaments which react with anti-mammalian vimentin antibodies. J Embryol Exp Morphol 83:169–184.
Godsave SF, Wylie CC, Lane EB, Anderton BH (1984b): Intermediate filaments in the *Xenopus* oocyte: The appearance and distribution of cytokeratin- containing filaments. J Embryol Exp Morphol 83:157–167.
HeidemannSR, Hambborg MA, Balasz JE, Lindley S (1985): Microtubules in immature oocytes of *Xenopus laevis*. J Cell Sci 77:129–141.
Hermann H, Foquet B, Franke WW (1989): Expression of intermediate filament proteins during development of *Xenopus laevis*: I. cDNA clones of vimentin. Development 105:279–298.
Hill DP, Strome S (1988): An analysis of the role of microfilaments in the establishment and maintenance of asymmetry in *Caenorhabditis elegans* embryos. Dev Biol 125:75–84.
Hill DP, Strome S (1990): Brief cytochalasin-induced disruption of microfilaments during a critical interval in 1-cell *C. elegans* embryos alters the partitioning of developmental instructions to the 2-cell embryo. Development 108:159–172.
Hoyle HD, Raff EC (1990): Two *Drosophila* beta tubulin isoforms are not functionally equivalent. J Cell Biol (in press).
Huchon D, Jessus C, Thibier C, Ozon R (1988): Presence of microtubules in isolated cortices of prophase I and metaphase II oocytes in *Xenopus laevis*. Cell Tis Res 254:415–420.
Iwao Y, Elinson RP (1990): Control of sperm nuclear behavior in physiologically polyspermic newt eggs: Possible involvement of MPF. Dev Biol 142:301–312.
Jeffery WR (1989): Localized mRNA and the egg cytoskeleton. Int Rev Cytol 119:151–196.
Jessus C, Thibier C, Ozon R (1987): Levels of microtubules during meiotic maturation of the *Xenopus* oocyte. J Cell Sci 87:705–712.
Jones P, Jackson P, Price GJ, Patel B, Ohanion V, Lear AL, Critchley DR (1989): Identification of a talin binding site in the cytoskeletal protein vinculin. J Cell Biol 109:2917–2927.
Kemphues KJ, Raff RA, Kaufman TC, Raff EC (1979): Mutation in a structural gene for a β-tubulin specific to testis in *Drosophila melanogaster*. Proc Natl Acad Sci USA 76:3991–3995.
Kemphues KJ, Kaufman TC, Raff RA, Raff EC (1982): The testes-specific β- tubulin subunit in *Drosophila melanogaster* has multiple functions in spermatogenesis. Cell 31:655–670.
Kemphues KJ, Priess JR, Morton DG, Cheng N (1988): Identification of genes required for cytoplasmic localization in *C. elegans*. Cell 52:311–320.
Kimble M, Incarona JP, Raff EC (1989): A variant β-tubulin isoform of *Drosophila melanogaster* (β3) is expressed primarily in tissues of mesodermal origin in embryos and pupae, and is utilized in population of transient microtubules. Dev Biol 131:415–429.
Kimble M, Dettman RW, Raff EC (1990): The β3-tubulin gene of *Drosophila melanogaster* is essential for viability and fertility. Genetics (in press).
Klymkowsky MW, Maynell LA, Poulson AG (1987): Polar asymmetry in the organization of the corticalcytokeratin systemof *Xenopus laevis* oocytes and embryos. Development100:543–557.
Klymkowsky MW, Maynell LA (1989): MPF-induced breakdown of cytokeratin filament organization during oocyte maturation in *Xenopus* depends upon the translation of materal mRNA. Dev Biol 134:479–485.
Klymkowsky MW, Maynell LA, Nislow C (1991): Cytokeratin phosphorylation, cytokeratin filament severing, and the solubilization of the maternal mRNA Vg1. J Cell Biol (in press).
Kobel HR, Du Pasquier L (1986): Genetics of polyploid *Xenopus*. TIG 2:310–315.

MacLean-Fletcher S, Pollard TD (1980): Mechanism of action of cytochalasin B on actin. Cell 20:329–341.

MacDonald PM, Struhl G (1988): *Cis*-acting sequence responsible for anterior localization of *bicoid* mRNA in *Drosophila* embryos. Nature 336:595–598.

Michielis F, Falkenburg D, Muller AM, Hinz U, Bellman R, Glatzer KH, Brand R, Biajolin S, Renkawitz-Pohl R (1987): Testes-specific β2 tubulins are identical in *Drosophila melanogaster* and *D. hydei* but differ from the ubiquitous β1 tubulin. Chromosoma 95:387–395.

Munro S, Pelham HRB (1984): Use of peptide tagging to detect proteins expresses from cloned genes: Deletion mapping of functional domains of *Drosophila* hsp70. EMBO J 3:3087–3093.

Nüsslein-Volhard C, Fröhnhofer HG, Lehmann R (1988): Determination anteroposterior polarity in *Drosophila*. Science 238:1675–1681.

Palecek J, Habrove V, Nedvidek J, Romanovsky A (1985): Dynamics of tubulin structures in *Xenopus laevis* oogenesis. J Embryol Exp Morphol 87:75–86.

Pondel M, King ML (1988): Localized maternal mRNA related to transforming growth factor β mRNA is concentrated in a cytokeratin-enriched fraction from *Xenopus* oocytes. Proc Natl Acad Sci USA 85:7612–7616.

Raff EC, Fuller MT, Kaufman TC, Kemphues KJ, Rudolph JE, Raff RA (1982): Regulation of tubulin gene expression during embryogenesis in *Drosophila melanogaster*. Cell 28:33–40.

Rebagliati MR, Weeks DL, Harvey RP, Melton DA (1985): Identification and cloning of localized maternal RNAs from *Xenopus* eggs. Cell 42:769–777.

Rudolph JE, Kimble M, Hoyle HD, Subler MA, Raff EC (1987): Three *Drosophila* beta tubulin sequences: A developmentally regulated isoform (β3), the testes-specific isoform (β2), and an assembly-defective mutation of the testes-specific isoform ($\beta 2t^8$) reveal both an ancient divergence in metazoan isotypes and structural constraints for beta-tubulin function. Mol Cell Biol 7:2231–2242.

Stephenson EC, Chao Y-C, Fackenthal JD (1988): Molecular analysis of the *swallow* gene of *Drosophila melanogaster*. Genes Dev 21:1655–1665.

Strome S, Wood WB (1982): Immunofluorescence visualization of germ-line-specific cytoplasmic granules in embryos, larvae, and adults of *Caenorhabditis elegans*. Proc Natl Acad Sci USA 79:1558–1562.

Strome S, Wood WB (1983): Generation of asymmetry and segregation of germ-line granules in early *C. elegans* embryos. Cell 35:15–25.

Tassin A-M, Maro B, Bornens M (1985): Fate of microtubule-organizing centers during myogenesis *in vitro*. J Cell Biol 100:35–46.

Thompson-Coffe C, Goffe G, Schatten H, Holy J, Mazia D, Schatten G (1990): Isolation of the mitotic centrosome of sea urchin embryos and generation of a novel monoclonal antibody. (Submitted).

Warren RH (1974): Microtubular organization in elongating myogenic cells. J Cell Biol 63:550–566.

Webster DR, Borisy GG (1989): Microtubules are acetylated in domains that turn over slowly. J Cell Sci 92:57–65.

Weeks DL, Melton DA (1987): A maternal mRNA localized to the vegetal hemisphere in *Xenopus* eggs codes for a growth factor related to TGF-β. Cell 51:861–867.

Wilson L, Bryan J (1974): Biochemical and pharmacological properties of microtubules. Adv Cell Mol Biol 3:22–72.

Wylie CC, Brown D, Godsave SF, Quarby J, Heasman J (1985): The cytoskeleton of *Xenopus* oocytes and its role in development. J Embryol Exp Morphol 89 (Suppl):1–15.

Yisraeli JK, Sokol S, Melton DA (1990): A two-step model for the localization of maternal mRNA in *Xenopus* oocytes: Translocation and anchoring of Vg1 mRNA. Development 108:289–298.

Index